Animal products from the Mediterranean area

The EAAP series is published under the direction of Dr. P. Rafai

EAAP – European Association for Animal Production

CIHEAM – International Centre for Advanced Mediterranean Agronomic Studies

EZN – Estação Zootécnica National

FAO – Food and Agriculture Organization of the United Nations

FCT Fundação para a Ciência e a Tecnologia
MINISTÉRIO DA CIÊNCIA, TECNOLOGIA E ENSINO SUPERIOR

FCT – Foundation for Science and Technology

The European Association for Animal Production wishes to express its appreciation to the *Ministero per le Politiche Agricole e Forestali* and the *Associazione Italiana Allevatori* for their valuable support of its activities

Animal products from the Mediterranean area

EAAP publication No. 119

Editors:

J.M.C. Ramalho Ribeiro, A.E.M. Horta, C. Mosconi and A. Rosati

Wageningen Academic
Publishers

Subject headings:
Animal production
Mediterranean basin

ISBN 10: 90-76998-86-8
ISBN 13: 978-90-76998-86-2
ISSN 0071-2477

First published, 2006

Wageningen Academic Publishers
The-Netherlands, 2006

The organisation of the Symposium was made possible by the kind support of the following scientists and structures, who are kindly acknowledged.

Local Organizing Committee
J.M.C. Ramalho Ribeiro
A.E.M. Horta
J.S. Pires da Costa
Luís Telo da Gama
O.C. Moreira;
J. Santos Silva
C.C. Belo

International Organizing Committee
D. Gabiña (IAMZ, CIHEAM)
F. Guessous (IAV Hassan II, Morocco)
A. Rosati (EAAP)

International Scientific Committee
J.C. Flamant (France)
D. Gabiña (Spain)
S. Galal (Egypt)
A. Georgoudis (Greece)
F. Guessous (Morocco)
C. Papachristoforou (Cyprus)
J.M.C. Ramalho Ribeiro (Portugal)
A. Rosati (Italy)
G. Zervas (Greece)

Table of content

Session 3. Possibilities for improving traditional systems
Main Papers

Session 4. Quality and traceability of typical products
Main papers

Session 5. Free communications and Posters

Preface

From 25 to 27 September 2005, EAAP, CIHEAM Zaragoza and FAO, in collaboration with the Estação Zootécnica Nacional of Santarem, Portugal, organised the Mediterranean Symposium, whose main topic concerned the "Comparative Advantages for Typical Animal Products from Mediterranean Area".

About 90 participants from countries of the Mediterranean basin took part to the symposium, presenting their personal experience, case studies on of typical local products and breeds, descriptions of the production systems and conservation techniques of endangered breeds/products.

The speakers presented the natural constraints of the area that, due to climate and geography, seems to be unfavourable to mass production at low cost. It was agreed that the exploration for ways of valorise the specificity of the products coming from extensive and traditional systems, involving local breeds, can meet the needs of the population requiring safe foods at a reasonable costs. These profit-related aspects were also discussed considering the different economic realities of the northern part of the basin compared to those of the southern part.

Participants presented the results of their investigations on the characteristics of typical animal production in the regions, being conscious of the positive and negative impact of the production systems on the environment, and aware of the necessity to adjust them to the climatic uncertainty and the seasonal variability of feed resources.

Four sessions and a round table were organised during the symposium, focusing the debate on the following areas:

- Animal production economy and social impact in the Mediterranean area.
- Utilisation of natural resources and environmental impact of the animal production systems.
- Possibilities for improving traditional systems.
- Quality and traceability of typical products.
- Round Table: "From traditional to certified animal products. Products perception by consumers, marketing systems and interaction with producers: what future role for organic animal farming in the Mediterranean?".

The Editors
J.M.C. Ramalho Ribeiro, A.E.M. Horta, C. Mosconi and A. Rosati

Session 1. Animal production economy and social impact in the Mediterranean area

Main papers

Mediterranean products: moderate consumption, diversified production

Vitor Barros[1] & Antonino Rodrigues[2]

[1]Estação Agronómica Nacional, INIA, Av. República, Quinta do Marquês, Nova Oeiras 2784-505 Oeiras, Portugal
[2]Lisboa, Portugal

Summary

After analysing where the EU lies in the international market, in terms of both production and consumption of food, defend the need for Southern European Countries to deepen their political and economical relations with the other Mediterranean countries. This would create a new market, which within the EU frame would have more chances to overcome the globalization threat.

The fact that all Mediterranean countries have common historical, cultural and civilisation roots enables a common start point, the Mediterranean diet. From there new and adjusted rural development policies can be developed, thus enabling a fostering environment for sustained production and rational consumption of this region's typical products.

Keywords: identity, Barcelona Agreements, new market, Mediterranean diet, redefining consumer habits, biodiversity, rural development, production systems.

Introduction

Portugal, a small, peripheral country at the westernmost point of Europe has the Atlantic at its feet and opened itself to the world, conquered an empire it did not know how to explore or develop and ended up losing it because of the lack of man power and technical means. Portugal turned to Europe and became part of the European Union in 1986 after having lost its colonies and it is thanks to this fact that it has been able to survive in the globalized world we live in.

Although the coastal waters of Portugal are not the Mediterranean sea, it is as much mediterranean as are Spain, Italy, Greece and southern France. When the European Union was made up of 15 countries, the tendency was to have northern countries impose policies on southern countries. In increasing the number of member countries to 25, the centre of power will be even more to the north and west.

With new and broader economic areas and greater international commerce, countries tend to dilute their influence and even loose their independence and sovereignty when they are unable to affirm themselves as States with a unique identity. This is a feat only Countries with History can achieve.

With such an accelerated and sometimes unrefrained globalization, the identity and sovereignty of a country are affirmed only through the knowledge, culture and state property of the country itself.

When we look at mediterranean countries, it is easy to find common denominators that should be the foundation on which to build a solid union that will allow us to defend our common interests.

The history, culture, climate, agriculture, rural development and gastronomic heritage of all of these countries have their roots in the Mediterranean sea. We should also add food quality and safety to these values, because it is always easier to control production in our own country or in neighbour countries than in far off countries.

Countries from southern Europe should use the power of these values to exercise their influence, not only moving the centre of Europe, but also involving all other Mediterranean countries. All aspects addressed in the Barcelona Agreements must be carefully analysed, because the more democratic and developed the societies surrounding us, the stronger the cohesion of the European Union.

The world market

There are an increasing number of fundamentalists who defend the total and unrestrained liberalization of the market worldwide. According to these fundamentalists, markets will work perfectly if there are no barriers, because demand equals supply, whether it is jobs or any other product or good. However, Europe has the duty to defend the values on which its social organization is set and should therefore try to pass the idea that liberalization should have people and environment in mind.

In order to exist, the market must have people, rules and must trigger purchasing power. This is the only way to have just, balanced and respectful values in the different phases of an acceptable model for the liberalization of world trade.

When Nobel Prize winner for Economy, Joseph Stiglitz, analysed the harmfulness of a globalization with no rules and no respect for people, he established a terrible and chilling analogy which is up-to-date and shows the brought on effects: *"Modern technological warfare is designed to ban physical contact – when someone drops a bomb from 50 000 feet, that person doesn't realize what he's done. Modern economic management is similar. It is easy to mercilessly impose politics from a luxurious hotel room when you can't see the people whose lives are about to be destroyed by them"*.

Although the foundations of the present EU were set in 1957 with the Rome Treaty, the true beginning dates 1952 with ECSC (European Coal and Steel Community). It was made up of six countries with a population of 157.9 million and covered an area of 1 170 846 km^2. Fifty three years later, the European Union is made up of 25 countries, has a population of 455 66 million and covers an area of 3 974 598 km^2.

With this population and area, the European Union is in a position where it can be an economic and social influence and set rules that will prevail in international relations.

However, in order to be more competitive and still be a just and solidarity society, the European Union must have more people with new and better jobs and must offer new professional training skills.

We need a new market

With life expectancy increasing and birth rate decreasing, Old Europe is unable to generate the necessary human resources to invert the tendency for decline. Even so, this situation is not as serious as expected, because there has been significant immigration from Eastern Europe, Middle East and Northern Africa.

The demographic situation is so worrying that all projections indicate that the workforce will begin to decrease in 2020 and consequently weaken the sustainability of social security systems.

Europe needs a greater workforce and increased productivity.

Since it is not possible to issue a decree for the increase of birth-rate or make elderly people work, immigration has become crucial for the future of Europe. However, foreigners who seek the

EU must find what they are looking for: better living conditions, peace, prosperity and trust. It is thus urgent to broadmindedly analyse the open and new relations with neighbour countries.

This is the case, for example, of southern Mediterranean countries which are not part of the UE and yet have the same historic and cultural references as EU countries. The southern region of the Iberian Peninsula, part of southern France and Italy and all of Greece have more in common with southern Mediterranean countries than with countries from northern Europe.

Commercial transactions show there are favourable conditions to increase the free trade, between these countries and the EU. In fact, when considering agricultural and processed food products, south and south-eastern Mediterranean countries (Morocco, Alger, Tunisia, Egypt, Israel, Lebanon and Turkey) import 32.6% from the EU, 24.1% from the USA and 43.3% from the rest of the world. In turn, the EU imports 46.3% from these countries, while the US imports only 6.6%.

Mediterranean diet

The culturally and productively influential Mediterranean has maintained common references throughout Time. Although the Mediterranean was the birthplace of western civilization, it was someone from the New World who called everyone's attention to the importance of the Mediterranean Diet in human health. In the 1960's, Dr. Ancel Keys, researcher at the University of Public Health in Minnesota, USA, published a paper entitled "Study of Seven Nations", in which he compared the incidence of heart disease in seven countries – Japan, Italy, Netherlands, Finland, USA, Greece and former Yugoslavia and showed that people from the mediterranean region had a lower rate of heart disease. This difference is observed because northern countries, USA and Japan have 50% more calorie intake per person than Mediterranean countries.

In 1996, Oldways Preservation, a non-profitable organization that jointly studies and promotes healthy eating habits with the WHO (World Organization for Health) European Office and the Harvard School of Public Health – USA, introduced the concept of the Traditional Mediterranean Food Pyramid. This pyramid was very useful in convincing people to change their eating habits by presenting correct food choices. This diet was inspired by that of the people from mid XX century Crete and includes some variants of the traditional diets from Spain, Portugal, Greece, Morocco, Tunisia, Turkey, Syria, southern France and southern Italy.

The base of the pyramid includes foods you should preferentially choose – cereals, whole rice and pastas, vegetables, fresh fruits and dry fruits and nuts, while the top of the pyramid shows foods you should eat moderately, such as meat and sweets.

Although Mediterranean countries have some differences in terms of diet, there are also common points:

- Abundance of vegetables, cereals, potatoes, legumes, green vegetables and fresh or dry fruits. Bread plays an important role and is both eaten with the meal and used as an ingredient in traditional dishes, such as "*açorda*", "*migas*", "*ensopado*" and "*gaspacho*".
- In general, foods were consumed fresh in the region of their production and were not subjected to industrial treatment. Legumes, green vegetables, herbs and fresh fruits (all rich in vitamins, minerals and antioxidant enzymes), as well as dry fruits (rich in polyunsaturated fatty acids) were consumed on a regular basis in all countries.
- Olive oil was the main fat used in the Mediterranean diet. It is a mono-unsaturated fat which increases HDL levels (known as good cholesterol) and avoids the deposition of LDL (bad cholesterol). Pig fat represents a small percentage and margarines/butters are seldom used.
- Cheese, milk and yoghurt intake was very low.

- Fish, poultry and egg intake was moderate. However, sardine, mackerel and later dry cod were the main source of protein for people living near the coast.
- Fresh fruit was the main dessert.
- Red meat was an option on festive occasions only.
- Wine was moderately consumed. Since grains are rich in anthocyanins (antioxidants), people who drank wine regularly and moderately had a lower risk of heart disease.
- Water intake was constant and its importance is shown by the number of public fountains and thermal water resorts in all Mediterranean countries.

The growing role and importance of the Mediterranean diet is confirmed by specialists. Recent studies show that if people with heart disease switch to the Mediterranean diet, they lower their risk of suffering a second attack by 70%. This is why nutritionists increasingly advise people from developed countries to change their eating habits so as to eat more cereals, fruits and vegetables. Suggestions contained in Table 1 should be followed as closely as possible by those who want to eat more sensibly.

Food in Portugal

Portuguese eating habits have evolved, although the essential characteristics of the Mediterranean diet persist.

Ancient sources state that in the II century a.c., Polibius made reference to "the prices of wheat, barley, wine, goats, hares, rams, pigs, figs and meat of wild animals in Lusitânia", and also to "ocean and estuary fish and the good quality of tuna".

At a later time, the geographer Estraban described the main food resources of Lusitânia: "the territory was rich in fruit and animals; rivers were abundant with fish; there were many olive groves and vineyards; people were sober and frugal; they drank water, beer and goat milk; in the mountains and during the greater part of the year oak acorns were used for food (later ground to make bread), while wine was saved for festive occasions only".

According to Mendes Correia, the people of Lusitânia survived in much the same way as many others around the world: they used available natural resources: fruits, legumes, vegetables, game or meat from herds and fish.

According to the same author, "these products and all of the changes that occurred in the agricultural sector defined the eating habits of Portugal until mid XIX century". These changes were also observed in other European countries of the Mediterranean, especially in the Middle Ages during which the role of monasteries and convents improved conditions for agriculture. Later, the discovery of other countries triggered intense commerce and consequently the introduction of new foods. Some examples

Table 1. Nutritional style as suggested by the Mediterranean diet.

Eat less meat	Choose recipes that have vegetables, pastas, dry fruits and legumes as main ingredients.
Try a greater variety of grains	Try recipes were cereals are fundamental, such as pastas, rye bread, etc.
Widen the variety of flavours	Use herbs, garlic, lemon and capers to enhance flavour.
Serve fruit in every meal	Fruit should be used in a variety of ways and not just as dessert. Take advantage of colour and flavour.
Use cheese as seasoning	Small portions of cheese bring out the flavour of foods.

are the potatoes (South America), corn (Central America), spices (India) and tropical fruits (Africa and Brazil).

However, the analysis of food using scientific methodology began with Ferreira Lapa (1873) and has continued into the XX century with laboratories for the analysis of food using chemical and analytical methods at the "Instituto Nacional Ricardo Jorge".

Ferreira Lapa was the first to define the basis and criteria for a Food Scale in Portugal. With available data, he designed a chart of foods consumed annually per person and defined the *per capita* proportion of nitrogen, nitrogen matter, salt and water for each food. He then divided these foods into 11 groups: bread cereals (corn, wheat and rye); meat (pork, beef, mutton, goat, poultry and game); starches (potatoes, rice and dry legumes); vegetables (string beans or similar); fresh produce; dry fruit; fish (fresh and dry); eggs and dairy products (eggs, milk, cheese and butter); oils (olive oil); fermented drinks and spirits; "goods from the colonies" (sugar, coffee, tea). The mediterranean influence is seen in the way food products are used in Portugal.

The conclusions for this study were precise, clear and very important for research that was carried out through the XX century and continues until present date.

According to Ferreira Lapa, "the great variety of foods in the average portuguese diet makes it possible to substitute those that are in shortage with other foods that have the same nutritional value". When he compared our diet with that of other countries, he concluded that despite of some differences, ours was the best and sufficient for the needs of the portuguese population.

The National Statistics Office has been working regularly on charts for the Food Scale of the portuguese population since mid XX century. These charts make it possible to analyse the eating habits of the portuguese and its changes, as well as study the best ways to define the most convenient foods for the health and well being of present and future societies.

As a consequence of the improvement in the standard of living in Portugal during the past decades, the tendencies for food intake, between 1980 and 1992, were as follows:

- Intake increase for the majority of foods, namely meats, fish, eggs, milk and dairy products, oils and fats with the exception of olive oil, potatoes, dry legumes, vegetables and fruits.
- The *per capita* intake of grains and sugar was stable until 1988 and decreased after 1992.
- Wheat was increasingly used as the main grain for bread making, while corn and rye decreased.
- The *per capita* intake of rice was stable.
- Pork became the most consumed meat, followed by beef and poultry.
- The tendency to use less olive oil and more vegetable oils was confirmed.
- Calories, lipids and proteins increased, while carbohydrates slightly decreased.
- There was a slight increase in the percentage of total calories supplied by lipids (fats), which went from 30.9% during 1980/84 to 34.6% during 1990/92. This increase is mainly due to the increase in vegetable oil and fat intake and to the greater intake of meat, milk and other dairy products.

Some tendencies of the last decade have worsened and we can clearly state that the eating habits of the portuguese have increasingly pushed away from the Mediterranean diet. There is an accentuated unbalance between the intake of animal origin products (meat, milk, butter, cheese and eggs), which has increased, and the intake of vegetable origin products (cereals and vegetables), which continues to decrease.

The food group with greater intake *per capita* was that of milk and dairy products (+41%), followed by meat and eggs (+32%). The fruit/vegetable and oil/vegetable fat groups had moderate increases (+8%), while grains, rice, legumes and potatoes presented a 12% decrease.

When analysing the eating habits of EU countries (15) with EUROSTAT (2000/2001), Greece is the country which most closely follows the Mediterranean diet. Greece is the greatest *per capita* (in kg/hab/year) consumer of grains (161.6 kg), vegetables and dry fruits (445.8 kg), oils and vegetable fats (45.9 kg – due to olive oil) and ranks last in the intake of milk and dairy products (66.7 kg). Greece is also one of the countries with lowest meat intake (86 kg).

Portugal ranks as follows in relation to each food group: second for grain intake (129.4 kg/hab/year), followed by Italy and The Netherlands - the latter with only 59.5 kg. For vegetable and dry fruit intake, Italy and Spain follow Greece and have 314 and 274 kg respectively, while Portugal ranks 5th with 190.8 kg and Sweden last with 107.9 kg. For dairy products, the greatest consumers are Finland with 190.3 kg and Ireland with 176.6 kg, while Portugal comes in last with 113.3 kg.

In relation to meat (all species included), the greatest consumer is Spain with 122.3 kg, followed by Denmark with 113.7 kg. Portugal has increased its meat intake and ranks 5th with 102.9 kg, while Luxemburg ranks last with 48 kg. In relation to oils and vegetable fat, Greece ranks first followed by Spain (31.6 kg) and Italy (26.4 kg), while Portugal comes in 6th with an intake of 21.3 kg and Sweden in last with only 3.2 kg.

Redefining consumer habits

The history of human feeding blends in with the history of humanity, because they are both about survival and are influenced by whatever changes have occurred in agriculture from the beginning of life until present day.

While it is true that the pleasures of good food have been conditioned by religious aspects (this still happens in some countries), it was the lack of food that triggered the majority of conflicts between people and nations.

In present day Europe where there is freedom and comfort, these problems do not have the same weight as before and the worries of Europeans are centred on dioxins, growth promoters, BSE and others. In other words, present day worries are related to Food Quality and Safety. In order to ensure the quality and safety of foods, new concepts must be introduced, such as identity, authenticity, diversity and certification of both products and production systems.

It is unthinkable that a current agricultural policy not be focused on the quality of products and rural areas.

It is not enough for food to look appealing and taste good. The consumer also wants new information, such as geographic origin and especially production systems. Although the protected designation of origin system marks a turning point, it must grow to include more products and ensure producers that they are the beneficiaries of this kind of system. However, added value is lost in certification systems, stamps and staff expenditure which are a result of new professions that emerge in this circuit.

Other problems that make the situation worse and must therefore be addressed are the age of producers and the limited and slow profit return of material and human investments. These aspects must be analysed if we are to guarantee biodiversity and satisfy the growing demands of consumers.

We have reached a point in which the contribution of agricultural production (GAP/GDP) is decreasing in EU countries. Even with 25 member countries, food production is loosing importance as an economic activity and being looked at as a sector with strong links to public health, physical planning and conservation of natural resources. This is why it is important to further analyse the strategy for rural development, using available support and producing foods that will satisfy consumer needs at reasonable prices.

Organic production (OP) is a good alternative if it is sustained by a simplified and efficient structure in which the distance between who produces and who consumes is as small as possible and where production costs are as low as possible. All of this may be achieved by increasing quantities and concentrating areas of production guaranteed by an economic, credible and feasible certification system.

Organic agriculture will only contribute to true and sustainable development when it becomes accessible to all consumers rather than being a niche product.

The UE has been trying to convince producers to use new production methodology and consumers to prefer these new products. Thus:

- A 1988 Commission's communication to the European Parliament and Council regarding the future of the rural world was the first turning point and established that "unless the diversification of rural economy should improve the potential of external development, the perspectives for the expansion of conventional agriculture will be reduced and there will be less foreign investments in rural areas".
- In 1991, LEADER I established "*...some supporting measures with agricultural and environmental objectives*".
- CEE Regulation N. 2092/91 marks the role that biological production may play in CAP in relation to environmental protection and management of rural areas.
- With PAC Reform in 1992, a series of regulations that clarified and supported these new alternatives of production were approved. EEC Regulation N. 2078/92 reinforced the environmental dimension of agriculture, as well as the idea of an institution with financial mechanisms to support MPB (Biological Production Model), and strengthened its development in the EU. In much the same way, EEC Regulations N. 2081/91 and 2082/92 (Protected Designations of Origin, Geographical Indications and Traditional Specialties) reinforced the trust of consumers in agricultural products that were addressed in this regulation.
- In 1996, the European Conference on Rural Development was held in Cork – Ireland. The following conclusions were drawn:
  - Alert public opinion to the importance of renewing rural development policy.
  - Transform rural areas into more attractive living and working spaces for people from all origins and ages.
  - Support the programme and cooperate as partners in the carrying out of objectives stated in the declaration.
  - Actively promote durable rural development on an international scale.
- With the Agenda 2000, Rural Development became the second pillar of CAP.
- In the following year of 2001, the Declaration of Copenhagen was approved as a guidance document and confirmed the role of Organic Agriculture in the resolution of problems regarding food production, environment, animal welfare and rural development.
- In early 2003, the European Commission elaborated and opened to discussion a document entitled "European Plan of Action regarding food and biological agriculture". The final version of this plan was proposed in 2004 and was submitted to discussion in the form of a statement to the Council and European Parliament. This plan aims to clarify and reinforce measures that support organic production and is known as the "European Plan of Action for Organic Agriculture and Animal Production".
- EC Regulation N. 692/2003 introduced some explanatory norms in the regulation regarding designations and protected geographic areas, namely "...important decisions are expected with the aim of promoting the system of designations of origin used in the EC as a model to be used in the rest of the World" or "...the possibility of producer organizations indicating that the product was handled in the area of geographic production, if the referred to handling ensured the preservation of quality, traceability and control", and "the registration of a homonymous designation must in the future have local and traditional aspects in mind, as well as the real risk of having confused consumers".

Conclusion

The dice are rolled. The potential of Mediterranean countries is well known regarding the number and distribution of PDOs, PGIs and ETGs (Guaranteed Traditional Speciality) already available on the market.

The Mediterranean diet uses traditional agricultural products. On the other hand, methods for traditional agricultural production are compatible with the preservation of the environment, which is a growing concern of european consumers.

Besides the intrinsic qualities of traditional foods, their link to their region of production and the curiosity in relation to production processes and culture which are at the bases of their origin may offer a new set of opportunities to farms and local communities, because they associate services to the production and selling of products. We have no doubts that this link will be the true contribution traditional products will give Rural Development.

In order to be more efficient, the relationship between countries on both sides of the Mediterranean must be reinforced and open.

We must control migratory flows both from the south to the north of the Mediterranean and from inland to the coast.

A true policy for Rural Development that is adequately applied to each country will create conditions for the settling of rural populations in agricultural regions, thus avoiding the continued and unsustainable concentration of people in big cities and a consequential loss in quality of life.

References

Barros, V., 2001. Produtos Tradicionais Portugueses. MADRP/DGDR
Campos, J. S., 1976. "A Balança Alimentar do Continente Português". INE
CIHEAM. "Rapport Annuel – 2004"
Comunidades Europeias, 2004. "Factos e Números Essenciais sobre a EU"
Correia, M., 1951. "A Alimentação do Povo Português". INE.
EUROSTAT, 2000/2001
INE. 1994. Balança Alimentar Portuguesa
INE. 2003. Destaque "Dia Mundial da Alimentação".
Keys, A. 1965. "Estudo das Sete Nações". Univ. Minesota – EUA
Stiglitz, J., 2002. "Globalização – A grande desilusão". Ed. Terramar.

The future of dairy products in the Mediterranean market: specialties or commodities?

D. Rama

Osservatorio sul mercato dei prodotti lattiero-caseari, Via Milano 24, 26100, Cremona, Italy

Summary

Dairy sector in the Mediterranean basin have some special features: it corresponds to an area of relative specialization, particularly for cheese and the sub-sector of small ruminants; at the same time consumption increase, sometimes supported by policymakers, led to the development of import flows of dairy commodities; in the last decade however production growth overcame consumption growth, partly moving the focus of the dairy policies from the demand to the supply side.

This has important implications for the relations between dairy specialties and commodities, which in this case must be defined not only in terms of purchasing and consumption behaviour, but also of production elements as tradition and localism.

The analysis of structural and commercial situation for countries of the North, East and South coast of the Mediterranean sea shows that for the first group specialties can play as an answer to lack of cost competitiveness on the dairy commodities market, in the last one market development should be based on commodities more than specialties while a complementarity can exist for Eastern Mediterranean countries.

Keywords: Mediterranean area, dairy specialties, traditional products.

Recent evolution of dairy market in the Mediterranean basin

Production of milk from all species in the Mediterranean area[1] marked a substantial progress during the last decade, increasing by over 9% from 1993 to 2003 (Table 1); this is the effect of different trends, where European Mediterranean countries show an almost stable production, that of Eastern Mediterranean countries rises at about the same rate than the average, and milk output from North African countries boomed by almost 6% per year. The single product with the strongest progression is buffalo milk, mostly produced in Egypt, Italy and Turkey; the weakest production trend is for sheep (-0.8% in ten years) and goat milk (+2.5% in the same period).

In fact, although about 90% of the milk in the region comes from cows, production of milk from small ruminants is a characteristic of the Mediterranean livestock sector: this covers about one third

[1]In our definition "Mediterranean area" includes twenty-four countries, twelve of which belonging to the North coast group, i.e. (from west to east): Portugal, Spain, France, Italy, Slovenia, Croatia, Bosnia-Herzegovina, Serbia-Montenegro, Macedonia, Greece, Malta; seven to the East coast group: Turkey, Syria, Lebanon, Israel, Palestinian territory, Jordan, Cyprus; five to the South coast: Egypt, Libya, Tunisia, Algeria, Morocco.

of the world sheep and goat milk production (Table 2), against a share of 13% for cow milk and slightly more than 7% in terms of population.

For almost all the considered species – cow, buffalo, sheep and goat, the only exception being camel – production increase slowed after 1998, so that the Mediterranean area share on world total has been reduced. In the eastern part of the region the trend of milk production from small ruminants has even been reversed due to a severe decrease in Turkey, by far the most important producing country of the sub-area.

Another specific feature of the dairy industry in the Mediterranean area is the good importance of cheese as a dairy product (Table 3). Indeed, although it can be said that the coagulation culture characterizes the North coast of the Mediterranean sea, while the fermentation culture is typical of the South coast, nonetheless in this part of the region the increase of cheese production has been

Table 1. Production of milk from all species in the Mediterranean area, from 1993 to 2003 (in 1 000 Mt).

	Year			Variation percent	
	1993	1998	2003	03/98	03/93
North	52 344	54 472	53 593	~1.6	+2.4
East	13 334	13 693	14 608	+6.7	+9.6
South	5 242	6 883	9 276	+34.8	+76.9
Total Mediterranean	70 921	75 047	77 477	+3.2	+9.2
% Medit. over world total	13.4%	13.4%	12.6%		

Source: Author's elaboration from Fao/Agrostat.

Table 2. Production of sheep and goat milk in the Mediterranean area, from 1993 to 2003 (in 1 000 Mt).

	Year			Variation percent	
	1993	1998	2003	03/98	03/93
North	3 924	4 110	4 052	–1.4	+3.3
East	2 112	1 954	1 965	+0.6	–6.9
South	581	593	625	+5.2	+7.4
Total Mediterranean	6 617	6 658	6 642	–0.2	+0.4
% Medit. over world total	35.6%	33.7%	32.6%		

Source: Author's elaboration from Fao/Agrostat.

Table 3. Production of cheese in the Mediterranean area, from 1993 to 2003 (in 1 000 Mt).

	Year			Variation percent	
	1993	1998	2003	03/98	03/93
North	2 942	3 240	3 540	+9.3	+20.3
East	465	593	588	–0.9	+26.3
South	142	206	446	+116.0	+214.9
Total Mediterranean	3 549	4 039	4 574	+13.2	+28.9
% Medit. over world total	25.2%	26.0%	26.2%		

Source: Author's elaboration from Fao/Agrostat.

particularly strong (more than tripled in a decade) and nowadays the ratio between cheese output and quantity of milk produced ranges from 0.04 and 0.07 in the three sub-areas.

Cheese production developed consistently in the last decade in the whole Mediterranean basin, so that the region's weight in world production have been gradually incremented.

In the past decades, growth of milk production, particularly in Maghreb and South-Eastern Mediterranean countries, has been slower than demand, carrying along a consistent increase in imports, mostly of milk powder to be reconstituted (Auriol, 1989). Nowadays the situation has changed, also because of the lower dynamics of demography: from 1998 to 2003 total population in the area increased "only" by 4.5%, lower than the world average (+5.3%). Therefore increase in domestic food utilization of dairy products, expressed in milk-equivalent quantity, has been lower than milk output increase in the decade 1993-2003, particularly during the second part of the period (+1.3% vs. +3.2% for milk production) (Table 4).

Difference in speed between production and domestic food use is even clearer when specifically observed for cheese: in the five years from 1998 and 2003, cheese production growth rate exceeded the one of domestic consumption by almost 1% (2.51% per year vs. 1.64%) (Table 5). While the Northern part of the area shows a persistent equilibrium between output and consumption, Eastern part of the Mediterranean basin – where a solid tradition exists for production both of white and yellow cheeses – results more and more as a net exporting area. As a consequence, in the final year the overall balance between production and domestic consumption is reversed: in 2003 the regional cheese output exceeds food use by 2.6%.

Table 4. Domestic food use of milk in the Mediterranean area, from 1993 to 2003 (in 1 000 Mt).

	Year			Variation percent	
	1993	1998	2003	03/98	03/93
North	44 097	47 134	47 674	+1.1	+8.1
East	11 311	11 943	10 919	–8.6	–3.5
South	6 703	7 281	8 633	+18.6	+28.8
Total Mediterranean	62 111	66 358	67 225	+1.3	+8.2
% Medit. over world total	15.3%	14.5%	13.7%		

Source: Author's elaboration from Fao/Agrostat.

Table 5. Domestic food use of cheese in the Mediterranean area, from 1993 to 2003 (in 1 000 Mt).

	Year			Variation percent	
	1993	1998	2003	03/98	03/93
North	2958	3263	3530	+8.2%	+19.3%
East	325	375	378	+0.7%	+16.4%
South	330	472	550	+16.6%	+66.5%
Total Mediterranean	3613	4110	4458	+8.5%	+23.4%
% Medit. over world total	25.9%	26.9%	26.0%		

Source: Author's elaboration from Fao/Agrostat.

The above outlined situation can be summarized in some stylized facts:

- the Mediterranean basin represents an area of relative specialization in milk production: although comprising no more than 7-8% of world population, in this area 12-13% of total milk is produced, which rises to 33-35% for small ruminants milk;
- the specialization is even stronger when focus is limited on cheese: from this region comes 25-26% of total cheese and 55-56% of sheep and goat cheese;
- despite this relative specialization, during past decades the increasing utilization of milk and dairy for human consumption led, especially in the South coast countries, to the development of import flows of dairy commodities;
- in the last decade production growth overcame consumption growth, especially in the South of Mediterranean basin, partly moving the focus of the dairy policies from the demand to the supply side.

Implications of the last point in terms of nature and role of dairy products in the regional markets, are worth a deeper investigation before going on with the analysis of market perspectives.

The concept of dairy specialty and commodity in the Mediterranean context

Generally speaking, a specialty product can be defined as a good which is considered somehow unique by the consumer, requiring from him a special effort and implication in its purchasing process (Stanton *et al.*, 1991), while commodities are standard products, generally low priced and having a high utilization frequency, requiring a very limited time for purchasing decision and action.

However, especially for food products where the distinctiveness of goods to consumer's eyes is more and more made up by being natural and genuine and the identification with local traditions (von Ansleben, 1997), this demand side concept of specialty should be implemented by considerations on the supply side: being geographically identified, part of a national or regional heritage and resulting from a locally integrated product chain are often among the distinctive aspects of a food specialty.

For traditional, highly differentiable products like cheese, wine or other foods in the Mediterranean area, demand and supply side components of the "specialty" concept naturally tend to integrate: specialties are distinctive products belonging to specific traditions of production and consumption, obtained by local raw material and manufactured in a traditional way, while commodities are standard, low cost products which can (and often are) originated by an import source.

Therefore in order to maintain and develop local food specialties, improve their valorisation and eventually reduce import dependence, actions must be taken on the whole *filière,* from the consumer to the farmer, using tools as traceability, standardization (which must be combined with the ability to valorise and organize the natural variability of heritage products) and actors coordination, aiming at the obtainment of a "common asset" (Vallerand, 1999), moving focus from "products quality" to "qualification process".

Perspectives, opportunities and threats for dairy specialties

All through the Mediterranean basin, dairy specialties coexist with commodities, either because they are different components of a common agricultural model, as it is in Northern coast countries, or due to the globalization process and a policy mostly orientated to consumer needs, which is often the case of Southern coast countries. An important question which arises is: what type of equilibrium (synergy or competition) exist among these two components of the dairy system? A connected question is: which perspectives exist for dairy specialties, how and to which extent should they be publicly stimulated or supported?

Because of the differences in products nature, production structure, consumption patterns and institutional arrangements, it is not possible to give single answers to these questions, which can usefully addressed separately for the main sub-areas of the Mediterranean basin, *i.e.* the Northern, Eastern and Southern coasts.

Northern Mediterranean countries

European Mediterranean countries, which include southern EU countries (Portugal, Spain, Italy and Greece), western Balkans (Former Yugoslavian republics and Albania) and a mixed country like France (partly Mediterranean and partly continental), represent by far the most important sub-area in terms of milk production as well as development of dairy specialties, mostly in form of cheese.

From 164 cheeses with a PDO (Protected denomination of origin) or a PGI (Protected geographic indication) into the European Union, 124 come from Mediterranean countries: 12 from Portugal, 19 from Spain, 42 from France, 31 from Italy and 20 from Greece. Their relative weight is highly variable, going from 2% of the total cheese volume in Portugal to 70% in Italy (representing about 40% of the total milk availability), passing through 5-6% of protected cheese in Spain and 16% in France.

Greece is a special case as the great majority of the cheese is produced from sheep or goat milk: only 37 thousand tons over the total 226 thousand are cheese from cow milk[2]. About 2/3 of the total is Feta (white cheese) while the remaining part is semi-hard yellow cheese (mostly Kasseri, Kefalotiri, Graviera, Kefalograviera). Theoretically almost all this cheese is covered by PDO, while actually only a small part of it enters into the controls system. The case of Feta cheese has been recently solved, as the Denomination originally recognized had been withdrawn from the PDO list in 1999 following a position of the European Court of Justice, but re-established in 2004.

Small ruminant milk utilization in cheese making is however common also to other south European countries: its share is around 25% both in Spain and Portugal, much lower in France (7-8%) and Italy (almost 10%), as both are large cow milk producing countries. The situation is different in the former Yugoslavia republics, where about 98% of 70 thousand tons of cheese produced is from cow milk: the most common varieties belong to the family of Kashkaval.

While it is a common thought that traditional products will "spontaneously" generate an added value representing the value of their exclusivity, in fact the real valorisation of PDO *filières* is highly variable. Portuguese protected cheeses represent an excellent test area, as after the recent introduction of the EU denomination system it is possible to find on the market both PDO cheeses and similar, undifferentiated products. In some cases the difference in price exceeds 30% (Rabaçal +94%, Serra da Estrela +55%, Queijo de Cabra Transmontano +36%, Castelo Branco +32%) while many of the remaining PDO cheeses show a price advantage over similar products ranging from 18% to 30%.

Even in Greece it can be observed that, although the average price of imported Edam or Gouda ranges from 4.45 to 4.75 euro per kg, Feta is priced at around 5 to 6 euro and Kefalograviera from

[2]Figures for 2003 supplied from CIRVAL, Centre international de ressources et de valorisation de l'information des filières laitières petits ruminants. Statistics concerning Greek cheese production are rather erratic: for the same year, cheese production is estimated from FAO at 244 thousand tons, of which 68 thousand from cow milk, while Eurostat "official" figure indicates total production at 155 thousand tons.

6.8 to 8.3 Euros. However, it is also true that pasteurized milk has a price margin over raw cow milk of 200%, while Feta price (with all the expenses due to cheese-making) corresponds, in terms of raw sheep milk price, to a margin of only 67%.

Recently the inter-professional federation of Roquefort (the well-known sheep milk cheese, representing the second French PDO cheese by volume), confronted to a decrease by 4.3% in 2003, decided to establish a supply management system; by now, part of the sheep milk in the Denomination area is not processed into Roquefort, being instead exported as frozen curd or generic sheep cheese to Italy and Spain. As a whole, PDO cheeses in France suffered from 1999 to 2003 a decline of 2.5% of final sales per year, while having on the average a 25% price premium over non-PDO prices.

In Italy Pecorino Romano suffered from the export refunds reduction of recent years. Grana Padano and Parmigiano Reggiano implemented, from 1978, a system of concerted supply management; it has been abolished in 1997 after a sentence of the Antitrust authority and since these two cheeses show serious declines in price. While Italy traditionally has one of the highest cow milk prices in the EU, Spain used to have much lower prices; after the gradual increase subsequent its adhesion to the EU and years of production growth by installing large, modern stables, nowadays Spanish dairy industry is complaining that milk cost is too high endangering its competitiveness. The valorisation of cheese *filières* appears as the only way to deal with the cost threat, but producers experienced that only small denominations have been price effective, while for large ones like Manchego (11% of the whole Spanish market), their impact on milk price keeps often low in absence of marketing innovation.

Eastern Mediterranean countries

Like at the North of the Mediterranean basin, also the countries situated at the East corner of this sea traditionally have an important cheese sector, while this is coupled with a remarkable production of fermented milk: in countries like Turkey, Syria and Jordan both yogurt (Laban, Labne - condensed yogurt - and Ayran - liquid yogurt -) and white cheese (Feta and Halloumi) are basic staples. In Turkey, the most important milk producer of the sub-area (almost 80%), 44% of milk is processed into cheese, 85% of which represented by Feta produced with cow milk. Cheese consumption, at 8.5 kg per person, is about at 50% of the EU average and can be considered high, given the level of per capita revenue; yogurt consumption, at 36 kg per capita, is the highest in the world.

While production of non-cow milk was 15% of the total in the early 90's, nowadays it is limited to 10%; it is mainly processed into Kashar (local version of Kashkaval, from sheep milk) and Tulum (goat cheese).

The reduction of traditional product lines and the development of western-type consumption is indeed the main trend in the Turkish dairy sector. This is partly due to the entry in the sector of European companies, namely Nestlé and Danone, which accompanied the privatization of the former state controlled dairy industry; also the rapid growth of modern retailing, with hyper- and supermarket, plays a role in change consumers habits. Main European retailers active in Turkey are Metro, Carrefour and Tesco. The highest sales trends are for mozzarella cheese, fruits flavoured yogurt and ice-cream.

Syria is experiencing an evolution similar to Turkey, although at a slower rate, while Jordan and, even more, Lebanon are traditionally much more open to imports from Europe. Israel represents a separate case: its cow milk output, around 1.2 million tons per year, is produced by 900 family farms, obtaining on average 500 thousand litres of milk per year, and by 200 cooperatives (kibbutz) which average size is 3.3 million litres per year. The dairy herd is almost totally constituted by "Israeli Holstein". The structure of the processing industry is also different from the neighbouring countries, as only a formal sector exists; however the product range is quite similar. About 50% of the total milk is processed into cheese, two thirds white cheese (Feta-type, but less thick and more spreadable than Greek Feta) and one third yellow cheese. Moreover, from sheep and goat milk (about 3% of the total), some two thousand tons per year of traditional cheese is produced.

Southern Mediterranean countries

Egypt is the most important milk producer country on the African coast of Mediterranean sea and the one where the self-sufficiency ratio is the highest: almost 90% of the 50 litres available per capita and per year come from domestic production. This availability is much lower than the recommendation of World Health Organization, suggesting a milk consumption of at least 90 litres per person. In the early 90's, 58% of milk was coming from buffalo and 38% from cow (sheep and goat milk together were covering the remaining 3%). Today the development both of local cow breeds and of Holstein carried the share of cow milk to 49%, equivalent to the buffalo milk one. Consumption is mostly as white cheese (about 7 kg per capita), both Feta-type (spreadable) and Domiaty, a cheese rather hard and salted, while liquid milk intake is very low, no more than 15 litres per year. In fact, due to the absence of a cold chain, 90% of liquid milk consumption is non treated milk, therefore possible only in production areas: coagulation of milk keeps here the original scope of food preservation. Less than 20% of cheese is a traditional semi-hard cheese called Roomi. However the faster consumption development is for processed cheese, nowadays 10% of total consumption, which is produced by local branches of foreign companies (Fromagéries Bel and Bongrain) or under licensing (Lactalis).

At the opposite corner of Northern Africa, Morocco has some common points with Egypt. Although dairy consumption is somehow traditional, it does not exceed 41 litres per capita; because of the weak consumption level, this is fulfilled by 90% with domestic production, also because of a 75% import tax. But here cheese consumption is very low, about 300 g per person, resulting from a level of 900 g in urban milieu and only 120 g in the countryside; it is mostly processed cheese and no cheese specialty exists. Milk is consumed either pasteurized or processed into yogurt by small artisan firms, or as sterilized or powdered milk for the lowest revenue families. During last years a production flow of processed cheese has been developed, mostly through foreign investment (Fromageries Bel), for export towards sub-saharian Africa and Middle East.

While Libya, with a very limited internal supply, is largely dependent from import of milk powder and its regeneration into liquid milk, curd, yogurt and cheese, both Tunisian and Algerian consumers are supplied by a mix, to different degrees, of domestic production and commodities import. Indeed Tunisia, by means of heavy public investments, tripled its milk production from 1987 to 2001, arriving at the level of 940 thousand tons from a starting point of 356 thousand; only 3% of milk is from sheep and goat. Meanwhile consumption arose from 760 to 990 thousand tons; domestic production is integrated by import of milk powder and milk fat (under specific authorization) and whey powder, casein and lactose (free). Liquid and fermented milk represent the large majority of consumption, while cheese and fresh products are developing slowly.

In the case of Algeria, the objective of increasing consumption level has been pursued through the reinforcement of the processing industry but without integration with local production. The goal of milk production intensification has largely failed: nowadays 7-8% of dairy herd belongs to intensive stables, supplying about 40% of milk. On the other side milk consumption is very high, equivalent to 115 litres per person, due to the low price imposed at the retail level: the price of a litre of pasteurized milk is 30 times lower than that of a kg of beef. Therefore the self-sufficiency ratio is lower than 40% but, due to the lack of mutual connection between production and processing, the milk produced mostly goes to the informal market and only 10% of the industry needs are supplied by local production. Symmetrically, only 10% of national production is collected, the remaining part being self-consumed, locally distributed as raw milk or processed at the farm into traditional products: Leben (a sort of buttermilk), Raeb (a coagulated curd obtained by spontaneous acidification of raw milk), Djben (traditional white cheese produced with sheep milk using cardoon flowers as coagulant factor) or farm butter.

Conclusions

Dairy specialties are, through the Mediterranean basin, mostly the matter of Northern coast countries and, partly, of the Eastern part of the area. For all European Mediterranean countries PDO cheeses represent a relevant part of milk utilization; in most cases – Italy being the main exception – they are obtained particularly by sheep and goat milk. Specific legislation helped in (partially) organizing the sector in Greece, Spain and Portugal following their adhesion to EU.

Clear opportunities emerge from the analysis of several cases: while cow milk production cost, higher than in continental Europe, represents a specific difficulty for dairy commodities development, differentiation based on local specificity and tradition can represent an important success factor. Generally speaking either specialties and standard products are independent components of the market, or a substitution relation emerges between them, although some cases have been reported where the existence of a Denomination covering part of the production in the specific area had positive effects even for the commercialization of non-denominated products (Fragata *et al.*, 1999).

However in cases where differentiation covers large quantities, is based on weak bases and/or is not supported by creative marketing efforts these specialties can be confronted with the same market difficulties of standard products, possibly made harder by differential costs and lower flexibility. A suggested tool for this kind of situation is the management of supply, which is however a typical tool of commodities market organization and can be sanctioned by Antitrust authorities.

For Southern coast countries dairy specialties are inexistent or have a limited impact only on the informal sector. In some cases a consumer oriented policy aimed at reaching recommended per capita consumption levels resulted in domestic production development, or in building a processing capacity based on raw material import; in other cases – Egypt and Morocco – consumption remains at a very low level. In order to increase consumption, in these cases, a commodity sector development must be planned, although ensuring an effective *filière* coordination.

Eastern Mediterranean countries are somehow in an intermediate position: dairy specialties are part of their heritage, although they are endangered by the vague of modernization and consumer behaviour globalization. An appropriate policy for their valorisation is needed, including precise specification of typical characteristics and partial normalization as well as building brand reputation and organizing the necessary partnership among actors. Having as a starting point consolidated consumption habits and in some cases an opening on export markets, the traditional specialty segment of these markets can contribute to increase sector sustainability and develop the competitive arena.

References

v. Ansleben, R., 1997. Consumer behaviour. In: Agro-food marketing, D. I. Padberg, C. Ritson & L. M. Albisu (Editors), Cab International, Oxon (UK), pp. 209-224.

Auriol, P., 1989. Situation laitière dans les pays du Maghreb et du sud-est de la Méditerranée. Options Méditerranéennes – Série Séminaires, 6: 51-72.

Fragata A., D. Alberto & I. Coelho, 1999. Social and economic impact of a PDO cheese ("Queijo de Nisa") on the local cheese production and processing (North of Alentejo, Portugal).

In: Livestock production in the European Less Favoured Areas, J.P. Laker & J.A. Milne (Editors), LSIRD network, Macaulay Land Use Research Institute, Aberdeen, pp. 175-178.

Laker, J.P. & J.A. Milne (Editors), 1999. Livestock systems in integrated rural development. LSIRD Network Macaulay Land Use Research Institute, Aberdeen, pp. 208.

Mijaèevic, Z, & S. Bulajic, 2004. Traditional manufacturing of hard cheese-kachkaval on Stara Planina mountain. Acta agriculturae slovenica, 84, 1: 11-15.

Oliveira, V., 2002. Produtos tradicionais com nomes protegidos: apresentação e análise de dados sobre produção, preços e comercialização. Ministério da Agricultura, do Desenvolvimento Rural e das Pescas, Direcção Geral de Desenvolvimento Rural, Lisboa, Portugal, pp. 71.

Rachid A., 1995. La filière lait en Algérie: entre l'objectif de la sécurité et la réalité de la dépendance. Options Méditerranéennes - Série B, 14: 229-238.

Soliman, M., 2001. Stratégie des acteurs et restructuration des marchés dans la filière lait en Egypte. Options Méditerranéennes - Série B, 32: 133-145.

Stanton, W.J., M.J. Etzer & B.J. Walker, 1991. Fundamentals of marketing. McGraw-Hill, New York, USA.

Vallerand, F., 1999. Heritage and innovation in Mediterranean animal products: some research questions. In: Livestock systems in integrated rural development, J.P. Laker & J.A. Milne (Editors), LSIRD Network Macaulay Land Use Research Institute, Aberdeen, pp. 69-75.

Voorbergen, M., 2004.The Turkish dairy sector. Gearing up for EU entry? Rabobank International, Amsterdam, pp. 20.

Pale meat from young ruminants in the Mediterranean. Diversity and common traits. Implications for trade

J.-P. Boutonnet

Institut National de la recherche Agronomique, Unité Mixte Innovation, département SAD, 2 place Viala, 34060 Montpellier Cedex 1, France

Summary

The main characteristics of ruminant meat in the Mediterranean area are its colour (pale), the age of animals (young), and the feeding (grain). This livestock systems contribute to maintain rural population in less favoured areas, but is not competitive in terms of price with commodity meats on the global market. Segmentation and promotion of Mediterranean specificity, is the strategy adopted in order to improve the competitiveness of Mediterranean meat producers. Several ways used by farmers and processors are discussed.

Keywords: beef, lamb, competitiveness, market segmentation, quality.

Introduction

Due to the bio-climatic characteristics of the Mediterranean area, livestock farming systems are strongly dependant on the environmental conditions. Livestock managers of these regions have had a particular know-how in taking advantage of the diverse landscape available to them: dry range land, fertile and irrigated arable land, forest, high altitude pastures, etc. In this framework, animal products, particularly ruminants meat, come from a range of very diverse systems: extensive systems based on permanent pasture, intensive feed lots, or a succession of extensive and extensive phases in the life of the same animal (Boyazoglu & Flamant, 1990).

As far as milk production is concerned, cattle have not had the same evolution than small ruminants. Local dairy cattle breeds have declined and cattle milk is produced by high potential international breeds such as Holstein, fed intensively. But most of the Mediterranean countries are not self sufficient and import significant quantities of dairy products. On the opposite, local dairy breeds of sheep and goat have been improved, keeping their traits of adaptation to the local conditions. Dairy sheep and goats are a characteristic of the Mediterranean area. Young animals from cattle, sheep, and goat dairy farms are slaughtered as young as possible in order to take milk from the mother.

Suckling livestock farming has kept traditional breeds, well adapted to their environment. They are kept in pastoral, harsh areas, where livestock farming is one of the few activities able to provide some income to rural people. The scarcity of good quality feed does not allow to meet the requirements of both adult females and fattening young animals.

As a result, traditional meat production in these areas comes from young animals, slaughtered as soon as possible. Consumers in these regions have therefore developed a strong preference for a pale colour of the meat, a good tenderness, not strong taste. This is obtained by meat industry from young, light animals. These animals are fattened in large feed lots near harbours or in fertile areas (source of grain), and near the big cities (proximity to the market. They have been bought as store animals, either in the Mediterranean countries, or in Northern part of Europe, in dairy or suckling herds (Boutonnet *et al.*, 2005a).

Recently, and after the crisis which have affected the meat markets, Northern European farming systems are re-orienting their production towards the production of lighter, younger animals, able to satisfy the requirements of Mediterranean consumers. The question is then: Is the Mediterranean meat industry (farmers, abattoirs, wholesalers, retailers, butchers) able to identify, characterise, and promote (at higher prices than the standard) specific meats from the Mediterranean area?

Quality as a mean of market segmentation

According to ISO, quality consists in "Properties and characteristics of a product, a service, or a process, that gives it an aptitude to satisfy needs explicitly or implicitly expressed by the user". Three consequences can be observed from this definition:

- There are several criteria of quality, dependant from the type of consumer, from the circumstances of consumption, from the season or the place, etc. Several types of the same commodity can be present on the same market. The market size can be increased by the availability of several qualities of product at the same time. On a same "commodity" market, there can be several "specialties" (Rama, 2005).
- There is not "good" and "bad" quality, but quality adapted to a particular market.
- The special product does not exist *per* se. For each type of special product, the concerned people have to negotiate the characteristics of the product, the process, the ways of guarantee for the customer. All this negotiation has to take into account the history, the natural conditions, the technical and organisational innovations in the area, etc.

Apart from very short channels, at a local level where consumers know personally small producers, in all other cases, the specific products have to be labelled or branded, in order to make the buyer sure that the product bought is really the expected one. This traceability is necessary to the buyer as a guarantee of the reality of the specificity of the product, and for the seller, as a mean of catching the rent, attached to the specialty product. This rent is the result of the building of a system of "monopoly competition" (Chamberlin, 1933).

In the case of Mediterranean meat, this rent allows higher prices than those of meat as a commodity on the global market. If "Mediterranean" meat is sold as different to the "International" meat, as a segment of the global market, then the concerned people have to agree on the characteristics of the product and of the production process. The *local commodity*, usually bought as "meat" by local consumers, has to become a *global specialty*.

Case studies: specific meat from Mediterranean France

Several strategies are used for building quality schemes, create the rent, and distribute it among the partners of the scheme.

One is the private, individual commercial brand. This brand's name can be sold or leased. The product can be made anywhere, and using any process (Nobody knows where and how Nestlé products are made nor packaged). The whole rent, if any, is caught by the owner of the brand's name: agro-food processing company or retailer. The farmers who sell the raw material find it very difficult to participate to the benefits of the market segmentation.

Our research (Boutonnet *et al.*, 2005b) has focussed on three collective particular schemes on meat products in Mediterranean France (Table 1). People concerned have chosen three different ways of building their system of identifying their product and catching the rent.

For "Taureau de Camargue" beef, they have chosen the European sign "Protected Denomination of Origin". PDO is managed by a group of producers and processors and controlled by a State

Table 1. Main traits of specialty meat products of Mediterranean France.

	Taureau de Camargue (Beef)	Agneau El Xai (Lamb)	Veau Rosée des Pyrénées (Veal)
Nature of the sign	PDO	Label Rouge	• Brand + Official certification • PGI applied
Production area	• Some districts around the delta of Rhône river	No	• Hills of East Pyrenees (France and Spain)
Starting year	1996	• Brand: 1992 • Label Rouge: 2004	• Brand 1991 (French part)
Starters	• Farmers and one meat processor	• Sheep farmers Co-operative	• Association of cattle farmers, members of a co-operative
Objectives	• Add value to a poor quality product (according current commercial standards)	• Protecting traditional production, specific of this area.	• Distinguish a specific product
Characteristics (different from standard)	• Dark, strong taste meat, tenderness versatile	• Marketed with head and fry (liver, heart, lung) • Age 60-90 days • Pale colour	• Age 5-8 months local breeds • Dark pink colour
Specific process	• Cattle bred and reared for race and games ("course camarguaise" and spanish "corrida") • Extensive farming in Camargue • Six months fed exclusively with humid natural pastures	• Suckling lamb, complemented with concentrate	• Suckling calve fed exclusively with mother's milk and natural pasture
Characteristics of farms	• Big size	• Small and medium, multiple job holdings	• Small and medium, multiple job holdings

body: INAO. Characteristics of the product and of the process are defined by law. PDO does not belong to anybody, and everyone who is in the delimited area, and using the requested processes, can use the name. The sign is used for a very special product, coming from a herd reared for races and not for meat. The meat is prized by local consumers. The use of PDO guarantees the higher price (+10 to 15%) returns until the processing company and the farmers. It seems retailers have few benefit from this product, so they are not enthusiastic at selling this product specially in the supermarkets.

For "Agneau El Xai" lamb, the farmers, grouped in a co-operative company, started several years ago by a commercial brand. But their very specific product, traditional in the region, was imitated by some meat-wholesale companies, buying lambs elsewhere. The threaten was a decreasing in the average quality of lambs on the regional market, and the loose of the particularity of this regional market. So, in association with two meat processors and wholesalers, and traditional butchers and some supermarkets, they applied for a "Label Rouge". This commercial brand belongs to the French ministry of agriculture. It can be used by collective groups only, for a high quality product, specified in a special document, and whose process and organoleptic characteristics are controlled regularly by an independent bureau. This quality product can be produced anywhere, once the official specifications are used. In this case, the higher price is distributed equally between all the partners (farmers, processors and retailers), but the controls are difficult and this constraint makes that some farmers and butchers went out of the scheme, in a period where the price of standard (commodity) lamb remains high.

For "Rosée des Pyrénées" veal, local farmers wanted to promote a very special product: a veal made only with its mother milk and grazed grass. This is not a traditional product, but very characteristic, nowadays, of the eastern part of Pyrenees. The simple commercial brand used at the beginning, was not sufficient for a good definition of the product. The members of the co-operative company decided to have a set of rules for this product, and to make them controlled by a third party. Then they applied for a Protected Geographical Indication (PGI) in order to be allowed to use the word "Pyrenees". This scheme is quite successful on the regional market, where they obtain high prices. But the retailers outside the region do not reach sufficient margins, and they do not want to sell this product regularly. As for "Taureau de Camargue" beef, it seems it difficult for the "local commodity" to become a "national specialty".

Conclusion

As it can be seen through these examples, the specificity of Mediterranean meats from ruminant (pale from young animals) can be promoted by different ways, according to the economic, cultural, and agronomic context of each region. However, the consequences of each way of promotion are very different as far as the repartition of the benefits is concerned. Also, as trade of animals, meat, processed products is more and more important among Mediterranean countries, it is unlikely that the different meat companies in the different countries have the same words for the same products. For the preservation or the promotion of typical meat products in the Mediterranean Basin, it is necessary to describe precisely what are the characteristics of each product, in order to make each trading company know exactly what they are buying. An official – or at least standard – certification of the main products and processes would help keeping a segment for these products on the Mediterranean meat market. It is a necessary condition to the preservation of livestock farming in this area.

References

Boutonnet, J.-P., Candau M., Alquier C., Bocquier F., Tchakerian E., Auréjac R., Bourbouze A., Molénat G., Fargeas E., (Editors.), 2005a. Productions animales en Méditerranée: viande jeune et claire. Actes du symposium Réseau Méditerranéen Elevage, Montpellier, 18-19/03/2004. Editions Agropolis International, pp. 102.

Boutonnet J.-P., Devautour H., Danflous J.-P., 2005b. Études de cas comparatives des démarches d'appellation en France. In: Produits du terroir méditerranéen: conditions d'émergence, d'efficacité et modes de gouvernance. Ilbert H., (Editor), CIHEAM-IAMM, pp. 298.

Boyazoglu, J., Flamant, J.-C., 1990. Mediterranean Systems of Animal Production. Dans The world of Pastoralism, Galaty, J.-G. and Johnson, D.-L. (Editors). The Guilford Press, New York, pp. 353-393.

Chamberlin, E.-H., 1933. The theory of Monopolistic competition, Cambridge: Haravard University Press.

Rama, D., 2005. The future of dairy products in the Mediterranean market: specialty or commodities? Communication to the International symposium on Comparative Advantages for Typical Animal Products from the Mediterranean Areas, 25-27 September 2005, Vale de Santarém, Portugal.

The role of livestock typical Mediterranean products in sustainable rural development: the case of small ruminant herdings in less favoured mountainous areas of Southeastern Spain

J. Calatrava & S. Sayadi

I.F.A.P.A., Dpt. Agricultural Ecomomics and Rural Studies, Apdo. 2027, 18080 Granada, Spain

Summary

The need to diversify rural economies by making use of local resources, particularly crops and livestock, which not only provide products but also serve territorial and environmental functions, calls for the consideration of the potential that traditional products from livestock, especially small ruminants, can have in the developmental processes in rural Mediterranean zones.

The present work examines the diversification of production in rural economies and the multi-functionality of agro-livestock systems, analysing the role of traditional livestock products in the local development of rural communities. The conditions surrounding of these types of products is studied in Spain, presenting two cases of handcrafted cheese in the Southeastern mountains of the country.

Keywords: typical livestock products, sustainable rural development, small ruminants, less favoured areas.

Introduction

Changes during the last two decades in the rural zones of Mediterranean countries of the European Union, still not thoroughly analysed, have involved drastic socio-economic and cultural changes. One of the most salient characteristics of this transformation of rural societies has been the loss of the relative importance of farming and livestock activities in terms of income and employment. This loss of the importance of agriculture at local level is, in some ways, desirable, as it was previously on a global scale, since it implies growth in non-agrarian economic activities and, in short, the modernization of the economic system.

As an example, in Spain today, agriculture occupies 5.5% of the active agrarian population of the country, representing hardly 3% of the GNP, while in 1960 these figures were 40.7% and 22.8%, respectively. At the local level, the trend is the same in rural areas, though later. The active agrarian population in towns of fewer than 10 000 inhabitants (which we can call rural) is today 16% of the active rural population (Calatrava, 2005), this percentage being 23.5% a decade ago (Calatrava and Sayadi, 1998) and 63.4% at the beginning of the 1970s. In the rural areas of the Betic Cordillera of Southeastern Spain, direct agricultural income scarcely reaches 15% of the total income of families in these mountainous regions, with most of the family income earned from tourism, construction, services, and social transfer of capital to these areas.

In the first stages of the current phenomenon of rural development, the decline in agricultural relative importance, was an aim to be pursued almost as a slogan, and it is reflected as such not only in the literature (scientific and non-scientific) on the subject, from the 1980s and beginning of the 1990s, but also in planning and political discourse. Agriculture, which until then identified with the rural setting, began to be considered as somewhat backward or at least not necessarily desirable. There was even a certain somewhat ironic voices, that identified the phenomenon as "rural-agrarian divorce" (Guigou and Hullo, 1996). This echoes a similar call two decades earlier in the period of rural development in the USA, proposing the basic objective of achieving a "non-farm rural America".

In reality, the true reason for this supposed anti-agrarianism, in the case of the European Union, arises, in our opinion, from the desire of rural communities to break their ancestral dependence on agriculture and take up industrial and service-oriented activities. A major obstacle to agriculture in the rural development processes was the major difficulty of transformation and conversion of agrarian systems with respect to the activities in more dynamic sectors, such as establishing the infrastructure for tourist lodging or recreational activities, for example, in which the investment, grants, or credits had a more visible short-term effect. Local managers of development policies, normally acting quite independently, without multi-disciplinary technical teams to provide proper support, and thus with little possibility of making detailed analyses to guide management, were unlikely to value agricultural potential in local development. Also, given that their management was often judged by its speed in executing investment plans that rapidly depleted the budgets allotted, they had clear limitations and lacked experience in promoting positive attitudes and changes with respect to agricultural activity in the rural population. At the second stage, now towards the second half of the 1990s, consideration began to be given to the importance of agriculture and livestock in the development of rural areas (Menéndez de Luarca, 1999, Bokeman, 1995, EAAP, 2005). Various key factors brought about this change:

- The dynamics of rural change, with the massive increase in non-agricultural activities, particularly services, in addition to the consequent alteration of values and socio-cultural traditions, have prompted rural societies to assume an enterprising capacity which could hardly disregard the local resources derived from agro-livestock activities.
- The growing appreciation of food quality on the part of consumers, which led to the revaluation of local products (traditional, natural, etc.), led to the consideration of agriculture as a source of income and added value.
- The growing consideration of the multi-functionality of agriculture, plus its inclusion in the planning and practice of rural development, caused agro-livestock activities to be considered according to their diverse functions rather than only their primary productive function.
- The goal of environmental sustainability has been set for developmental processes. The absence of agro-livestock activities would enormously determine the lack of sustainability in a rural area.
- Recent experiences in European Mediterranean zones where agricultural activity has reached the brink of extinction have shown how the absence of certain agricultural functions which are not directly productive (aesthetic, territorial, recreational, etc.) have brought negative consequences, sometimes serious, to the current model of development, based, often excessively, on tourism (Sayadi and Calatrava, 2002).

This context of reconsidering the role of the agro-livestock systems in rural development should include an analysis of the potential that local quality livestock products have in the sustainable development of rural settings. In the present work, after examining the need to diversify rural economic activities in relation to the multi-functionality of agriculture, we explore the role that traditional livestock products can play in such areas. We analyse the example of products derived from the raising of goats and sheep, presenting some case studies on the handcrafted production of cheese in mountainous zones of the Spanish Mediterranean and we draw from these a number of conclusions.

The present work was conducted within the framework of the Andalusian Regional Project "PIA.8.2" *Livestock systems in rural development of less favoured and mountainous areas* financed by the Regional Programmes of The Andalusian Institute if Research (IFAPA).

Diversification of the rural economy and agricultural multi-functionality

Development implies a diversification of the economic activity of rural zones. This is simultaneously a need and a strategy, justified by several reasons as follows (Calatrava, 2005):

- Economic diversification is the response to growing multiple demands for rural goods and services.
- This diversification tends to make maximum use of the local resources in an intent to meet these demands.
- It favours the creation and integration of income at the level of the family unit and of the local community.
- It helps lower the levels of underemployment and disguised unemployment of traditional rural societies.
- It tends to ameliorate the negative environmental signs of excessive spatial concentration of productive activities.
- In this way, it favours economic and environmental sustainability in the developmental process.
- It minimizes risks, uncertainty, and excessive dependence on outside influences.
- It makes use of the multiplicative and synergetic effect of multiple activities.
- It is, in short, the answer to adapt to rural areas the current post-fordist model of flexible production functioning in the world economic system.
- It is explicit in, and corresponds to, the institutional support strategies of current policies for rural European development.
- It tends to make maximum use of the aid from these policies.

The diversification strategy affects the occupation and use of space and thus local crops and livestock in two ways:

- Diversification of agro-livestock activities can improve food quality as well as the environment, thereby responding to new market demands.
- Diversification makes direct or indirect economic use of different agricultural functions. In this sense, one more reason for economic diversification is to favour the internalization of the greatest possible part of the multi-functionality of local agro-livestock systems.

In both contexts, diversification strengthens livestock activities as well as the manufacture and transformation of their products. Agrarian multi-functionality is specified in a number of agricultural functions, both productive as well as territorial and environmental. The main functions include:

1. Primary production of food from crops.
2. Primary production of food from livestock.
3. Primary non-food crop and livestock production.
4. Secondary production of food from crops.
5. Secondary production of food from livestock.
6. Secondary non-food crop and livestock production.
7. Repercussion on the quality and peculiarity of the local supply of tertiary activities (leisure related activities such as farm-tourism, accommodations, etc., recreation, other services).
8. Environmental impact: positive and negative externalities.
9. Balance of development and sustainability of the system.
10. Territorial occupation.
11. Aesthetic contribution to the landscape.

12. Water control in high-mountain farming systems.
13. Others (therapeutic, etc.).

Of all the productive functions, we will explore the points (2) and (6).

Livestock in Mediterranean rural development: towards local quality

Within the above scheme of rural economic diversification and of the multifunctionality of agro-ecosystems, and within a context of environmental sustainability, livestock activities have to contribute to rural development. This has special relevance in the case of European Mediterranean mountain zones, and particularly as far as the raising of small ruminants is concerned. The reasons explaining the importance and potential of including livestock production in this rural developmental processes are:

- In rural Mediterranean areas, tradition and experience in livestock production, particularly sheep and goats, persists, though drastically diminished and in some cases almost having vanished due to the rural exodus and the consequent agro-livestock crisis (i.e. abandoned land, scarcity of shepherds.
- Also, local knowledge of the handicraft techniques survives for transforming livestock products (particularly cheese production), although in many zones this has never been commercial but rather for family consumption and local trade.
- Both advantages - tradition and local knowledge - constitute local resources not sufficiently taken advantage of or valued, which could increase income as well as employment for livestock raising and could have an appreciable impact on the local economic system.
- The inclusion of livestock activities in the rural developmental processes is consistent with the aforementioned principles of economic diversification and incomes integration.
- Furthermore this would favour the sustainability of the local agro-ecosystem by bolstering the integration of agriculture and livestock (EAAP, 2005; Boyazuglu, 2002; FAO-CIHEAM, 2002).
- In general, this would favour environmental sustainability throughout the developmental process by fostering the sustainable occupation of large spaces, contributing to the fertility of the lands, and participating in the landscape.
- The demand for handcrafted livestock products, local and/or with some official denomination of quality, is growing both in local markets (boosting income and tourism in rural areas) as well as regional, national and international ones.
- Livestock production presents a strong potential to increase in added value, by transformation, which in most cases is feasible at the local level (Le Jaouen, 2004). In this sense, the possible raw materials for products (cheese, yoghurts, sausages, etc.) enables, if processed locally (by the producers themselves or not) participate in rural industrialization, the highly desirable objective throughout the developmental process.
- Also, there is clear synergy between livestock and tourist activities, not only with respect to demand for local farm products, but also with regard to the tourist use of what Rahmann (1997) defined as "indirect goods" produced by livestock, which tourists "consume" in diverse ways: viewing, touching (animal stroking by children: lambs, calves, chicks), using (horse riding) and good feeling (living landscape, nature protection, animal welfare).
- In relation to the above, it should be added that the consumption of gastronomical specialities based on handcrafted livestock products (e.g. goat and lamb stews, roasts as well as cheeses) have become a part of traditional hospitality of many Mediterranean mountain areas, attracting tourism (Flamant *et al.*, 1995; Sayadi and Calatrava, 2001).

- In addition, the maintenance and strengthening of livestock activities can make use of ancient fields and pasture lands, which otherwise would be marginalized.
- The rearing of small ruminants adapts well to the conditions of Mediterranean mountain and valley areas.
- On the other hand, the development of handcrafted livestock production, with higher than conventional prices could help in the recovery of local strains of animals which have peculiar characteristics and in some cases could not survive under low profitability, rescuing a source of livestock biodiversity. Good examples are breeds such as the sheep Payoya en the sierras of Cadiz province or the goat Blanca Serrana in the north-east of Malaga province.
- Finally, and no less important, is the policy of the European Union, both specific and direct support for the rearing of small ruminants, as well as more general assistance that might include aid to transform livestock products (Leader, Proder, etc.).
- Related to the above, consideration should be give to the recent interest on the part of the young, who are not necessarily members of a farmers family nor were previously linked to the land but who dedicate themselves to livestock activity. The reasons are the few alternatives of work for youth, and the fact that to rear livestock is relatively easy in the sense of not requiring land investment. With little money (normally provided by the family) to buy the first animals, a profession can be begun, and in a few years the herd can be easily doubled, supported by a substantial subsidy from the European Union.

In contrast to these positive aspects that explain the interest of craftsman livestock production, there are negative aspects that tend to limit its development, such as:

- Excessive climatic dependence, particularly for those extensive livestock systems, that use to be more linked to local livestock quality products.
- Higher costs and a scarcity of qualified labour, both in production, particularly of shepherds, and in transformation and industrialization (e.g. making of cheeses and yoghurts).
- A certain resistance among farmers to become associated not only for production but also for marketing and services supply.
- Linked to the above, difficulties in taking advantage of the existing scale economies in these activities, particularly in relation to manipulation and transformation of products.
- Low social consideration of labour linked to livestock sector. This constitutes a limiting factor for the development of livestock activities despite that a recent relative improvement in social consideration can be appreciated in rural societies regarding herding, livestock rearing, etc. Calatrava & Sayadi (1999) showed that among the possible occupations in rural society in the 1990s, the population of the Alpujarra (south-eastern mountains of Spain) negatively evaluated agrarian occupations, ranking livestock rearing last.
- Little knowledge or interest in management quality control systems, or the methods of evaluation of the products subject to this control. In addition, these systems represent additional excessive costs, and more so if the farmer afterwards does not easily recover the expenses in the market.
- Limited knowledge by the livestock raisers regarding different market demands that could affect both the level of types and range of products, new products, etc., such as at the level of the use of non-commodity concerns.
- Need by many livestock growers to increase their technical knowledge both in handling and management, sanitation, etc., of the flock, as well as the manipulation and transformation of products.
- In relation to the above, a need for professional training in both aspects.
- Deficient infrastructures in rural settings that limit both the access to farms for collecting livestock raw materials as well as for transportation and distribution of the finished products.

- It would be mentioned, finally, aspects related to institutional constraints that limit the more adequate size of herd, and the profitability of scale effect already mentioned.

The handcrafted product has much more likelihood of enhancing the added value to the producer in two ways: the price is usually higher than for industrial products; and the possibility of demand provides a more direct commercial channel for trade with fewer intermediaries.

For the consumer, the handcrafted product is linked to the perception of a product made with techniques derived from personalized or traditional know-how an original product, with local taste and distinction, as opposed to the mass-produced industrial product. The consumer deems something handcrafted which is made from raw materials, transformed under craftsman conditions, in the same place as it was produced. For the consumer, the demand for handcrafted conditions ensures organoleptic characteristics and food health that approximate those of tradition.

In the urban environment the consumer acquiring the product demands that the origin and production methods be certified with controls and labels, and from this information the consumer assigns the product a quality level and thus a value (always subjective), with the subsequent willingness to pay more. When the consumer, on the other hand, acquires a product in its place of origin, and even more so at the production site, the demand for controls and labels diminished, as the very site and production techniques can be directly viewed and evaluated, with the corresponding increasing in willingness to pay.

In an attempt to rank the livestock growers with respect to the possible level of revaluation of handcrafted products, the following elements should be considered:

1. Processing site. A product can be manipulated and/or processed at the production site, within the productive and accounting structure of this site, also the livestock raiser may not process the product alone but through some type of farmer cooperative, which may act as a separate and professional processing enterprise. Also, the product may be processed by some local company not necessarily comprised of livestock growers, which buys the raw materials from the growers and/or transforms them at the site of origin. Finally, the product may be processed far from the livestock-raising area. Even in this latter case, the product could maintain certain handcrafted characteristics in the production methods.
2. Site and manner of selling to the consumer. The sale of the product can be direct or indirect. Direct sales are considered those in which the producer or group of producers sell directly to a consumer or a group of consumers or association of consumers, to institutional consumers, or to catering services. Some authors (Wirthgen & Kuhnert, 1997; Bokeman, 1997) apply the concept of direct sales also to the case in which the product is sold to retailers. In this context, direct sales are also applied to products sold to retailers in the local market or to large outlets. The first definition implies no intermediaries between the producers and consumers, while the second signifies no intermediaries between the producer and the retailer. The direct sales of livestock products is in general not recent, though now it is sometimes acquiring modern commercial logistics. The preparation of animal products (sausages, yoghurt, cheeses, etc.) and the sale of the surplus from home consumption to neighbours or even nearby towns has been traditional in rural Mediterranean regions to complement the income from livestock raising. In some countries, direct selling by the producer is quite developed. In Germany, for example, 4.5% of the producers engage in direct selling, for 7% of the total agrarian sales at prices set by the producer (Wirthgen & Kuhnert,1997). Direct sales are widespread among ecological producers. Thus, according to the same authors, 80% of ecological growers participate in direct sales. In Spain, few, although steadily more livestock raisers, undertake individual direct sales, though most sales are through groups or cooperatives of producers. In northern Spain, direct sales is more common than in the south, where it is beginning, particularly for some types of handcrafted cheese.

3. Degree of product diversity. Another important aspect that characterizes livestock raising is the diversity of the products and services of marketing to be offered. Kuhnert & Wirthgen, (1997) reported that in Germany 80% of the livestock growers having direct-sales systems offered more than one product, or at least sold the raw material and the product transformed at the farm (e.g. milk, and homemade cheese). The mono-product is more usual at sites that sell through conventional networks of intermediaries. Within direct sales, normally the larger the distance between the production site and the consumer or retailer, the more frequent is the provision of a mono-product. When the market is local, diversification is more common, because, except in the case of areas with special tourist attractions, the local market is very limited and the diversification offers broader possibilities for income from the livestock.
4. Systems of certified quality. One important aspect to consider in livestock raising is whether the products have a system of quality certification. This does not necessarily imply a high degree of craftsmanship in the products, though there is a certain relationship between the two. In the following section, quality-certification systems are discussed for Spanish livestock products. It should be borne in mind that to obtain a quality certification implies to accomplish some requirements and assume additional costs for the producer, a fact often not sufficiently valued in the market. We shall return to this subject below.

Goat and sheep production in Spain

The official statistics from December 2003 (MAPA, 2003) list some 23.48 million head of sheep in Spain (19.5 million begin adult females) and 3.16 million goats (2.6 being females more than 2 years old). The sheep are spread throughout Spain, with half of the national total concentrated in Castilla y León, Extremadura, and Castilla-La Mancha, an (also with high numbers in Andalusia (3.26 million) and Aragon (2.86 million). Almost 43% of the country's goats are found in Andalusia (1.35 million), followed far behind by Castilla-La Mancha (0.46 million) and the Canary Islands (0.33 million). Recent statistics (December, 2004) from the same source indicate a dip of 6% in sheep and almost 8% in goats. The reason for the recent decline in small ruminants in Spain is owed to lower consumption of products (except in cheeses, which is often imported, particularly goat cheese), to the rise in production costs, to administrative obligations (obligatory electronic identification of individual livestock since July 2005), and to lower sale prices to the grower in contrast to the high margin of distributor chains. Only a plan that foments the consumption of sheep and goat meat and milk products, together with the increase in high-quality handcrafted production marketed in some form of direct sales, can stop this decline in the rearing of small ruminants in Spain.

In the context of the European Union, which has 10% of the world's sheep and almost 15% of all goats, the United Kingdom and Spain are the leading European sheep producers (35.8 million and 23.5 million, respectively), followed at a great distance by France (9.3), Greece (9.2), and Italy (8.1). In goats, Greece with 5 million and Spain with 3.1 lead production, followed by France (1.2), and Italy (1 million).

In terms of meat, Spain produces 240 million kg of mutton, of which 40 million is suckling lamb, 180 million spring lamb, and 20 million from older sheep. The production of suckling lamb has had a strong upward trend, doubling in the last two decades (18.7 million kg in 1985). This increase, while other types of mutton decreased in consumption, is notable because the suckling lamb lends itself most to local quality production. Also, seasonality is great, slaughter being concentrated in December to March and April.

The situation with goat is similar, with a total production of 15 million kg, although in this case the increase in production of suckling kids (less than a month and a half old) has been somewhat lower, going from 5.1 million kg in 1985 to slightly less than 7 million kg at present.

In terms of milk, Spain produces some 400 million litres of sheep milk, which is all used for industrial transformation, human consumption of this milk being almost nil despite that almost 6 million litres were consumed in the mid-1980s. Some 515 million litres of goat milk are produced, of which 15 million are used for direct human consumption while 500 are used for industrial transformation.

With respect to cheeses, Spain (in 2003) produced 25 million kg of pure sheep pure cheese, 8.8 million kg of pure goat cheese, and 110 million kg of cheese from mixed milk. Some 20% of the pure sheep cheese is produced by the livestock growers themselves at their farm or in a group. Goat cheese, given that the cheese made by the farmers themselves is pure, some 62% of pure goat cheese is made by the farmers themselves. The cheeses made from mixed milk (these being the most numerous cheeses) are generally made in industries independent of the farming sector that sell the milk. A comparison of these figures with those recorded a quarter of a century ago (Calatrava, 1982) shows that in 1979 the cheese production by the farmers themselves represented 18.3% of the sheep pure cheese, and thus the situation has not appreciably changed. However, in terms of goat cheese, 33.5% was produced by the farmers and 26.9% by local industry unrelated to the producer. Thus, 60.4% was produced at the local level, by the farmers themselves or by others. Comparing this figure with the 62.0% produced today by farmers, and although exact current statistics are lacking for local production (which can be estimated at more than 80% of all pure goat cheese), it can be concluded that there has been an increase in local goat-cheese production, and within this increase the producers themselves have taken the greater part of the initiative. Pure goat cheese is seen in Spain mainly as local production linked to the rural economy. This concurs with the poll of Spanish consumers who identified goat cheese has "rural" and "home-made", while sheep cheese was identified more with "quality" and "strong flavour" (MAPA, 2002).

Quality livestock products in Spain

The European systems of quality control certification of Protected Denomination of Origin (PDO), Protected Geographical Indication (PGI), Guaranteed Traditional Specialty (GTS), and ecological agricultural products (AE) are used in Spain. Of the 169 products under PDO and PGI, there are 14 fresh meats, 3 sausages, 4 hams, 1 butter, 2 honeys, and 19 cheeses. Livestock products represent 25% of all products protected by PDO and PGI. The meats in general are protected by PGI and the cheeses by PDO. The denomination GTS has been used little in Spain to date, with only 3 types of products using this system.

In terms of ecological livestock, Spain (as of 2003) has 1751 producers, predominantly of cattle (766) and sheep (438) for meat, followed by goats for meat (120) and hogs (71). Ecological producers of milk are scarce, with only 398 for cows, 32 for goats, and 18 for sheep in all of Spain. There are 89 honey producers and 113 poultry growers certified as AE, and of these latter some 2/3 produce ecological eggs.

Ecological livestock is much less developed than ecological agriculture, which, with nearly 800 000 ha represents almost 2% of the SAU, perhaps because of the ease of converting certain crops to ecological methods, compared to the greater difficulty and risk of switching conventional herds to ecological management.

With regard specifically to small ruminants, only 3 of the existing denomination of origins for fresh meat apply to sheep, and there is none for goats. The protected sheep meat are not in depressed zones, except for lamb in La Mancha, nor in typical Mediterranean areas. In Andalusia, no meat of small ruminants is protected by PDO or PGI.

For cheeses, there are 5 PDO for goat cheese and 7 for sheep. The former are located in the Canary Islands (2), Murcia (2) and Extremadura (1), while the latter are found in Cantabria (3), Zamora (1), Extremadura (2), and La Mancha (1). There are no PDO or PGI for cheeses in Andalusia, nor in the Betic Cordillera, despite the great number of goats and sheep there (Sayadi & Calatrava, 2005) and in spite of the high-quality goat cheeses such as Ronda, Alba, Filabres, Grazalema, Aracena as well as sheep cheeses such as Calahorra and Grazalema (from Payoya sheep).

Ecological sheep meat is produced in 438 sites, with 151 in Extremadura, 126 in Andalusia, the 65 in the Balearic Islands, and 44 in Catalonia, the rest being scattered throughout other regions. Regarding ecological sheep milk, of the total number of producers (18), half are in the Basque Country (9). In Andalusia, there is no ecological herd used for sheep milk. In goats, the 120 ecological growers of meat are concentrated mainly in Andalusia (72), Catalonia (27) and Extremadura (8). In goat milk, the few growers (32) in Spain are distributed throughout Castilla-La Mancha (6), Andalusia (5), the Canary Islands (5), Catalonia (4) and other regions.

Apart from these official certifications of origin and quality, there are many local instances of non-certified craftsman production, some of which has a certain commercial prestige and good demand in the local market and at times in markets beyond the local level. This is mainly the case of more Mediterranean local livestock products.

Case studies

Local production of handcrafted cheese in the Alpujarra (mountainous regions south of the city of Granada)

This case, which has been treated in detail by Calatrava & Sayadi, (2003), concerns the transformation of goat milk into handcrafted cheese by a small local company unaffiliated with milk producers and comprised of 4 people who were locally born. The company, founded in 1991 was called Central Lechera Alpujarreña (CLA).

In the mid of the1980s local goat milk was sold by breeders to a large dairy located far from the district of the Alpujarra, which used the district's name as a sign of local distinction in the production of a cheese that combined goat and cow milk, the latter never existing in the area, except for business purposes.

The CLA began in 1991 to buy the milk of almost 5 000 goats in the municipality of Valor and its surroundings (see Calatrava & Sayadi, 2003, for details) to manufacture handcrafted cheese according to local experience and knowledge. A pure goat cheese was produced with different degrees of ageing, with an aged crust and a crude, hand-made aspect following tradition. The cheese thus carried on the traditional types produced on a domestic scale in the area since ancient times.

At first, the market was local but spread afterwards throughout the provinces of Granada and Almeria (the provinces to either side of the locality), with a very positive market response. Currently, part of the 100 000 kg of annual cheese produced can be found also in the large national supermarkets (particularly in Madrid) and small portions have been exported to Japan, Canada, and Germany, though the Consorcio de Queso Tradicional de España (Spanish Traditional Cheese Consortium). All this occurred without losing the local traditional character of the cheese.

The impact in the area has been notable. The local livestock raisers sell their milk at a higher price and its collection is rapid and immediate. The existence of the CLA has been decisive in maintaining the goat production in the zone after a sharp decline up to the 1990s. The socio-economic impact of the CLA can be appreciated through the income that it brings to the village of Valor, amounting to roughly 8% of the total income of local families (Calatrava & Sayadi, 2003). This income is derived basically from tourism and related services and from social capital transfers. If the latter are not taken into account, but only the income generated by the local productive system, the economic impact of the CLA in terms of income has almost doubled, approaching 15%.

Production projects of milk and handcrafted cheese in the north-eastern area of Malaga province

The study area is composed of 7 municipalities of the north-eastern area of Malaga province, the main town being Archidona. The population centres are under the jurisdiction of the Proder[1] Nororma (north-eastern Malaga). Among other actions, the Proder is supporting 2 projects of handcrafted cheese manufacture of different characteristics and dimensions, which are represented in Table 1.

These initiatives differ sharply both in their conception as well as in the nature of their impact on the zone.

Project I was established by two young people, children of farmers, without practical experience in cheese making, but with university studies in veterinarian medicine and with recent courses in cheese making. They did not previously know the market but had considered the types of cheeses after consulting experts and studying demand. They had one herd, very large for the average in the area, and they had sufficient land. With this, they supplied the raw material at a reduced cost. The cheese factory has a large dimension within the context of artisan activities. They maintain a careful balance of agriculture and livestock on their farm and employ local workers, but their impact on the maintenance of goats in the zone is limited as they supply themselves with milk.

Project II, on the contrary, though also a local project, was not impelled by a livestock herder but by one person with prior knowledge of handcrafted cheese making and of the market. This was a very small factory that works in a very handcrafted manner. No salaried workers are employed (outside the family) but the acquisition of goat milk from local goatherds helps maintain goat raising in the area, despite the small size of the operation.

Conclusions

- In Mediterranean areas, local traditional products derived from livestock constitute a resource which, though not always put to good use and re-evaluated, have a clear potential for development.
- Furthermore, the traditional livestock products present externalities that can have very positive synergetic effects, both for recreation and leisure as well as for environmental sustainability and developmental equilibrium.
- In the case of small ruminants, many aspects of different nature confirm this potential, not only related to the primary productive potential but above all to the secondary productive one.

[1]The Proder-2 programme (2000-2006) (Projects of Economic Development and Diversification in Rural Areas) has very similar perspectives to those of the Leader programme and responds to a national Spanish policy of rural development with a territorial and integrated focus. These projects are supported by the Spanish Ministry of Agriculture and Fishing (MAPA) in areas where generally the Leader programme does not apply. The financing is mainly European (63%: FEOGA-Orientation and FEDER in objective 1 regions and FEOGA-Garantia in the other regions) and the rest Spanish (central government 13%, and regional and local governments 24% (MAPA, 2003).

Table 1. Summary of 2 projects on the handcrafted cheese manufacture of different characteristics and dimensions.

	Cheese Factory I	Cheese Factory II
Location	- Archidona	- Cuevas de San Marcos
Previous experience in farming	- Promoter: None - Father: Always been a farmer: 540 ha of mountain	- None
Previous experience in livestock rearing	- Promoter: 6 years rearing goats	- None
Motivation	- Looking for better returns from the goat farms through cheese making - Trust in market potential	- Experience of 15 years making handcrafted goat cheese at the family level - Trust in market potential
Legal (fiscal) nature	- Anonymous Society (S.A.)	- Limited Responsibility Society (S.L.).
Date project began	2001	2001
Total amount of investment declared (€)	525.27	103.35
Percentage of PRODER subsidy (% total capital)	21.28	34.89
Existence of own livestock	Yes	No
Size of herd	- 900 head (600 Malagueña breed and 300 Blanca Serrana)	
Previous experience in cheese making.	- No - Prior courses in cheese making	- Yes, 15 years making handcrafted cheese at the family level. - Prior courses in cheese making
Milk supply	- Own production	- Local goat herders
Products: nature, composition and type.	- Handcrafted cheese, 100% goat - Fresh: 20%; soft: 40% and aged: 40%	- Handcrafted cheese, 100% goat - Fresh 100%
Total maximum capacity of cheese farm	- 2000 Kg/day maximum	- 100 Kg/day maximum
Markets size	- Regional: Small shops - Local: individual and local small shops	- Local: individual and local small shops
Sales channels	- Direct sales	- Direct sales based on previous market knowledge
Labour	- Promoter (self employed) - 3 full-time employees, to work in the cheese farm and with the goats	- Promoter (self employed)

- Also some constraints to livestock production have been identified, namely physical, socio-economical, institutional and linked to infrastructure level in rural areas.
- There is a broad casuistic of initiatives in revaluing livestock products at the local level, having identified such differentiating elements as: the processing site, the site and manner of selling to consumers, the degree of product diversity, and the fact of having or not having certified quality.
- This diversity is reflected in the study of the three cases presented concerning local artisan cheese in the mountain zones of Mediterranean Spain: commercial production linked to the livestock producer, commercial production not linked to the livestock producer, and small-scale family production not linked to the producer. In the second case, local, national, and even incipient international trade is involved. In each case presented, added value was found, as well as the creation of employment, and the influence on the maintenance of local livestock systems.
- In view of the above, in the planning of developmental processes at a local level in Mediterranean areas, consideration should be given to the implementation of projects and activities based on traditional livestock production in a triple line of craftsman production, high-level quality, with possibility of certification, and participation of the livestock growers themselves in the transformation and commercialization of the products, making maximum use both of potential demand of the local markets as well as the possibility of trade beyond the local level.

References

Bokerman, R. 1997. Direct marketing of livestock products in Germany. Part II. Contribution of on-farm-processing and direct marketing to safeguarding agricultural incomes. In: Laker, J.P. & J.A. Milne (Editors). Livestock systems in European rural development. Proceedings of the 1st Conference of the LSIRD network. Nafplio, Greece, 37-41.

Boyazoglu, J., 2002. Livestock research and environment sustainability with special reference to the Mediterranean Basin. Samll Ruminant Research 45: 193-200. Proceedings of the 1st Conference of the LSIRD network. Nafplio, Greece. 162: 43-54.

Calatrava, J. 1982. La producción de queso de cabra y oveja como factor de desarrollo en zonas de agricultura marginal". ITEA. 1: 64-109.

Calatrava, J. 2005. Non-food crops and rural development in Mediterranean Europe: some considerations about their potentiality under the scope of the CAP. International Conference on Industrial Crops and Rural Development. 17-21 September. Murcia (Spain).

Calatrava, J. & S. Sayadi, 1998. De l'exploitation agricole à l'exploitation rurale: Nouveaux regards sur l'agriculture des zones de montagne du Sud-Est español. Revue Etudes Recherches sur les Systèmes Agraires et le Développement; Gestion des exploitations et des ressources rurales: entreprendre, négosier, évaluer, 31: 387-396. INRA, France.

Calatrava, J. & S. Sayadi, 1999. Agrarian Crisis, Farming Abandonment and Social Regard for Agriculture in Depressed Mountainous Areas of Southeastern Spain. IX European Congress of the Agricultural Economists "European Agriculture Facing the 21st Century in a Global Context". 24 -28 August. Warsaw, Poland.

Calatrava, J. & S. Sayadi, 2003. Milk production systems in rural development: the case of goat cheese making at the eastern Alpujarras. In: Proceedings of the joint EAAP-CIHEAM-FAO symposium, Prospect for a sustainable dairy sector in the Mediterranean, Hammamet, Tunisia, October, Publication 99. Wageningen Academic Publishers, 34-43.

EAAP, 2005. Animal production and natural resources utilisation in the Mediterranean mountain areas, EAAP Scientific Series 115, pp. 600.

FAO-CIHEAM, 2003. Sustainable grazing, nutritional utilization and quality of sheep and Goat Products. Proceedings First joint Seminar of the FAO-CIHEAM Sheep and Goats Nutrition and Mountain and Mediterranean Pastures Sub-Networks. 2-4 October.

Flammant J.-C, J, Boyazoglu & A. Nardone, 1995. Animal Production and Rural tourism in Mediterranean Regions. EEAP. Wageningnen Academic Publisher, 74, pp. 302.

Guigou J & P. Hullo, 1996. L'agriculture et le rural: un divorce en perspective". Colloque "Nouvelles Fonctions de l'agriculture et de l'espace rural. Enjeux et défis identifiés par la recherche". INRA. Second Volume. Exposés Pléniers, interventions des rapporteurs et tables ronde finale. 17-18 December. Allaire G., B. Hubert & A, Langlet (Editors): 113-118.

Le Jaouen, J.C, 2004. Réalités et avenir des produits fermiers. La chevre. 265: 8-10 November/December.

Ministerio de Agricultura, Pesca y Alimentación (MAPA), 2003. Libro Blanco de la Agricultura y el Desarrollo Rural. Madrid, Spain.

Ministerio de Agricultura, Pesca y Alimentación (MAPA), 2002. Estudio sobre hábitos, actitudes e imagen del queso en el mercado español. November, pp. 40.

Menéndez De Luarca, S., 1999. Integrated economic development of rural communities engaged in livestock production. Newsleter of the LSIRD Network issues. November. FAUNUS, Issue 6.

Rahman, G., 1997. Contribution of rural tourism to the market for livestock products in LFAs in Germany. In: Laker, J.P. & J. A. Milne (Editors). Livestock systems in European rural development. Proceedings of the 1st Conference of the LSIRD network. Nafplio, Greece. 162: 55–60.

Sayadi, S. & J. Calatrava, 2001. Análisis de la potencialidad agroturística para el desarrollo rural: el caso de la montaña Penibética. IV Congreso de la Asociación Española de Economía Agraria (A.E.E.A) "Economía Agraria y Recursos Naturales: Nuevos Enfoques y Perspectivas". 19-21 September, Pamplona, Spain.

Sayadi, S. & J. Calatrava, 2002. Análisis Funcional de los Sistemas Agrarios para el Desarrollo Rural Sostenible : las Funciones Productiva, Recreativa y Estetica de la Agricultura en La Alta Alpujarra. Ministerio de Agricultura, Pesca y Alimentación (Madrid). Secretaría General Técnica. Serie Estudios. 148, pp. 328.

Sayadi, S. & C. Calatrava, 2005. Small ruminants projects in LEADER-PRODER European Rural Development Programs in the South of Spain: A survey analysis. International Symposium on Comparative Advantages for Typical Animal Products from Mediteranean Area. 25-27 September, Santarém, Portugal.

Wirthgen, B. & H. Kuhnert, 1997. Direct marketing of livestock products in Genrmany. Part I. The direct marketing sector in Germany. In: Laker, J.P. & J.A. Milne (Editors.). Livestock systems in European rural development. Proceedings of the 1st Conference of the LSIRD network. Nafplio, Greece, 35-36.

Session 2. Utilisation of natural resources on the animal production systems

Main papers

Annual legume species for the improvement of pastures

N.M. Simões[1], M.S. Pereira[2], A. Quintas[2], A.M. Barradas[1], C. Vicente[1], C.C Belo[2] & M.M. Tavares de Sousa[1]

[1]Estação Nacional de Melhoramento de Plantas, Apartado 6, 7351-951 Elvas, Portugal
[2]Estação Zootécnica Nacional, Fonte Boa, 2005-048 Vale de Santarém, Portugal

Summary

In Portugal, between the general agricultural census (GAC) of 1989 and 1999, the area of permanent pasture increased by 575 208 hectares, 72% of this area in Alentejo. These pastures, composed by spontaneous species with low feeding value, allowed grazing by about 0.40 of normal cattle heads (NCH) per hectare. Due to traditional agriculture practices with successive soil mobilisation and herbicide application, the majority of the interesting autochthonous grass species disappeared from pastures. The most efficient solution to recuperate will be the introduction of greater potential species namely annual legumes.

Keywords: agricultural census, permanent pastures, annual legumes.

Introduction

Portugal comprises extensive areas where agriculture is not profitable. Areas with low depth soils, and sometimes steep slopes, have a better utilization throughout productive activities that include conservation and improvement of natural conditions.

On these areas, mainly the inland regions and southern Portugal, animal production systems have been developed, based on existing plant resources and animal landraces, originating food products which attained reputation for their unique characteristics.

In the last years, due to the abandonment of traditional crops such as cereal production, permanent pasture areas increased over the entire country, especially in Alentejo and areas under holm and cork oak groves with low grazing capacity. The recuperation of this important ecosystem, the oak, may be achieved through extensive animal production due to the increase in soil fertility.

Therefore, it is urgent to increase the feeding potential of permanent pastures in order to profit from the country's capabilities, reflected on the production of animal products recognized by their characteristics.

The importance of permanent pastures on forage area

In Continental Portugal, the area of permanent pastures has significantly increased in recent years to 1 331 000 ha, as a result of a decrease in agricultural activities, such as cereal production, which were relevant just a few years ago.

The decrease in what is known as "ploughable clear land" (PCL), between the 1989 and 1999 general agricultural census (GAC) was 393 879 ha, meaning that the area of PCL in the 1999 GAC represented 79% of that indicated in the 1989 GAC. Curiously enough, in the 1999 GCA, the area of PCL represented 40% of the land used for agricultural purposes (LUAP) for all the regions in continental Portugal.

Between census, the area of land used for permanent pastures was increased by 576 208 ha. However, Alentejo absorbed 72% of that increase, which corresponded to 416 703 ha, amounting to 818 302 ha. In the 1999 GAC, permanent pastures in Alentejo represented 42% of LUAP, while the mean for other regions was 25% and the general mean for continental Portugal was 36%.

In Alentejo, permanent pastures are mainly found under holm and cork oak groves. In fact, 63% of the 818 302 ha of permanent pastures in this area were observed on these groves.

As previously indicated, permanent pasture areas greatly increased in the Alentejo region, especially in holm and cork oak groves. These trees are essential in the "montado" ecosystem and their sustainability should be ensured.

In order to study the evolution of holm and cork oak groves utilization in the Alentejo, data from 239 civil parishes was divided in 4 groups according to the percentage of area under holm and cork oak groves (AUG) of LUAP: AUG1 (0%<AUG<25%), AUG2 (26%<<AUG<50%), AUG3 (51%<AUG<75%) and AUG4 (AUG>76%).

Concerning the evolution of soil utilization, there was a clear increase of permanent pastures in PCL, especially for the area classes with greater "montado" area (AUG2, AUG3 and AUG4) where the increase triplicated. These facts show clearly the decline of crop production in favour of animal production, especially in the civil parishes with greater AUG.

Considering only the evolution of the forage production area, i.e. temporary and permanent pastures as well as forage crops, a positive variation essentially due to the increase in the area of grassland and permanent pastures was observed. This area increased in all classes of AUG, especially for the civil parishes with greater AUG, where the variation was higher than 100%. In the civil parishes with AUG1, there was only a 20% increase. The variation for the temporary pastures and forage crops corresponded to a 38% increase in AUG2 civil parishes, and a 32% decrease in AUG4 civil parishes, demonstrating the greater investment on forage crops for farms in civil parishes with smaller AUG.

However, the majority of permanent pastures are classified as spontaneous of low feeding value, due to the plant species present, with a low grazing capacity. They gradually occupy the soil after the abandonment of cereal production since the area of improved and sown pasture decreased by 53% in the civil parishes with smaller AUG and by 15% in AUG4 civil parishes.

The area occupied by permanent pastures in the Alentejo region has made possible the increase in extensive animal production. In fact, the number of grazing animals was greater in 1999, cattle to be the only ones increased for all classes of AUG, although at a higher rate for the intermediate classes. In the AUG civil parishes, the number of sheep varied in a peculiar way since it decreased by 19% in the civil parishes with smaller AUG and increased by 28% in AUG4 civil parishes, where there was a 16% increase in bovine cattle. It should also be referred that, during this period of time and throughout Alentejo, the ovine and caprine numbers decreased 5 and 30% respectively while the cattle population increased by 55%.

Grazing capacity of permanent pastures

The ruminant production in the Alentejo region is an expanding activity since 84% of the farms with "montado", considering the 1999 GAC census, reported it. With the assistance of the Agriculture Regional Direction of Alentejo, data included in the 1999 GAC census was organized in view of the water availability during the raining season, to better characterize the distribution of permanent pastures and the way in which they are articulated with other crops used as supplement feeding for grazing animals, at the farm level. All forms of evaluation of the agricultural activities that follow, refer to farms located in areas according to three real evapotranspiration zones (EZ), which limits were set using charts from the Ministry of the Environment: from 400 to 450 ml (EZ1), from 450 to 500 ml (EZ2) and above 500 ml (EZ3).

The farms were also organized according to classes of total area (CA), for a better understanding of the partition of the agricultural activities. Thus, 5 classes were considered: 100 ha<CA1<200 ha; 201 ha<CA2<400 ha; 401 ha<CA3<600 ha; 601 ha<CA4<800 ha; and CA5>801 ha.

The study of the forms of soil utilization in the "montado" from Alentejo comprised 932 farms with a total area greater that 100 ha and, at least, 1ha of AUG, chosen from a total of 3 000.

Overall, the area reserved for each type of LUAP utilization was independent from CA, standing out the importance of the AUG for any of the CA and EZ considered. However, there was a tendency for a greater utilization of the available AUG for EZ1 farmers, especially for smaller CA.

Permanent pastures and grassland were present mainly in areas under holm and cork oak groves (AUG) (78.1%), followed by clear land (18.5%) and, only 3.3% were under permanent crops.

Animal production is the main activity for this type of agriculture, the "montado", based on the utilization of poor soil, with scarce agricultural aptitude, and favourable to direct grazing where the quantity of acorn in the AUG may has a decisive role due to its availability during the year. The animal production activity was presented by analysing the distribution of the bovine, ovine, goats, swine and equine number of heads. The five animal species referred were found on all CA and EZ, except for the CA5 (EZ2), where the farms did not have goats. The association of different species was common but the proportion of ruminants, namely sheep and cattle, was greater than the other species. Pigs was present on all situations. The equine species had some significance on CA4 (EZ3).

The number of normal cattle heads (NCH) per hectare corresponding to cattle was greater than that for sheep on all CA and EZ, with mean values of 54.9 and 39.1% respectively. The exceptions were CA1 farms on EZ1 and EZ3 and CA2 farms (EZ1), where the ovine species predominated.

When evaluating the NCH according to areas used for animal production – temporary pastures and forage crops, fallow land and permanent pastures – the stocking rate was lower than 0.40 NCH, but presenting higher values for the EZs where cereal crops were sown in a greater proportion.

The relative importance of the forage production area components was evaluated by analysing the stocking rate on farms with permanent pastures and fallow land and other farms where, besides these components, there was also the production of forage crops. The stocking rate, allowed by the situation with the lowest investment, was 0.35 NCH/ha and the mean values for the EZs were 0.46, 0.34 and 0.22 NCH/ha (EZ1, EZ2 and EZ3, respectively). The stocking rate was independent from farm areas except for farms located on EZ1 with total areas of 200 ha, with a stocking rate of 0.62 NCH/ha. If farms producing forage crops were also included, the stocking rate increased to 0.41 NCH/ha, due to the rise observed for EZ2 and EZ3, 0.43 and 0.38 NCH/ha of foraging area respectively.

Finally, considering the stocking rate in relation to CA and percent of AUG, there was a decrease in NCH as farm area increased and, for all CAs, a decrease as the percent AUG increased on the farms. The greater stocking rates, 0.64 NCH/ha, are observed on the farms with smaller area (CA1 and CA2) and less than 33% of AUG. These farms, as previously described, have 100-400 ha, with less fallow land and AUG (60 to 70% with crops), thus having greater proportions of LUAP used for the production of wheat and oat for grain, the stubble being used for animal feeding, and annual forage consociations.

On the contrary the lowest stocking rate, 0.32 NCH/ha, was found on farms with an area greater than 400 ha with more than 66% of AUG, which forage area is mostly permanent pastures.

Natural pastures versus sown pastures

Due to substantial increase of extensive animal production in oak areas, it is imperative to evaluate the productivity and quality of natural pastures versus improved pastures, not only in the fertilization practices of the natural pastures but also the introduction of sown pastures with fertilization in the system.

The production of dry matter in five soil types submitted to three treatments (natural pasture - NP; fertilized natural pasture - FNP; fertilized sown pasture - FSP) can be seen in figure 1. These results correspond to the year following installation of the sown pasture (Barradas, 2003). It can be stated that in the majority of the cases there wasn't a huge increment of the natural pasture's production as a result of fertilization. On the other hand, there was a great increase in the production of dry matter between natural and sown pasture. This fact results from the inexistence, in almost all natural pastures, of adapted species with high production potential.

Successive soil mobilizations and herbicide applications during cereal production resulted in a decrease of biodiversity, with loses of interesting species for animal feeding. This fact associated to soil degradation, namely by erosion and loss of organic matter, resulted nowadays in a reduction of natural pasture's production.

As result of the dry matter's production improvement, as well as the introduction of better quality species, there was an increase in normal cattle heads (Figure 2), three fold in some cases. In the sandy soil, the cattle rejected the natural pasture due essentially to its poor quality, inclusively breaking the fence more than one time seeking the fertilized sown pasture on the other side.

Annual legume species for pasture improvement

In the Forages, Pastures and Alternative Crops Department of the National Plant Breeding Station (Estação Nacional de Melhoramento de Plantas), based on the idea that to improve natural pastures it is necessary to introduce species with a good quality, high yield potential and well adapted to the

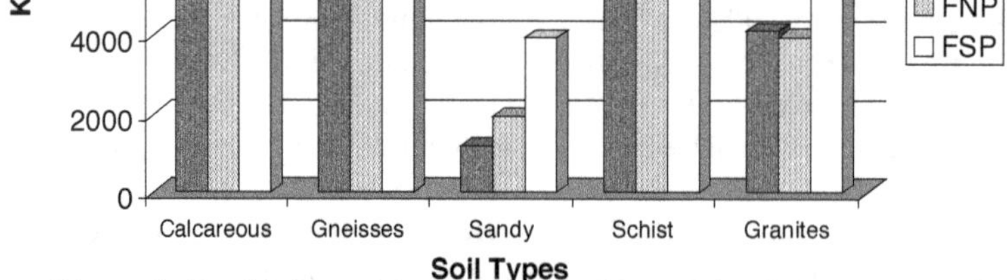

Figure 1. Production of dry matter per hectare for 5 soil types with 3 different treatments (natural pasture – NP; fertilized natural pasture – FNP; fertilized sown pasture – FSP).

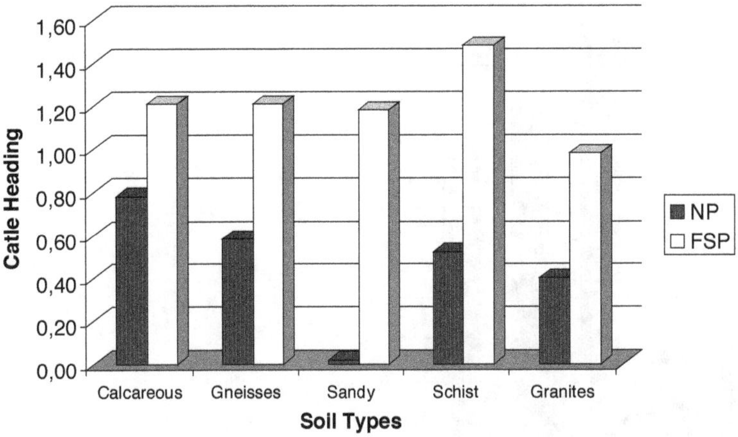

Figure 2. Normal cattle heads (NCH) in 5 soil types with 2 different treatments (natural pasture - NP; fertilized sown pasture – FSP).

environment where they will be established, it was developed a selection programme of annual legumes species for pasture, specially directed to soils with some limitations (acid pH; low content in organic matter; low depth soil; lower capacity of cationic exchange; low capacity of water retention, etc.). These species with high rusticity and good adaptation are to be used as pioneer species, promoting the improvement of these soils fertility in short periods of time, and promoting a better development of others more productive species, such as subterranean clover (Simões, 2003).

Seven ecotypes of annual legumes species of natural re-sowing capacity were pre-selected (*Biserrula pelecinu (Bp)*; *Ornithopus compressus (Oc)*, *Trifolium cherleri(Tc)*; *Trifolium glomeratum (Tg)*; *Trifolium hirtum (Th)*; *Trifolium lappaceum (Tl)* and *Trifolium resupinatum (Tr)*) considering the collections of germplasm made across Continental Portugal (Simões *et al.*, 2005a, b).

The average production of dry matter and seed for these pre-selected species can be seen on figure 3. The higher average production of dry matter per plant was observed for *Trifolium hirtum*, *Trifolium lappaceum* and *Trifolium resupinatum*. Regarding the production of seed, *Ornithopus compressus* produced the best results, in average, followed by *Trifolium hirtum* and *Biserrula pelecinus*.

The number of days since emergence till full bloom (when more than 50% of plants present flowers) and the period of blooming are presented on figure 4. *Ornithopus compressus* was the species with a greater number of days till full bloom; on the other hand, the ecotypes of species *Trifolium cherleri* and *Trifolium hirtum* reached this phase earlier.

The ecotypes belonging to the species *Ornithopus compressus* presented, in average, the longest blooming period, being the *Trifolium lappaceum* ecotypes the ones with the lowest values for this characteristic.

Conclusions

The area of permanent pastures increased considerably in Portugal, mainly in the Alentejo region, in areas under holm and cork oak groves. Most of the area of permanent pastures corresponds to soils

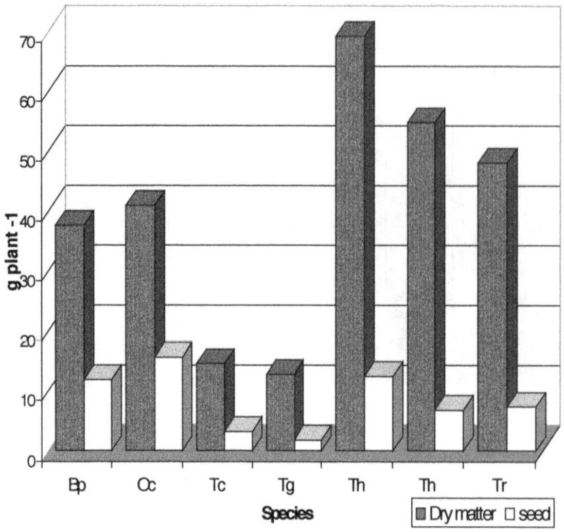

Figure 3. Average production of dry matter and seed per plant of the ENMP's pre-selected species.

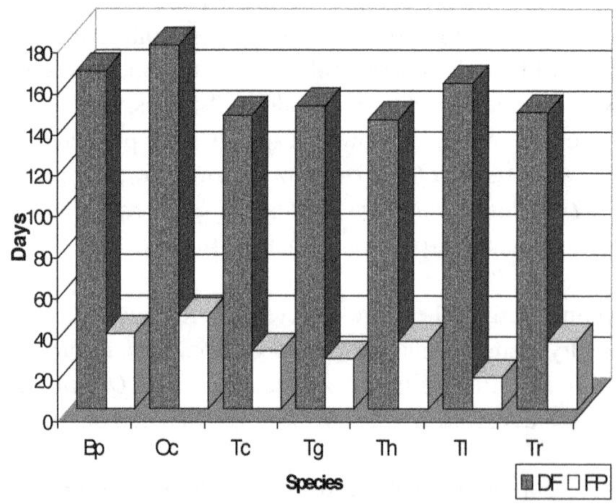

Figure 4. Average number of days since emergence till full bloom (DF) and blooming period per plant (FP) of the ENMP's pre-selected species.

used previously for cereal crop production where plant species of lower feeding value started to develop, corresponding to low grazing capacities. In these areas fertilisation may not be sufficient to increase production and improve the feeding value quality. The most efficient solution will be the introduction of high potential species, in a biodiversed mix, well adapted to the environment, with great persistence. This biodiversity will guarantee a better capacity of the pasture to cover the different soil variations, forming in each one a niche and minimising the effect in total dry matter production, of different inter-annual climate fluctuations according to the typical weather of Mediterranean areas.

References

Barradas, A.M. 2003. Resultados preliminares de la mejora de pastos en suelos mediterráneos. Thesis for the obtention of High Studies of 3rd cycle in the Universidad de Extremadura, Spain, 32 pp.

Simões, N.M. 2003. Caracterización de vários ecotipos de especies pratenses – Leguminosas anuales. Thesis for the obtention of High Studies of 3rd cycle in the Universidad de Extremadura, Spain, 38 pp.

Simões, N.M.; Tavares-de-Sousa, M. M.; Costa, A. R. & Vicente, C., 2005. Estudo comparativo de leguminosas anuais pratenses. I – *Biserrula pelecinus*; *Ornithopus compressus* on Revista da Sociedade Portuguesa de Pastagens e Forragens. 25, Portugal, 12 pp.

Simões, N.M.; Tavares-de-Sousa, M. M.; Costa, A. R. & Vicente, C., 2005. Estudo comparativo de leguminosas anuais pratenses. II – *Trifolium cherleri*; *Trifolium glomeratum*; *Trifolium hirtum*; *Trifolium lappaceum*; *Trifolium resupinatum* on Revista da Sociedade Portuguesa de Pastagens e Forragens. 25, Portugal, 14 pp.

Effects of climate variability on animal productions

B. Ronchi & A. Nardone

Università degli Studi della Tuscia, Dipartimento di Produzioni Animali,
Via S. Camillo de Lellis, 01100 Viterbo, Italy

Summary

The performance, health, and well being of livestock animals are strongly affected, directly or/and indirectly, by climate change and variability. Impacts of climate change and variability on agriculture and livestock production systems are expected to be more severe in developing countries, also due to poor access to technologies and financial supports. Future animal husbandry management systems are requested to incorporate information and guidelines regarding responses to weather challenges. Adaptation to current climate variability can potentially offset some adverse climate change impacts, improving food security for the global population.

Keywords: animal diseases, land use, adaptation.

Introduction

Since the beginning of animal domestication livestock production systems were forced to adapt to climate conditions. Nomadism and transhumance are typical examples of adaptation to annual and interannual variability of climate and to its effects on pasture availability. Continuous movements and seasonal migrations in search of pastures were in the past, and still now are in many regions, the main land use strategies enabling people to cope with climate constraints and variability (Manderscheid, 2001; Boyazoglu & Hatziminaoglou, 2002). Geographical distribution of livestock species and breeds is largely conditioned by adaptability to climatic constraints (Nardone, 2000).

Vulnerability to environmental changes is strictly related to frequency, duration, and intensity of climate adversities, but is also dependent on economic and policy factors (Galvin *et al.*, 2001). According to definitions in U.S. Climate Change Science Program (2001), "climate variability" can be considered as "the variations in the mean state and other statistics of climate features on temporal and spatial scales beyond those of individual weather events"; "climate change" represents " a statistically significant variation in either the mean state of the climate or in its variability, persisting for an extended period (typically decades or longer)"; "global change" is used to indicate "changes in the global environment that may alter the capacity of the Earth to sustain life". Climate variability refers to anomalies observed in climate records in periods when climate system is not showing changes (Salinger *et al.*, 2000). The analysis of climatic variability focus on temporal and spatial fluctuations of parameters like temperature, precipitation, evapotranspiration (Ni & Zhang, 2000). Examples of climate changes can be also detected in the frequency and severity of extreme weather events. Characteristics of variability are transient behaviours of climate factors (trends, persistence, pseudo-cycles, extremes weather events). An example of anomalous climate tendency is represented by "heat waves", defined as an increase in air temperature more intensive than normal daily mean for one or more standard deviations (Baldi *et al.*, 2004). In the last decade Europe experienced several summer heath waves, with a strongly increasing tendency comparing to the past. The one occurred in 2003 was the most extreme in 500 years, both in terms of intensity and length. Figure 1 shows variations of daily temperature-humidity index (THI) registered on dairy cow farms in Central Italy during the year 2003.

The problem of climate change represents one of the most important topic worldwide, for its scientific, social, and economical implications. Climate data show evidences for a global warming at the surface of 0.6 degrees Celsius since 1856, with six of the warmest years occurring in the last decade and highest sea temperature in the tropics in 1998 (Hulme, 2001). The warming process of earth's surface is strictly related to the increasing concentration of greenhouse gases, including water vapor, CO_2, methane, nitrous oxide and some other gases from anthropogenic activities. The global warming measured during the 20th century results from a general decrease in daily temperature range and an increase in nighttime temperature (Karl *et al.*, 1993). The impacts of changes in temperature extremes on natural ecosystems are widely recognised to be more important than changes in average values (Katz & Brown, 1992). Global warming is not always associated to an increased climate variability. Data from studies conducted in Europe indicate on the contrary that in some areas warming is accompanied by a reduction of variability, both in daily temperature range and in the monthly day-to-day variability (Rebetez, 2001).

Climate variability is a strong determinant of year-to-year variations in production for pastoral agriculture. Impacts of climate changes and variability, and of extreme weather events would be negative for many regions of the world and an emerging risk of future large scale discontinuities can be predicted.

The broad question the review aims to explore are:

- What are the foreseeable impacts of climate change and variability on animal husbandry?
- Which livestock production system is currently more vulnerable to climate change and variability?
- What is the potential for adaptation to climate change and variability?

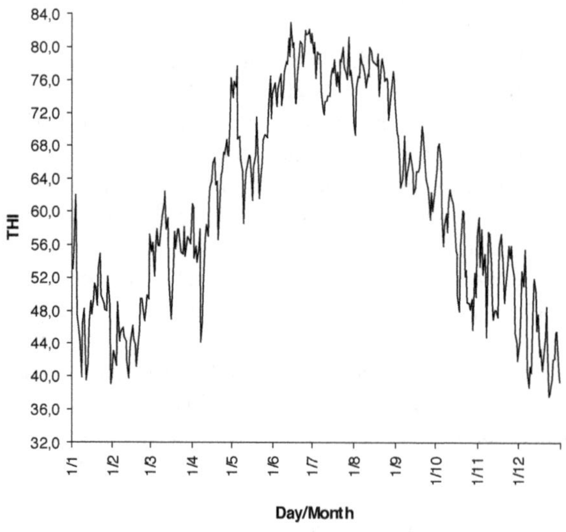

Figure 1. Changes of daily temperature-humidity index (THI) registered on dairy cow farms in Central Italy (Viterbo province) during the year 2003 (Bernabucci et al.*, unpublished).*

The dynamic responses are discussed of establishing criteria for increase sustainability of livestock production systems in the context of climate change and variability.

Climate variability and land use

Climate and weather influence land-use and land-cover in a complex way. Key links include for example the exchanges in greenhouse gases between the land surface and the atmosphere, the radiation balance, the availability of water resources, changes in biological chemical and physical processes in the soil. Most researches have focused on the impacts of long term climate change on agriculture. More recently also the impact of climate variability and of extreme climatic events have been taken into consideration (Rosenzweig & Hille, 1998).

Climate variability is responsible for altering land-use practices and land-cover, reducing availability and quality of natural resources. On the other hand changes in land-use and land cover can contribute to climate change and variability. The impact of climate variability on land-use practices and livestock farming systems may be substantially different, relating to the geographical position and characteristics of different regions. Some researchers consider that, on a global basis, climate variability and change may have an overall negligible effect on agriculture production; but on a regional level, some areas could benefit from an altered climate, whereas other could be adversely affected (Parry & Carter, 1989). Subtropical and tropical areas are expected to be more sensitive to climate variability and more negatively affected; on the contrary developed countries may benefit, due to availability of technologies and to possibilities to employ adaptive adjustments. The 1998 El Niño events were responsible in East Africa for an estimated five-fold increase in rainfall (Galvin *et al.*, 2001). On the contrary 1997 and, particularly, 1999 were drought years. Climate analyses suggest that part of East Africa in the next decades will became drier, with strong reduction of the length of the growing season and crop production. Other areas will become wetter, with positive effects on agriculture and animal productions (Thornton *et al.*, 2002).

Future climate scenario for European countries differ according to different agricultural regions. At middle and higher latitudes global warming will extend the length of growing season and will play a positive role on production (Olesen & Bindi, 2002). In regions like Scandinavian countries, cropping area may expand northwards, at an estimated shift of 120-150 km °C^{-1} increase in annual mean temperature (Carter *et al.*, 1996). In warmer, lower latitude of Europe an increase of temperature will negatively affect crop yield and will increase the turnover rate of organic matter and the demand for water for irrigation (Olesen & Bindi, 2002). In the west Mediterranean zone climatic change, associated with socio-economic factors, may be responsible for land degradation and desertification (Puigdefabregas & Mendizabal, 1998).

The effects of climate variability and extreme events on crops, depend on crop type, level and extent of exposure. Corn represents one of the most sensitive crops to climate variability, especially to maximum temperatures; a strong reduction on corn yield has been highlighted in U.S. as a consequence of summer water stress due to El Niño events (Phillips *et al.*, 1999).

An open question on the debate about consequences of climate variability on agriculture regards possible effects on pest population and, in turn, on pest treatment. If it is easy to predict that climate change will affect crop pests and weeds, the impact these change may have on the complex of pest and weeds management for specific crops in specific regions need to be assessed (Strand, 2000). Projected warming will increase the ability of some pests to survive winter and relative risks for plant diseases (CSIRO, 2001). Plant pathogens will also be affected by irregular weather conditions, such as alternation of dry periods and rainfall events. Climatic factors have a well known influence on growth, survival, dissemination and incidence of several fungi diseases, which are also responsible for transmission of toxic metabolite to vegetable and animal products for human consumption. In the case of *Fusarium* fungi, both a direct and an indirect effect of climatic conditions on the incidence of diseases has been proposed (Doohan *et al.*, 2003).

As a consequence of higher carbon dioxide levels in the atmosphere, pasture production is expected to increase in some regions due to an extended growing season, although accompanied by a reduction in forage digestibility (Olesen & Bindi, 2002). Data from laboratory studies (Newton *et al.*, 1994) demonstrated that exposure of ryegrass and white clover to 700 ppm (level expected in New Zealand by 2100, determines 8% higher growth than those grown at 355 ppm (current level). Other authors evidenced that the positive effect of rise in carbon dioxide on pasture growth can be detected only with adequacy of rainfall. On the contrary, in water–limited environments a double concentration of carbon dioxide may reduce pasture productivity by about 15% (Howden *et al.*, 1999). At the same time the increased forage availability may be accompanied by a reduction in feed quality, not only as a consequence of reduced digestibility, but also for an higher incidence of subtropical species, especially in areas where there will be an increase in summer temperature and rainfall. Also arable crops may benefit from warmer conditions and higher carbon dioxide, but potential yield increase will require higher fertilisers input and higher availability of water for irrigation. Water resources, both in terms of quantity and quality, are critical to a successful livestock production system. In many areas of the world water resources availability is becoming a serious problem, as the consumption levels are exceeding availability. In addition, in areas such as Mediterranean, fresh water resources are under pressure both in terms of quantity and quality, negatively affecting agriculture, livestock, ecosystems and necessity for human.

Climate variability and animal diseases

Relationships between climate and disease are well established both in human and in animals, but the possible effects of climate change and variability on disease appearance or spreading are controversial. This is due to the scarcity of health databases over long periods and to the presence of many confounding factors. Climate change is only one of many areas of environmental change that can impact on health. Demographic and societal changes occurred during the past decades are also consider associated to changing epidemiology and emergence of epidemic diseases (Gubler, 2002). The features of climate change that deserve most attention in the context of diseases appearance and spread are changes in frequency and severity of extreme weather events (Sutherst, 2004). Recently a relationships between El Niño events and cholera prevalence in Bangladesh has been found by Rodó *et al.* (2002). The study represents the first concrete evidence that climate variability, connected to intensification of El Niño Southern Oscillation (ENSO), is responsible for temporal dynamics and incidence of diseases. In the last three decades a strong intensification of the ENSO has been highlighted. The 1997-98 El Niño event was one of the strongest of the past century and responsible for extremely dry conditions in Southeast Asia. On the contrary in East Africa El Niño caused severe flooding, connected with a mosquito-borne Rift Valley Fever (RVF) epidemic (Linthicum *et al.*, 1999). RVF is a virus disease affecting ruminant animals and humans with severe consequences. Months of heavy rains and floods in 1997 favoured explosion of mosquito population, with spreading of RVF from Kenya to Somalia and death of hundreds of humans and thousands of animals. The potential health impacts of global climate change and variability include effects due to exposure to thermal stress and weather disasters (e.g. heatwaves, cold spells, droughts), and to ecological changes (vector-borne, feed-borne, and water-borne diseases; feeds availability and composition; water resources availability and quality).

The recent spread of Blue-tongue disease in most Mediterranean countries, particularly after 1998, raises a number of questions about possible relationships between Blue-tongue epidemiology and global warming. The Mediterranean basin is at the limit of the disease's distribution range. Insects from the genus *Culicoides*, responsible for the transmission of the Blue-tongue virus from one animal to another, require temperature higher than 18°C; only rarely Blue-tongue has been reported north of the 40° line of North latitude (Hendrikx, 2005). On the basis of Mellor's model (Mellor and Boorman, 1995), which forecasts that the increase of 1°C of ambient temperature would permit a movement of *Culicoides* toward north by 90 km, it can be hypothesized that the arrival of Blue-tongue in the European continent may be a consequence of climate change.

Climate change and variability influence the ecological balance and context in which parasites population breed, and develop (Patz *et al.*, 2000). Climate changes will alter patterns of parasitic abundance and disease occurrence in farm animals. Climate models for northern latitude forecast, due to warmer seasons, better conditions for nematode parasites and an amplification of parasite in the host population (Dobson *et al.*, 2003).

Climate variability and animal productions

The expected effects of climate change on animal productions are extremely variable, depending on a complex interaction of climatic, agricultural, social, and political factors. A strong influence on animal husbandry is exerted by availability and quality of feeds from land cultivation, since livestock in many agricultural systems are integrated with cropping activities.

In many developing countries climatic changes may worsen climatic condition with severe consequences on local economy and food security. The collective effect of climate changes on crop yields, forage yields, and animal welfare will result in many African countries, without the adoption of adequate policy based adaptations, in: decrease on animal products, increase in price level of meat and milk, increased risk of hunger and human diseases, abandonment of rural areas (Butt *et al.*, 2005). Extensive pastoral systems may be further marginalized by frequent climatic adversities, with economic and environmental consequences. In fact, pastoralists make a substantial contribution to the economy of developing countries, providing meat, milk, cheese, and other products from animal husbandry. On addition, pastoralism contribute to the maintenance of a great reserve of biodiversity, connected to the survival of genetic resources (vegetables and animals) disappeared in specialised livestock production systems. High climate variability complicate decisions for stocking rate and stocking policy and can aggravate existing problems of soil erosion and pasture degradation in developing countries (White *et al.*, 1996).

Livestock production may be negatively affected by climate variability also in temperate regions, where during summer animals are exposed to heat stress, like in the Mediterranean basin (Nardone, 2000) and in the south of USA (Klinedienst *et al.*, 1993). Heat stress represents one of the most important limiting factors particularly for dairy cow's farming. Warming trends will increase the extent of exposure to heat stress, reducing productivity and reproductive efficiency of dairy cows (Lacetera *et al.*, 1996; Ronchi *et al.* 2001). Genetic selection of dairy cows in the last decades has been mainly devoted to increase of productivity, with extraordinary results (69% increase of average milk production of Holstein in Italy between 1970 and 1995). On the contrary little attention has been paid to the adaptive capacity to environmental constraints. As a result, current dairy cows show high performance and concurrent high metabolic heat production, resulting more sensitive to heat stress. According to predicted climate change scenarios, in hot-humid regions of USA in dairy cows farms milk production will decline by 5-14% and conception rate will decrease by 36% (IPCC, 2001). Extreme weather events may be responsible for consequences on animal production systems more severe than those related to climate change. A typical example comes from USA, where in 1995 in Nebraska-Iowa a single heat wave caused over 5,000 head of feedlot cattle died (Hahn & Mader, 1997).

A minor impact of climate variability is predicted for very intensive livestock systems, such as confined poultry and pig farming systems, where animals are usually reared under a controlled climate. Hahn *et al.* (2002) reported 1.2 % increase in days to market for the 2030 scenarios. A warmer condition will however require higher housing expenses for insulation and air-conditioning (Olesen & Bindi, 2002).

On the other hand in northern regions climate warming may be beneficial for animal husbandry, expanding the grazing period, forages and concentrate availability, improving survival of newborn animals, and reducing energy costs. Results from studies to assess the potential impact of climate change on British grazing livestock systems indicate that ruminants at grass or in natural ventilated buildings will be able to adapt easily, and that a modest increase in grass production and stocking rates are expected (Parson *et al.*, 2001).

Adaptation to climate variability

Referring to climate, the term "adaptability" is defined as the complexity of adjustments made to reduce the vulnerability of agricultural systems when faced with climate variability and changes (Smit & Pilisofova, 2001).

The adaptability of livestock production systems to climate change and variability are not simply related to meteorological and climatic factors. Other than these, other factors exert a strong influence on adaptability, such as: social, political and economical factors, internal components and function of each system, interactions with other systems and resultant integrated behaviours (Basher, 1999).

Adaptation strategies can be distinguish into two categories (Table 1): "short-term strategies", such as crop and livestock diversification; "long-term strategies", such as development of new technologies (Kokic *et al.*, 2005). Short term adaptations are commonly associated with on farm responses to climate variability, rather than to climate change. Short-term adaptation strategies include also diversification in timing and/or intensity of agricultural production, regulating the length of the growing season to changing environmental conditions. In areas affected by hot and dry climate, the adoption of soil and water conservation measures may be effective in reducing soil erosion and preventing nutrient retention. Long-term strategies include changing crop types and location, and adoption of new technologies, conditioned by local factors (e.g. soil type). Adaptation measures usually implicate a considerable economic impact, both in terms of less immediate farm income opportunities, both in term of on-farm investment and capacity building. Adaptation options also need to be acceptable within the context of farm management (Kokic *et al.*, 2005).

Biotechnology and precision farming are likely to improve adaptability to climate variability and changes. Biotechnologies can help in developing plant resistance to heat and drought, and in providing feeds for animal husbandry. Biotechnologies applied to animal resources can help in selecting animals more resistant to infectious and parasitic diseases, and more tolerant to climate constraints. The application of precision farming methods, through the adoption of information technologies, will help farmers in the management of resources and in the adaptation to changing conditions.

Adaptations of agriculture and livestock farming systems to climate change and variability is considered a complex process, involving both a government policy response, both decision-making by agri-business and producers at the farm level (Smit & Skinner, 2002).

For each livestock production system there is a need to understand what types of adaptation are feasible and effective, who can be involved in their implementation, what is required to facilitate their

Table 1. Adaptation options in livestock farming systems to climate change and variability.

Short term options	Purposes
- Crop diversification. - Multicropping. - Mixed farming. - Adjusting cropping sequence, fertilizer, - Timing of irrigation. - Changing land use and tillage practices.	- Risk spreading, increasing productivity, defence against plant disease, soil protection.
- Change in grazing management (timing duration, location, stocking rate). - Change in feed conservation techniques. - Varying supplemental feeding. - Temporary migration. - Altering the breeding management program.	- Meet nutritional requirements, soil protection, feed quality and safety.
- Providing shades. - Adoption of technologies (e.g. Use of sprinklers).	- Withstand climate shocks and seasonal effects.
Long term options	**Purposes**
- Migration. - Promoting irrigation efficiency and water recycling.	- Overcome long-lasting climate impacts. - Water conservation.
- Chose of more adaptable species and breeds. - Animal selection for climate and disease resistance	- Withstand climate effects and increase productivity.
- Modernisation of farm operations. - Adoption of technologies (e.g. Precision farming). - Weather and climate information systems.	- Increase farmer's perception, sensitivity, professional capabilities.
- Develop market efficiency. - Policy measures. - Private insurance. - Extension services.	- Reduction of costs, increase of profitability. - Control of barriers, property rights, land use regulation. - Subsidies, pricing incentives, investment promotion.

adoption and assure their success. In cropping and improved pasture there is a wide scope for adaptation in response to climate variability and changes, such as the chose of new plant varieties more adaptable to rising temperature and drier climate (CSIRO, 2001). On the contrary, in rangelands there are few adaptation options to improve natural vegetation and production systems. Adaptation of livestock farming systems to climate constraints must include also the possibility of animal adaptation, by means of physiological and genetic changes (Khalifa, 2003). More precisely, the term "adaptation" include genotypic components of the interaction between the animals and the environment, whereas "acclimation" is related to phenotypic response (Collier *et al.*, 2003). The most part of breeds of domesticate animals reared in subtropical and tropical areas have acquired the ability to survive to climate variability, being more resistant to heat stress, to parasites, to underfeeding, to water deprivation, to prolonged physical exercise. One of the most effective component of adaptation of ruminant domesticated animals to hot and arid conditions is considered the degree of efficiency in urea recycling (Webster, 1991). A long-term approach to support strategies for livestock production in regions with high climatic constraints is the study and identification of genes associated with acclimation of domestic animals to thermal stress, and activation of breeding programs to increase resistance particularly to heat stress (Collier *et al.*, 2003).

The adoption of precise agriculture and livestock gives the farmers the capability to more effectively practice crop production and animal husbandry, and to adapt their activities to oncoming environmental constraints. Precision farming include a more effective use of crop inputs (fertilizers, pesticides, tillage and irrigation water), a more rational use of grazing resources, a correct approach to animal feeding and supplementation, the adoption of optimal technical solution to mitigate climatic stress, and to prevent infectious and parasitic diseases. Technology for precision farming include recent advances in electronics and computers, such as remote sensing (RS), geographic positioning system (GPS) and geographical information system (GIS).

Advances in climate science are providing information for decision makers to improve predictions of climate variability and projections of climate change, as a basis to develop integrated projects for agriculture adaptability.

The acquisition of precise and preventive information on spatial and temporal variability can be used for variability management through decision support systems (Mandal & Ghosh, 2000). Since climate can have devastating effects especially among traditional agricultural systems of poor populations, climate forecast information may have a strong potential for reducing poverty and vulnerability of rural population such as pastoralists (Luseno *et al.*, 2003). But the practical value of climate forecast will depend on many factors, such as: predictive accuracy and the spatial-temporal resolution of climate forecast, the confidence of indigenous population in external forecast, the educational level, the ability to forecast impact variables of direct interest, and finally, the production system. For traditional pastoralists, who have elaborated complex and flexible livestock management strategies, based on regular opportunistic migration, the meaning of external forecast information seems quite low. On the contrary, for agropastoral cultivators and for those who practice sedentarized animal husbandry, with low flexibility and possibilities for corrective adjustments, climate forecast may result very useful.

Conclusions

Agricultural and livestock production systems are strongly influenced by current and predicted scenarios of climate variability and change. The expected effects of climate conditions on animal productions are extremely variable, depending on a complex interaction of climatic, agricultural, social, and political factors. The impact of climate variability on land-use practices and livestock farming systems may be substantially different in type, timing and magnitude, relating to the geographical position and characteristics of different regions. Improved surveillance and monitoring is needed to identify consequences of global change and climate variability on distribution of agents of animals disease, especially in case of zoonosis diseases, to develop predictive models and to develop the required actions to minimise negative effects. There is a scarcity of information regarding ecology and transmission biology, as well as high-quality epidemiologic data from a surveillance system. Future climate scenarios will have devastating effects in many developing countries, with severe consequences on local economy and food security: increased risk of hunger and human diseases, abandonment of rural areas are predicted. An improvement of capabilities to assess climate information and to elaborate projection for the future is needed, as such as to provide needed information to decision makers at local, regional and national level. The development of new models linking climate with socio-economic aspects of land-use is strongly required, together with the prevision of different land-use decision. Agrometeorology and agroclimatology should play an important role in furnishing readily available inputs to integrate the impacts of weather, climate and their variability into the decision packages for short and long-term adaptation strategies.

References

Baldi, M., M. Pasqui, F. Cesarone & G. De Chiara, 2004. Heat-wave events in the Mediterranean: a recurrent feature or a global warming effect? EMS Annual meeting Abstracts, Vol. 1: 268.

Basher, R.E., 1999. Data required for developing adaptations to climate variability and change. Mitigation and Adaptation Strategies for Global Change, 4: 227-237.

Boyazoglu, J. & Y. Hatziminaoglou, 2002. Livestock genetic resources and production systems a Mediterranean overview. Arch. Latinoam. Prod. Anim., 10: 62-67.

Butt, T.A., B.A. McCarl, J. Angerer, T.P. Dyke & J.W. Stuth, 2005. The economic and food security implications of climate change in Mali. Climatic change, 68: 355-378.

Carter, T.R., R.A. Saarikko & K.J. Niemi, 1996. Assessing the risks and uncertainties of regional crop potential under a changing climate in Finland. Agric. Food Sci. Finland, 3: 329-349.

Climate Change Science Program, 2001. Strategic Plan for the U.S.; www.climatescience.gov.

Collier, R.J., C. Coppola, & A. Wolfgram, 2003. Novel approach for the alleviation of climatic stress in farm animals. In: Interactions between climate and animal production. EAAP Technical series, 7: 61-71.

CSIRO, 2001. Climate change projections for Australia. Climate Impact Group, CSIRO Atmospheric Research, Melbourne, 8 pp.

Dobson, A., S. Kutz, M. Pascual, & R. Winfree, 2003. Pathogens and parasites in a changing climate. Advances in Applied Biodiversity Science, 4: 33-38.

Doohan, F.M., J. Brennan, & B.M. Cooke, 2003. Influence of climatic factors on *Fusarium* species pathogenic to cereals. European Journal of Plant Pathology, 109, 755-768.

Galvin, K.A., R.B. Boone, N.M. Smith & S.J. Lynn, 2001. Impacts of climate variability on East African pastoralists: linking social science and remote sensing. Climate research, 19: 161-172.

Gubler, D.J. 2002. The global emergence/resurgence of arboviral diseases as public health problem. Archives of Medical Research, 33: 330-342.

Hahn, G.L. & T.L. Mader, 1997. Heat waves in relation to thermoregulation, feeding behaviour and mortality in feedlot cattle. Proc. 5 Int. Livestock Environ. Symp., Amer. Soc. Agric. Eng., St. Joseph, MI, pp. 288-295.

Hahn, G.L., T.L. Mader, J.A. Harrington, & J.A. Nienaber, 2002. Living with climatic variability and potential global change: Climatological analysis of impacts on livestock performance. In: Proc. 15th Conf. Biometeorol. Aerobiol.: 45-49.

Howden, S.M., P.J. Reyenga, H. Meinke, & G.M. McKeon, 1999. Integrated global change impact assessment on Australian terrestrial ecosystems: overview report. Working Australia, 51pp.

Hulme, M., 2001. Climate change and extreme weather events: what do we know? In: Adaptation to climate change, N Adfer, S. Huq & S. Toroq (Editors), Kohn Centre London: 3-6.

Hendrikx, P., 2005. Climate Change: fantasy or reality. Web site www.eid-med.org.

IPCC, 2001. Ecosystems and their goods and services. In: Climate change 2001. Impacts, adaptation and vulnerability. UNEP/WMO 257.

Karl, T., P. Jones, R. Knight, G. Kukla, N. Plummer, V. Razuvayev, K. Gallo, J. Lindseay, R. Charlon, & T. Peterson, 1993. A new perspective on recent global warming: asymmetric trends of daily maximum and minimum temperature. Bull AMS 74/6: 1007-1023.

Katz, R.W. & B.G. Brown, 1992. Extreme events in a changing climate: variability is more important than average. Clim. Change, 21: 289-302.

Khalifa, H.H., 2003. Bioclimatology and adaptation of farm animals in a changing climate. In: Interactions between climate and animal production. EAAP Technical series, 7: 15-29.

Klinedienst, P.L., D.A. Wilhite, G.L. Hahn & K.G. Hubbard, 1993. The potential effects of climate change on summer season dairy cattle milk production and reproduction. Clim. Change, 23: 21-36.

Kokic, P., A. Heaney, L. Pechey, S. Crimb, & B.S. Fischer, 2005. Climate change. Predicting the impacts on agriculture: a case study. Australian Commodities, 12: 161-170.

Lacetera, N., U. Bernabucci, B. Ronchi, & A. Nardone, 1996. Body condition score, metabolic status and milk production of early lactating dairy cows exposed to warm environment. Rivista Agric. Subtrop. E Trop., 90: 43-55.

Linthicum, K.J., A. Anyamba, C.J. Tucker, P.W. Kelley, M.F. Myers, & C.J. Peters, 1999. Climate and satellite indicators to forecast Rift Valley fever epidemics in Kenya. Science 285: 397-400.

Luseno, W.K., J.G. McPeak, C.B. Barrett, & P.D. Little, 2003. Assessing the value of climate forecast information for pastoralists: evidence from southern Ethiopia and Northern Kenya. World Development, 31: 1477-1449.

Mandal, D. & S.K. Ghosh, 2000. Precision farming - The emerging concept of agriculture for today and tomorrow. Current Science, 79: 1644-1647.

Mandersheid, A., 2001. Decline and re-emergence of nomadism: Tibetan pastoralist revive a nomadic way of life and production. GeoJournal, 53: 173-182.

Mellor, P.S. & J.P.T. Boorman, 1995. The transmission and geographical spread of African horse sickness and bluetongue virus. Annals of tropical Medicine and Parasitology, 89: 1-15.

Nardone, A., 2000. Weather conditions and genetics of breeding systems in the Mediterranean area. Proc. Intern Symp. "Produzioni animali di qualità ed impatto ambientale", Ragusa, Italy: 67-91.

Newton, P.C.D., H. Clark, C.C. Bell, E.M. Glasgow, & B.D. Campbell, 1994. Effects of elevated CO_2 and simulated seasonal changes in temperature on the species composition and growth rate of pasture curves. Annals of Botany, 73: 53-59.

Ni, J. & X.S. Zhang, 2000. Climate variability, ecological gradient and the Northeast China Transect (NECT).

Olesen, J.E. & M. Bindi, 2002. Consequences of climate change for European agricultural productivity, land use and policy. European Journal of Agronomy, 16: 239-262.

Parry, M.L. & T.R. Carter, 1989. An assessment of the effects of climate change on agriculture. Climatic change, 15: 95-116.

Parson, D., A.C. Armstrong, J.R. Turnpenny, A.M. Matthews, K. Cooper & J.A. Clark, 2001. Integrated models of livestock systems for climate change studies. 1. Grazing systems. Global Change Biology, 7: 93-112.

Patz, J.A., T.K. Graczyk, N. Geller, & A.Y. Vittor, 2000. Effects of environmental change on emerging parasitic diseases. International Journal for Parasitology, 30: 1395-1405.

Phillips, J., B. Rajagopalan, & C. Rosenzweig, 1999. The role of ENSO in determining climate and maize yield variability in the U.S. Cornbelt. International Journal of Climatology, 19: 877-888.

Puigdefabregas, J. & T. Mendizabal, 1998. Perspectives on desertification: western Mediterranean. Journal of Arid Environment, 39: 209-224.

Rebetez, M., 2001. Changes in daily and nightly day-to-day temperaure variability during the twentieth century for two stations in Switzerland. Theor. Appl. Climatol., 69: 13-21.

Rosenzweig, C. & Hille, D., 1998. Climate change and the global harvest. Oxford U. Press, New York.

Rodó, X., M. Pasqual, G. Fuchs, & A.S.G. Faruque, 2002. ENSO and colera. A non stationary link related to climate change. Proc. Natl. Acad. Sci. USA 99: 12901-12906.

Ronchi, B., G. Stradaioli, A. Verini Supplizi, U. Bernabucci, N. Lacetera, P.A. Accorsi, A. Nardone, & E. Seren, 2001. Influence of heat stress or feed restriction on plasma progesterone, oestradiol-17b, LH, FSH, prolactin and cortisol in holstein heifers. Livest. Prod. Sci., 68: 231-241.

Salinger, M.J., C.J. Stigter & H.P. Das, 2000. Agrometeorogical adaptation strategies to increasing climate variability and climate change. Agricultural and Forest Meteorology, 103: 167-184.

Smit, B. & M.W. Skinner, 2002. Adaptation options in agriculture to climate change: a typology. Mitigation and Adaptation Strategies for Global Change, 7: 85-114.

Smit, B. & O. Pilisofova, 2001. Adaptation to climate change in the context of sustainable development and equity. In: Climate Change: Impacts, Adaptation, and Vulnerability. Intergovernmental Panel on Climate Change. Cambridge University Press, Cambridge UK.

Strand, J.F., 2000. Some agrometeorogical aspects of pest and disease management for the 21st century. Agricultural and Forest Meteorology, 103: 73-82.

Sutherst, W.R., 2004. Global change and human vulnerability to vector-borne diseases. Clinical Microbiology Reviews, 17: 136-173.

Thornton, P.K., R.L. Kruska, N. Henninger, P.M. Kristjanson, R.S. Reid, F. Atieno, A. Odero & Ndegwa, 2002. Mapping povertà and livestock in the developing world. International Livestock Research Institute, Nairobi, Kenia.

Webster, A.J.F., 1991. Metabolic responses of farm animals to high temperature. EAAP Publ. no. 55, 15-22. Pudoc Pub., Wageningen The Nederlands.

White, D.H., S.M. Howden & H.A. Nix, 1996. Modelling agricultural and pastoral systems under environmental change. Ecological modelling, 86: 213-217.

Importance of biological diversity in the foraging of herbivores on Mediterranean rangelands

J. Rogosic

University of Split, Department of Natural Resources, Livanjska 5, 21000 Split, Croatia

Summary

Maquis and garrigues are one of the most extensive vegetation types in the Mediterranean region an important habitat for wild and domestic herbivores. Although the majority of these shrubs are nutritious, secondary compounds, like tannins and terpenes, reduce their forage value. A better understanding of how secondary compounds limit intake is vital for increasing consumption of shrubs and for enhancing and maintaining biodiversity in the Mediterranean shrublands.

In four trials, we examined the effect of polyethylene glycol (PEG) and number of species offered on intake of Mediterranean shrubs by sheep and goats. We offered 6 shrubs in Trial 1, three shrubs in Trial 2, two shrubs in Trial 3, and one shrub in Trial 4. Trials were conducted with 12 sheep and 12 goats (6 PEG vs. 6 control). In trial 1, goats ate more total shrub biomass than did sheep (60.7 vs. 45.9±2.6 g/kg BW, respectively). There was a trend (P=0.08) towards a positive PEG effect on total shrub intake, with PEG-supplemented animals consuming more total shrubs than controls (56.7 vs. 50.0±2.6 g/kg BW). In trial 2, both species of animals showed a numerical decrease in total shrub intake with or without supplemental PEG. Sheep receiving PEG ate more (P=0.002) total shrubs than did controls, but no such treatment effect was found for goats. In trial 3, supplemental PEG had a positive effect (P<0.001) on total shrub intake for both sheep and goats when only two shrubs (*Arbutus* and *Pistacia*) of lower palatability were offered. In trial 4, supplemental PEG had a marked positive effect (P<0.001) on intake of *Pistacia lentiscus* in both sheep and goats. PEG supplemented goats ate more *Pistacia* (39.6 g/kg) than did PEG supplemented sheep (28.1 g/kg), whereas control sheep and goats ate similar amounts (12.2 and 15.3 g/kg, respectively).

Our study suggest that biodiversity plays a very important role in herbivore's diet selection, enabling animals to better meet their nutritional needs and avoid toxicity. In addition, as the number of shrubs in the diet decreased, the impact of PEG on intake of shrubs increased. PEG alone had a greater influence on sheep than on goats, and it had the most influence on both sheep and goats when only one or two foods were available.

Keywords: Mediterranean shrubs, biodiversity, sheep, goats, diet selection, secondary compounds, polyethylene glycol.

Introduction

Sheep and goats are primary consumers of the Mediterranean vegetation maquis and garigues, and thereby shape the diversity, structure and dynamics of these extensive ecosystems (Naveh, 1972; Hodgson & Illius; 1996). Vegetation of these plant communities usually includes between 20 and 25 shrubby species (Rogosic, 2000), and its diversity, function and stability depends on biological/biochemical links between plants and animals in these ecosystems (Dziba *et al.*, 2004).

Plants possess a variety of flavor intensities and nutritional qualities that influence diet selection by herbivores (Villalba & Provenza, 2000). On the other side olfactory, gustatory, visual and tactile stimuli are important sources of information that herbivores use to detect potentially nutritious or toxic plants (Augner & Bernays, 1998). Traditionally, food selection has been attributed to the innate ability of animals to sense, through taste and smell, specific flavors, nutrients and toxins in plants (Rhoades, 1979), but food preferences are better understood as a learned process involving a complex interrelationship between food's flavor and its post ingestive effects (Provenza, 1995; Provenza *et al.*, 1998).

Given the diversity of plant cues herbivores typically encounter, their ability to learn to recognize food and to generalize over food cues are important traits (Augner *et al.*, 1998). Herbivores learn to prefer flavors of foods associated with macronutrients (Ralphs *et al.*, 1995; Villalba & Provenza, 1999), and to avoid the flavors of food paired with toxins (Ralphs, 1992). Once flavor-post ingesive feedback associations have been learned, herbivores use visual and olfactory cues to recognize a specific plant (Launchbaugh & Provenza, 1993), and to generalize preference and aversion across familiar cues in nutrition and toxic foods (Launchbaugh *et al.*, 1993).

Influence of herbivores on plant species richness has attracted more attention than the complex interactions governing plant biochemical diversity and herbivores (Landsberg *et al.*, 1999). Understanding how herbivores influence plant biochemical diversity and reciprocal influence of plant biochemical diversity on herbivores is critical for conceptualizing basic ecological processes and for managing ecosystems. Foraging is a dynamic process that is influenced by an animal's evolutionary and cultural histories and its ongoing experience of the environment (Provenza *et al.*, 1998).

Sheep and goats foraging on Mediterranean shrubs that contain different kind and amount of nutrient, different flavor and secondary compounds (toxin) may consume a variety of food to avoid toxicities (Freeland & Janzen, 1974). A variety of plant species with different kind of flavor, nutrients and toxins increase food intake and animal production by enabling individuals to select from that are biochemical complementary because nutrient-toxin, as well as toxin-toxin interaction increase food ingestion (Provenza *et al.*, 2003, Rogosic *et al.*, 2003).

In this paper, we discuss reciprocal interrelationships between Mediterranean shrubs and herbivores and show how plant biochemical diversity influences shrub intake by sheep and goats. We studied how different combinations (number) of the Mediterranean shrubs, which contain different amount of nutrient, toxins and flavor influence foliage intake by sheep and goats.

Materials and methods

Shrubs, treatments, animal management, and experimental protocol

Four trials were conducted at the Ecology Experimental Station 25 km from Split in the central part of the Croatian Adriatic coast (43° 52' N; 16° 23' E). In the first trial, six shrubs were offered simultaneously: *Quercus ilex* L. (Fagaceae), *Erica multiflora* L. (Ericaceae), *Arbutus unedo* L. (Ericaceae), *Juniperus phoeniceae* L. (Cupressaceae), *Viburnum tinus* L. (Caprifoliaceae) and *Pistacia lentiscus* L. (Anacardiaceae). In the second trial three shrubs were offered: *Quercus*, *Arbutus* and *Pistacia*, while in the third trial *Arbutus* and *Pistacia* were offered. In the fourth trial only *Pistacia lentiscus* was offered. Shrubs were harvested each week from the vicinity of the feeding trials. Shrub leaves and current season's growth (i.e., twigs) were clipped and ground to 1 cm length with a chipper, mixed for uniformity, placed in woven, polyethylene feed sacks, and stored at 4°C. Every day before the trial, sufficient bags of shrubs to feed the animals were removed from cold storage and offered to the animals.

The sheep (n=12) were crossbred hair-type 8 months of age, whereas the goats (n=12) were purebred Alpines 6 months of age. Each group comprised equal mix of both sexes. Sheep weighed 23.0 kg and goats 21.2 kg at the beginning of the trial and 24.0 and 21.5 kg, respectively, at the end. All animals were raised on the same farm on the island of Brac (Central Dalmatia) and were adapted to the shrubby vegetation of the Mediterranean maquis. All animals had previous browsing experience with these shrubs in a free-grazing setting, and this experience probably influenced the choices made by individual animals during these trials (Distel and Provenza, 1994).

To reduce neophobia, the experimental animals were offered a barley/PEG mix for 30 minutes/day for 5 days before the trials. Similarly, they were also offered each of the 6 shrubs for 120 minutes/day for 5 days before the trials commenced. Throughout the experiments, animals had free access to trace mineral blocks and fresh water. To establish a baseline, *ad libitum* intake of alfalfa was measured for 5 days. After the baseline was established, all animals had a 5-day preconditioning period where 25 g of PEG (molecular weight = 3 350; Spectrum Chemicals, Los Angeles, Ca) mixed with 175 g barley was given from 0800 to 0830, and all 6 shrubs were offered in individual feeding boxes and continually replenished from 0830 to 1400. Shrub intake was monitored, and animals divided into two treatments groups (PEG and controls) based on total shrub intake, by ranking animals and using all odd ranks as one treatment. Animals remained in their respective treatment groups for all 4 trials.

All trials lasted for 10 days and the trials ran consecutively. Sheep and goats in the PEG group received 25 g of PEG and 175 g of barley, and controls received 200 g of barley alone during Trials 1 and 2. During Trials 3 and 4, the amount of barley given to PEG animals was reduced to 100 g/animal mixed with 50 g PEG, and controls were not given barley. In all trials, animals were fed the PEG/barley mixture at 0800, and the ground shrubs at 0830. In all trials, sheep and goats in the treatment groups ate all the PEG-grain mixture within 30 minutes. Shrubs were fed (200 g) in individual boxes to each animal, and the amounts replenished as necessary during the day. Any uneaten amounts were weighed each afternoon at 1400.

During Trial 1, alfalfa pellets in quantities covering 70% of their baseline intake were fed to all animals at 1400. During trials 2, 3 and 4 the amount of alfalfa fed was reduced to 50% of their baseline intake.

Statistical analysis

The total amount consumed of all shrubs offered in each trial was used in the analysis because consumption of each shrub was not independent of the other choices. The experimental design for the PEG trials was a completely random design. Animals were a random factor in the mixed model analysis (SAS, 2000). The model included treatment (PEG vs. control), species of animal (i.e., goats vs. sheep), the species x treatment interaction, with individual animals nested within treatment and species. The model also used days as a repeated measure with all other interactions included. All analyses on shrub intake were adjusted for body weight (g/kg B.W.).

Results and discussion

Importance of biodiversity for small ruminants on Mediterranean rangelands

Total shrub intake increased as the number of shrubs offered increased. Although the 4 trials were not compared statistically, the results clearly showed that combinations of shrubs promoted greater intake in both goats and sheep. In general, treatment and control animals showed a tendency for increasing intake from the first to the last day in Trial 1. Even though sheep and goats differed in total amounts of shrubs eaten (Figure 1, Trial 1 with 6 shrubs), the rank order of the amount eaten for each shrub species was essentially the same for sheep and goats. The mean amounts of the shrubs eaten across all treatments and animal species were: *Viburnum tinus* (15.6 g/kg), *Erica multiflora* (14.4 g/kg), *Arbutus unedo* (8.84 g/kg), *Quercus ilex* (8.67 g/kg), *Pistacia lentiscus* (5.4 g/kg) and *Juniperus phoeniceae* (0.46 g/kg).

When 3 shrubs were offered to sheep and goats, either with or without supplemental PEG, both treatment groups and both species of animals showed a numerical decrease in total shrub intake (Figure 1) from the previous trial. Sheep receiving PEG ate more (P=0.002) total shrubs than did controls, but no such treatment effect was found for goats (Figure 1). Regardless of treatment, sheep and goats ate almost the same amount of total shrubs (34.4 g/kg BW for sheep vs. 41.4 g/kg BW for goats). Sheep and goats preferred *Quercus ilex* (18.58 g/kg) and *Arbutus unedo* (15.95 g/kg) over *Pistacia lentiscus* (3.34 g/kg).

Supplemental PEG positively affected (P<0.001) total shrub intake when only 2 shrubs (*Arbutus* and *Pistacia*) of lower palatability were offered (Figure 1). Both sheep and goats fed supplemental PEG ate more *Pistacia* (17.54 g/kg BW) compared to non-supplemented sheep or goats (7.57 g/kg BW; Figure 1).

When one shrub (*Pistacia lentiscus*) was offered to sheep and goats supplemental PEG had a marked positive (P<0.001) but differential effect on intake depending on animal species (Figure 1). PEG-supplemented goats ate more *Pistacia* (39.6 g/kg BW) than did PEG supplemented sheep (28.1 g/kg BW), whereas control sheep and goats ate similar amounts (12.2 and 15.3 g/kg BW, respectively).

Sheep eat more when offered several foods (3 > 2 > 1) that contain complementary toxins (Burritt & Provenza, 2000; Villalba *et al.*, 2004). These results are consistent with the satiety hypothesis (Provenza *et al.*, 2003), which contends diets and habitats that allow animals to select among alternatives enable individuals to better meet needs for nutrients and to better cope with toxins. All plants contain toxins, and the amount of toxin an animal can ingest depends on the kind and amount of nutrients and toxins in the forages offered. Nutrients and toxins both cause animals to satiate, and excesses of nutrients, nutrient imbalances, and toxins all limit food intake. Thus, individuals can better meet their needs for nutrients and regulate their intake of toxins when offered a variety of foods that differ in nutrients and toxins than when constrained to a single food, even if the food is "nutritionally balanced." Thus, feeding and grazing practices that allow producers to capitalize on biochemical diversity, as opposed to merely taxonomic diversity, are likely to improve performance of the herd. Under this hypothetical framework, goats and sheep should eat small amounts of poorer quality feeds even though other palatable feeds are available for consumption (Provenza *et al.*, 2003). Depending on prevailing conditions, livestock may be able to mix their diets in such a way as to provide sufficient energy and protein while reducing toxin loads.

The shrubs offered in this trial, contained variable types and amounts of secondary compounds, including tannins (*Arbutus, Quercus* and *Pistacia*), terpenes (*Juniper*) and iridoid glycosides and terpenes (*Viburnum tinus*) (Tomassini *et al.*, 1995). The efficacy of plant defenses varies with the mix of plants, and the chemical attributes of a single plant species must be considered within a larger context of the plant community (Bryant *et al.*, 1991). In a mix of plant species, an animal's preference may range along a continuum from strongly aversive, if nutrients and toxins are not complementary, to strongly positive if nutrients and toxins are complementary. Animals are likely to eat more if species differ in toxins, macronutrient concentrations, and flavors (Provenza *et al.*, 2003). Some toxins may be less aversive in plant communities containing high level of nutrients (macronutrient-rich species)

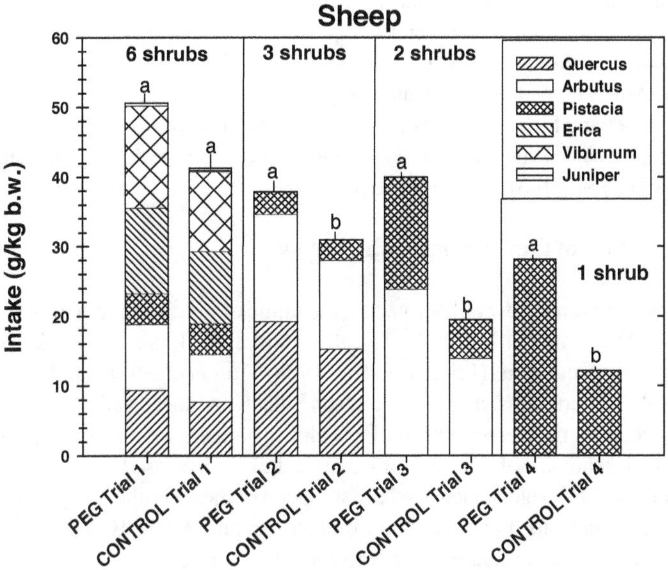

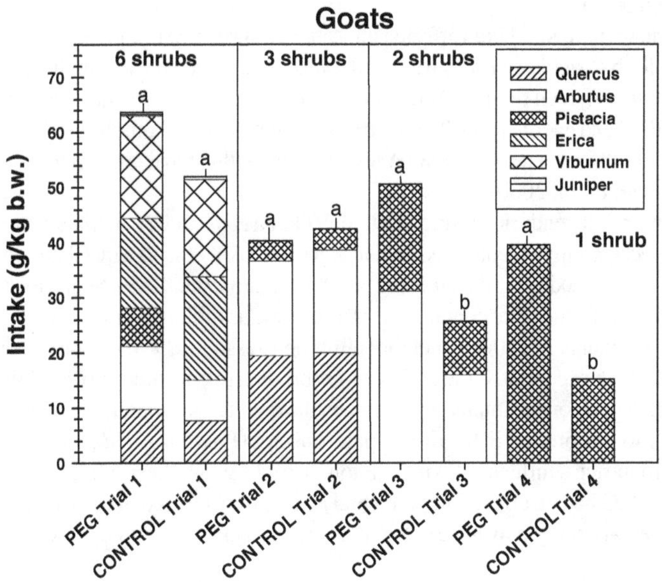

*Figure 1. Experiment with sheep and goats: Trial 1. six shrubs offered together; Trial 2. Three shrubs offered together; Trial 3. Two shrubs offered together, and Trial 4. One shrub (*Pistacia lentiscus*) offered to sheep and goats.*

needed for detoxication (nutrient-toxin complementary; Villalba & Provenza, 1999), or containing complementary toxins (toxin-toxin complementary; Burritt & Provenza, 2000).

We have failed to recognize that meaningful diversity requires species that are *biochemically complementary*, not just taxonomically different (Provenza *et al.*, 2003). Therefore, it is important to identify complementary mixtures in Mediterranean rangelands. Analysis of secondary compounds in Mediterranean shrubs, grasses and forbs will allow managers to elucidate the complementarity among toxins and macronutrients. Understanding such relationships may assist in the design of grazing systems that increase productivity of shrub rangelands and herbivores.

The effect PEG on forage intake of the Mediterranean shrubs

PEG is a polymer that binds to tannins irreversibly over a wide range of pH, thus alleviating the negative effects of tannins (Landau *et al.*, 2000). Supplemental PEG increases intake of tannin-containing plants by sheep and goats (Pritchard *et al.*, 1988; Titus *et al.*, 2000) and cattle (Hanningan & McNeil, 1998). Nevertheless, it is very interesting that when sheep and goats were offered 6 shrubs, and when goats were offered 3 shrubs, PEG alone (both groups had barley) had no beneficial effect on intake. In general, as the number of shrubs in the diet decreased, the impact of PEG on intake of shrubs increased. PEG alone increased consumption of the 3-shrub mix by sheep. The effects of PEG and energy in trials 3 and 4 are confounded because controls had neither supplement. Even so, the results of PEG and energy were substantial. When 3 species were offered to sheep (Trial 2), and 2 or 1 species offered to goats, PEG and energy substantially increased intake of shrubs high in tannin (Figures 1).

Interactions among nutrients, toxins and PEG affected utilisation of shrubs. For example, in Trial 3 the relatively unpalatable shrub *Pistacia lentiscus* was not avoided when offered with another more palatable shrub (*Arbutus unedo*) even though animals were not forced to eat *Pistacia*. Further, goats and sheep receiving PEG and energy in the 2-shrub trial ate much more of both shrubs than did control animals, indicating the effectiveness of PEG/barley in reducing the impacts of tannins from *Pistacia* and *Arbutus* (Banner *et al.*, 2000).

Food selection by herbivores depends on nutritional state (Villalba *et al.*, 2002). In trials 3 and 4, the control animals did not receive supplemental feed. When ingesting toxins entails a metabolic cost, herbivores are likely to respond by modifying their feeding behavior (Foley *et al.*, 1999). Supplemental macronutrients increase intake of foods that contain toxins as diverse as lithium chloride (Wang & Provenza, 1996), terpens (Banner *et al.*, 2000), menthol (Illus and Jessop, 1996), and quebracho tannin (Villalba *et al.*, 2002). Consequently, the ability of herbivores to consume shrubs high in secondary compounds will depend on the interaction between the quantity and quality of nutrients available and the classes of toxins present in the mix of plant species (Provenza *et al.*, 2003). The kinds and amounts of macronutrients influence toxin satiation. In trial 1 when sheep and goats were given supplemental barley, PEG had no effect. Supplemental energy may be used in detoxification pathways, and the supplemental energy may have attenuated any PEG effect on shrub intake.

Comparative responses of sheep and goats

In all 4 trials, goats ate more total shrubs than did sheep, suggesting goats have a higher tolerance for secondary compounds. Further, supplemental PEG and energy only affected shrub intake by goats when either 2 or 1 high-tannin shrub(s) were offered. Sheep, on the other hand, showed a positive response to PEG alone when 3 shrubs were fed in Trial 2. This suggests that the threshold for toxic (i.e., tannin) effects is higher in goats than in sheep.

Goats typically eat more browse than sheep, and utilize tannin-rich foods better than do sheep (Landau *et al.*, 2000). Food intake and dry matter digestibility of tannin-containing forages are often higher for goats than for sheep (Silanikove *et al.*, 1996), and goats often use protein more efficiently than sheep (Kronberg & Malechek, 1997). Difference in ruminal fermentation and adaptation of rumen microbes to tannins may also enable goats to use more efficiently tannin-rich foods (Landau *et al.*, 2000). The ruminal degradation is a primary mechanism for neutralizing the anti-nutritional effect of tannins in goats.

Some animals also have adaptative mechanisms such as production of proline-rich salivary proteins that bind tannins and minimize their potential adverse effects (Robbins *et al.*, 1987). Sheep and goats do not produce proline-rich salivary proteins (Austin *et al.*, 1989), but goats secrete more saliva containing a higher level of nitrogen than do sheep (Domingue *et al.*, 1991). A 50% reduction in tannins in extrusa samples from the esophagus of goats consuming blackbrush, a tannin-containing shrub (Provenza & Malechek, 1984), also suggest that even though praline-rich proteins may not be present in the saliva of goats, other salivary proteins contribute to forming complexes with tannins thereby alleviating their negative effects.

Management implications

Conservation of biodiversity is an important aspect of Mediterranean natural resource management. Many factors affect biodiversity including herbivory, which influences the structure and dynamics of plant and animal communities. Herbivory has the potential to either increase or decrease species diversity and abundance. The implications of our studies for managing Mediterranean shrublands for increased biodiversity are promising. Increasing use of these shrubs by livestock would likely enhance the production of grasses and forbs and create a more diverse mix of plants.

Most shrubs contain large quantities of secondary compounds that limit intake and cause animals to eat a variety of foods. In our studies, as the number of shrub species in the diet increased, so did intake. Indeed, when sheep and goats were offered 6 shrubs, and when goats were offered 3 shrubs, PEG alone had no beneficial effect on intake. Our findings suggest that biochemical diversity plays a very important role in herbivore's diet selection enabling animals to better meet their nutritional needs and avoid toxicity. Finally, PEG alone had a greater influence on sheep than on goats, and it had the most influence on both sheep and goats when only one or two foods were available.

References

Augner, M., F.D. Provenza & J.J. Villalba, 1998. A rule of thumb in mammalian herbivores? Anim. Behav. 56: 337 - 345.

Augner, M. & E.A. Bernays, 1998. Plant defence signals and Batesian mimicry. Evol. Ecol. 12: 667-679.

Austin, P.J., L.A. Suchar, C.T. Robbins & A.E. Hagerman, 1989. Tannin-binding proteins in salvia of deer and their absence in saliva of sheep and cattle. J. Chem. Ecol. 15: 1335-1347.

Banner, R.E., J. Rogosic, E.A. Burritt & F.D. Provenza, 2000. Supplemental barley and activated charcoal increase intake of sagebrush by lambs. J. Range Manage. 53: 415-420.

Bryant, J.P., F.D. Provenza, J. Pastor, P.B. Reichardt, T.P. Clausen & J.T. Du Toit, 1991. Interaction between woody plants and browsing mammals mediated by secondary metabolites. Annu. Rev. Ecol. Syst. 22: 431-446.

Burrows, G.E., & R.J. Tyrl, 2001. Toxic plants of North America. Iowa State University Press, Ames, IA, pp. 120-122

Burritt, E.A. & F.D. Provenza, 2000. Role of Toxins in Intake o Varied Diets by Sheep. J. Chem. Ecol., Vol. 26, No. 8: 1992-2005.

Distel, R.A.& F.D. Provenza, 1994. Effect of early experience on voluntary intake of low-quality roughage by sheep. J. Anim. Sci. 72: 1191-1195

Domingue, B.M., D.W. Dellow & T.N. Barry, 1991. The efficiency of chewing during eating and ruminating in goats and sheep. Br. J. Nutr. 65: 355-363.

Dziba, L.E., F.D. Provenza, J.J. Villalba & S.B. Atwood, 2004. Influence of terpenes and nutritional supplementation on intake of sagebrush by sheep. Range Ecology and Management. (In press).

Foley, W.J., G.R. Iason & C. McArthur, 1999. Role of plant secondary metabolites in the nutritional ecology of mammalian herbivores: How far have come in 25 years? In H.G. Fahey, Jr. (Ed.), Nutritional Ecology of Herbivores. Proc. Vth Int. Symp. Nutr. Herb. Am. Soc. Anim. Sci., II, pp. 130-209

Freeland, W.J. & D.H. Janzen, 1974. Strategies in herbivory mammals: The role of plant secondary compounds. Am. Nat. 108: 269-288.

Hanningan, N.A. & D.M. McNeil, 1998. Cattle preference for two genotypes of fresh leucaena folloving the manipulation of their tannin content with polyethylene glycol. Proc. Aust. Soc. Anim. Prod. 22: 401 (Abstr.).

Harshberger, J.W., 1920. Pastoral and Agricultural Botany. Blakiston's Son & Co, Philadelphia, p. 87.

Hodgson, J. & A.W. Illius, editors. 1996. The ecology and management of grazing systems. CAB International, Wallingford, UK.

Horvatic, S., 1958. Typological division of the Adriatic littoral vegetation of the garrigues and pine forest. Acta. Bot. Croat. 17: 7-78. (In Croatian)

Illus, A.W. & N.S. Jessop, 1996. Metabolic constraints on voluntary intake in ruminants. J. Anim. Sci. 74: 3052-3062.

Kronberg, S.L. & J.C. Malechek, 1997. Relationship between nutrition and foraging behavior of free-ranging sheep and goats. J. Anim. Sci. 75: 1756-1763.

Landau, S., A. Perevolotsky, D. Bonfil, D. Barkai & N. Silanikove, 2000. Utilization of low-quality resources by small ruminants in Mediterranean agro-pastoral systems: the case of browse and aftermath cereal stubble. Livest. Prod. Sci. 64: 39-49.

Landsberg, J., T. O'Conner & D. Freudenberger, 1999. The impact of livestock on biodiversity in natural systems, pp. 752-777, in Jung H.-J.G. and G.C. Fahey, Jr. (Eds). Nutritional Ecology of Herbivores, Proceeding of the 5th International Symposium on the Nutrition of Herbivores. Am. Soc. Anim. Sci. Savoy, Illinois.

Launchbaugh, K.L. & F.D. Provenza, 1993. Can plant practice mimicry to avoid grazing by mammalian herbivores? Oikos 66: 501-504.

Launchbaugh, K.L., F.D. Provenza & E.A. Burrit, 1993. How herbivores track variable enviroments: Response to variability of phytotoxins. J. Chem. Ecol. 19: 1047-1056.

Naveh Z., 1972. The Role of Shrubs Ecosystem in Present and Future Mediterranean Land Use. In "Wildland Shrubs, Their Biology and Utilization", pp. 414-427.

Odum, E.P., 1969. The Strategy of Ecosystems Development. Science 164, 262-270.

Pritchard, D.A., D.C. Stocks, B.M. O'Sullivan, P.R.Martins, I.S.Hurwood, & P.K. O'Rourke. 1988. The effect of polyethylene glycol (PEG) on wool growth and live weight of sheep consuming a Mulga (*Acacia aneura*) diet. Peoc. Aust. Soc. Anim. Prod. 17: 290-293.

Provenza, F.D. & J.C. Malechek, 1984. Diet selection of domestic goats in relation to blackbrush twig chemistry. J. Appl. Ecol. 21: 831-841.

Provenza, F.D. 1995. Postingestive feedback as an elementary determinant of food 21 preference and intake in ruminants. J. Range Manage. 48: 2-17

Provenza, F.D.1996. Acquired Aversion as the Basis for Varied Diets of Ruminants Foraging on Rangelands. J. Anim. Sci. 74: 2010-2020.

Provenza, F.D., J.J. Villalba, C.D. Cheney, & S.J. Werner. 1998. Self-organization of foraging behavior: From simplicity to complexity without goals. Nut. Res. Rev. 11: 199-222.

Provenza, F.D., J.J. Villalba, L.E. Dziba, S.B. Atwood & R.E. Banner, 2003. Linking herbivore experience, varied diets, and plant biochemical diversity. Small Rum. Res. 49: 257-274.

Ralphs, M.H., F.D. Provenza, W.D. Wiedemeier & F.B. Bundersom, 1995. Effect of energy source and food flavor on condition preference in sheep. J. Anim. Sci. 73: 1651 - 1657.

Ralphs, M.H. 1992. Conditioned food aversion: Training livestock to avoid poisonous plants. J. Range Manage. 45: 46-51.

Rhoades, D.F., 1979. Evolution of plants chemical defense against herbivores, pp. 3 - 53, in G.A. Rosenthal and D.H. Janezen (Eds), Herbivores: Their Interaction with Secondary Plant Metabolites. Academic Press. New York.

Robbins, C.T., S. Mole, A.E. Hagerman & T.A. Hanley. 1987. Role of tannins in defending plants against ruminants: Reduction in dry matter digestion. Ecology 68: 1606-1615.

Rogosic J. 2000. Management of the Mediterranean Natural Resources. Skolska naklada Mostar. ISBN 9958-774-00-, COBISS/BiH – ID 8755974, pp. 1-352. (In Croatian).

Rogosic, J., J.A. Pfister & F.D. Provenza. 2003. Interaction of Tannins and Saponin in Herbivore Diet. VII International Rangelands congress: Rangelands in the New Millennium. Durban, South Africa 26 July-1 Aug., pp. 104-105.

Silanikove, N., N. Gilboa, I. Nir, A. Perelovsky & Z. Nistan, 1996. Effect of a daily supplementation of polyethylene glycol on intake and digestion of tannins-containing leaves (*Quercus calliprinos, Pistacia lentiscus* and *Ceratonia siliqua*) by goats. J. Agric. Food Chem. 44: 199-205.

Tomassini, L., M.F. Cometa, S. Foddai & M. Niciletti, 1995. Iridoid glycosides from *Viburnum tinus*. Phytochemistry, 38, 423-425.

Titus, C.H., F.D. Provenza, A. Perelovsky, N. Silanikove & J. Rogosic, 2001. Supplemental polyethylene glycol influences preferences of goats browsing blackbrush. J. Range Manage. 54: 161-165.

Villalba, J.J. & F.D. Provenza, 1999. Nutrient -specific preference by lambs conditioned with intraruminal infusions of starch, casein and water. J. Anim. Sci 77: 378-387.

Villalba, J.J. & F.D. Provenza, 2000. Roles of flavor and rewards intensities in acquisition and generalization of food preference: Do strong plant signals always deter herbivory. Journal of Chemical Ecology, Vol. 26, No. 8, pp. 1911-1921.

Villalba, J.J., F.D. Provenza & R.E. Banner. 2002 . Influence of macronutrients and polyethylene glycol on utilization of toxic-containing food by sheep and goats. II. Responses to quebracho tannin. J. Anim. Sci. 80: 3154-3164.

Villalba, J.J., F.D. Provenza, & H. GouDong, 2004. Experience influences diet mixing by herbivores: Implications for plant biochemical diversity. Oikos 107: 100-109.

Wang, J. & F.D. Provenza, 1996. Food deprivation affect preference of sheep for foods varying in nutrients, and a toxin. J. Chem. Ecol. 23: 275-288.

The conservation of natural resources as an extrinsic quality attribute of lamb in Mediterranean areas

A. Bernués[1], J.L. Riedel[1], I. Casasús[1] & A. Olaizola[2]

[1]Centro de Investigación y Tecnología Agroalimentaria, Gobierno de Aragón, P.O. Box 727, 50080 Zaragoza, Spain
[2]Departamento de Agricultura y Economía Agraria, Universidad de Zaragoza, Miguel Servet 177, 50013, Zaragoza, Spain

Summary

This paper focus on the role that extensive sheep systems play on the conservation of grazing resources and landscape and on how producers can translate this condition into lamb quality products that satisfy better new consumer demands. First, the environmental benefits of extensive sheep farms on the natural resources of the "Sierra de Guara" Natural Park (Aragón, Spain) are demonstrated. Second, results of a consumer's survey in the same region, showing the increasing importance of environmental friendly production as a credence quality attribute of lamb and the advantages of specific quality lamb products, are presented. The paper concludes that extensive livestock production systems can have comparative advantages if farmers' organisations and small meat companies are able to translate the concerns of consumers in terms of environmental conservation into consumer-led quality products.

Keywords: sheep grazing systems, environment, credence quality, lamb, consumers.

Introduction

Extensive livestock systems, normally located in mountain and other less-favoured areas, have suffered a strong decline in the last decades. Concurrently, a process of intensification of sheep farming management could be observed in Spain in the last decades. As a consequence, there has been a decrease in the utilization, or even abandonment, of large pastoral areas (Bernués *et al.*, 2005). This process has originated changes in vegetation dynamics (invasion of shrub and forest vegetation) (Riedel *et al.*, 2004) and an increase in the risk of environmental hazards, of which the most important one in Mediterranean areas is forest fires. Changes in landscape conformation can also be observed.

Currently, there is an increasing interest in using and preserving grazing resources, especially those located in protected natural areas. Livestock systems based on grazing are considered as a cost-effective instrument to modulate the strong dynamic of vegetation towards shrub invasion (Casasús *et al.*, 2003). If adequately managed, grazing livestock can be a suitable tool to maintain traditional landscapes in protected areas and prevent forest fires (Kramer *et al.*, 2003).

Multifunctionality of grazing farming systems implies that, together with the traditional productive and economic objectives, wider social objectives need to be taken into account.

Together with the environmental interest described above, the new social concerns in terms of food quality and food safety are increasingly important. Food quality, as perceived by consumers, is a subjective, multi-dimensional and dynamic concept (Bernués *et al.*, 2003b). Meat credence quality attributes, those that cannot be ascertained even after consumption (Steenkamp, 1990) and which mainly focus on the production process rather than the product itself (extrinsic attributes) (Becker, 2000), are more and more important, in response to growing consumer demands about safety, health and ethical concerns (environmental friendly production, animal welfare, etc.) (Harrington, 1994; Wandel & Bugge, 1996). Hence, the demand for meat and meat products that deliver specific quality attributes is increasing across the EU.

In this context, extensive livestock production systems could have comparative advantages because their products are often perceived to be of higher quality or even 'safer' (credence quality) by a increasing number of consumers. The translation of extrinsic quality attributes (those that depend on the production process) into consumer-led quality lamb can constitute a useful strategy for farmer associations and small meat companies which can target better certain consumer segments.

In the case of protected natural areas, like the "Sierra de Guara" Natural Park (SGNP), the conservation of natural resources by livestock is a characteristic inherent to the production system. Therefore, this feature of the production process could become an extrinsic quality attribute of the lamb produced in the Park.

The role of sheep systems in preserving natural resources and landscape

Husbandry in Mediterranean areas has been traditionally considered as an "environmentally friendly" activity, in the sense that the production systems and techniques do not cause significant damage to the environment. Nevertheless, is necessary to expand the scope of this vision and to recognize the role of extensive livestock as an essential and effective tool for conservation of natural resources and landscape in many natural areas.

Sustainability of these areas depends on their capacity to reproduce their characteristics in terms of soil fertility, wildlife, bio-diversity, landscape and human activities (Thompson and Nardone, 1999). Grazing livestock plays an essential role to maintain the equilibrium among these components in agro-silvo-pastoral ecosystems.

The 'Sierra de Guara' Natural Park (SGNP) (81 491 ha) is a typical Mediterranean karstic mountain area located in the Aragón, north-east of Spain. The grassland vegetation of the Park is very diverse (as consequence of great variation in topography, soil and climatic conditions), ranging from typical Mediterranean pastures to Alpine and Atlantic pastures in more humid and higher areas. The Park has suffered a strong decrement of farming activities, with the consequent abandonment of pastoral areas that has originated changes in vegetation dynamics (invasion of shrub and forest vegetation) and, therefore, in landscape composition (Riedel, 2004).

An investigation started in 2000 aiming to analyse the evolution of sheep farming systems in the SGNP and to measure the real effect that domestic grazing animals had on vegetation and landscape. The main results from this second objective are presented now.

Six representative grazing areas were selected according to geographical location, grazing management and history of use. Herbaceous and shrub vegetation were compared inside (non-grazed) and outside (grazed) 12 fenced areas (2 per grazing site). Control measurements took place before (spring) and after grazing (autumn) in three consecutive years (2000-2003). Species composition, herbaceous and shrub biomass, dead:green ratio of herbage and chemical composition were analysed [See Riedel (2004) for details].

A general pattern of vegetation change was observed in all grazing areas under control. The grazing activity of animals maintained herbaceous biomass throughout the period of the study in grazed areas, while it accumulated in non-grazed areas (500 kg DM/ha/year). Differences between grazed and non-grazed areas were highly significant ($P<0.001$) from the second control measurement onwards (autumn 2001) (Figure 1).

When considering the green and dead fractions separately, no significant differences existed between grazed and non-grazed areas for green biomass. However, senescent material accumulated ($P<0.001$) in non-grazed areas, due to the lack of removal by herbivores (Figure 2). Therefore, accumulation of herbage biomass was only due to a strong increase in the amount of dead fraction, whereas green biomass followed the same pattern in all areas.

For shrub vegetation, accumulation of biomass could be observed in both grazed and non-grazed areas (Figure 3). However, when considering the whole period (2001-2003), significant differences in the increment rates were observed in non grazed vs. grazed areas (80.0% and 42.2% respectively; $P<0.01$). It could be observed that grazing activity reduced but did not stop the invasion of shrub vegetation.

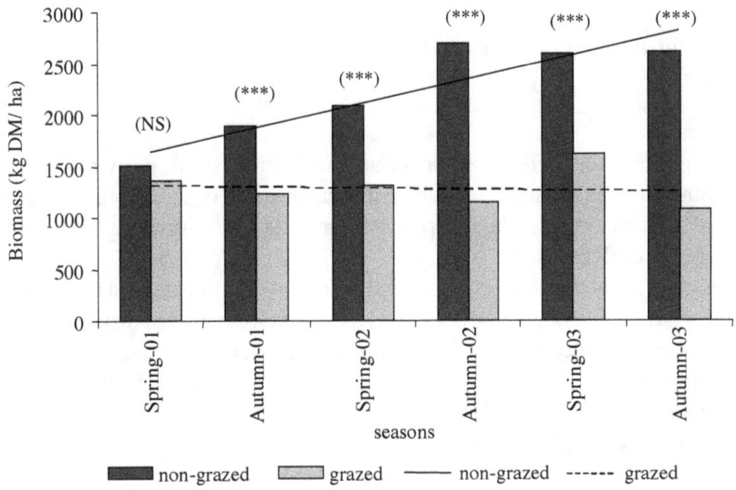

Figure 1. Changes in herbaceous biomass in the SGNP (2001-2003). Source: Riedel (2004).

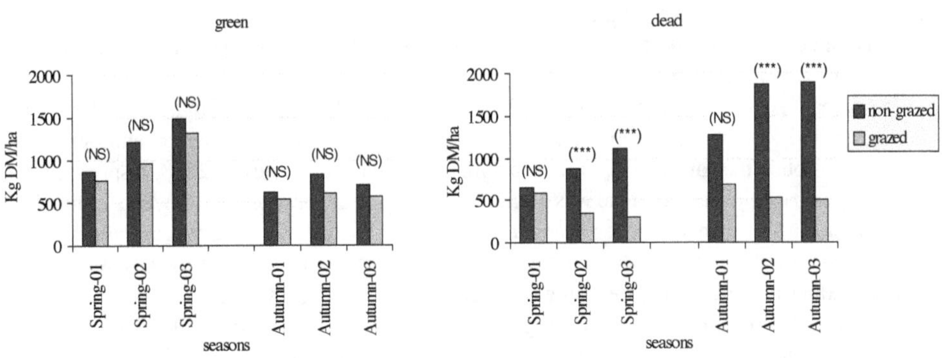

Figure 2. Changes in herbaceous green and dead biomass in the SGNP (2001-2003). Source: Riedel (2004).

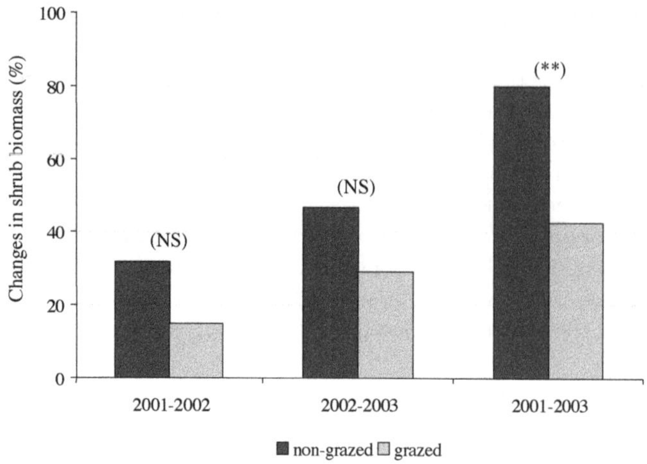

Figure 3. Changes in shrub biomass in grazed and non-grazed areas (2001-2003). Source: Riedel (2004).

The processes of change described above have important implications from an environmental point of view. The accumulation of dead herbaceous biomass and invasion of shrubs produce deep changes in the traditional "green short grass" landscape (Monserrat, 2001) and increases notably the risk and virulence of forest fires (Kramer *et al.*, 2003).

Although longer periods of study need to be considered to assess long-term changes in vegetation, specially for shrubs, the study demonstrated and quantified the role that extensive grazing animals play as regulators of the vegetation succession in the areas under control. From the point of view of the authorities in charge of managing the Park, the authors concluded that the conservation of large pastoral areas would require the establishment of specific policies aiming at maintaining or promoting the use of these areas with ruminants (Bernués *et al.*, 2005).

Preservation of the environment as a quality attribute of lamb

The role that extrinsic meat attributes, such as environmentally friendly production, respect for the animal welfare, etc., play in the consumer quality evaluation process is a controversial topic in consumer research at the present time.

In a research carried out in 5 EU regions located in France, Spain, Italy, England and Scotland, Bernués *et al.* (2003a) studied the importance of extrinsic attributes in the consumers' quality evaluation process. To obtain the data, a direct questionnaire in a sample of people (N=2288) responsible for the meat purchases in the household was designed. Among other questions, respondents were asked to report on the importance of seven characteristics (extrinsic attributes) to achieve quality in beef/ lamb: origin of meat/region of production, environmentally friendly production, animal welfare concerns, animal feeding, animal breed, processing and packaging and storage of meat.

For the lamb consumer sample in 3 regions (Cotswold, South-west of England; Languedoc-Roussillon, South-east of France; Aragón, North-east of Spain), the extrinsic attributes of meat that were more appreciated were animal feeding (83.2% of respondents rated this attribute as 'important' or 'very important') and origin of meat (85.7%) (Figure 4). Environmentally friendly production and

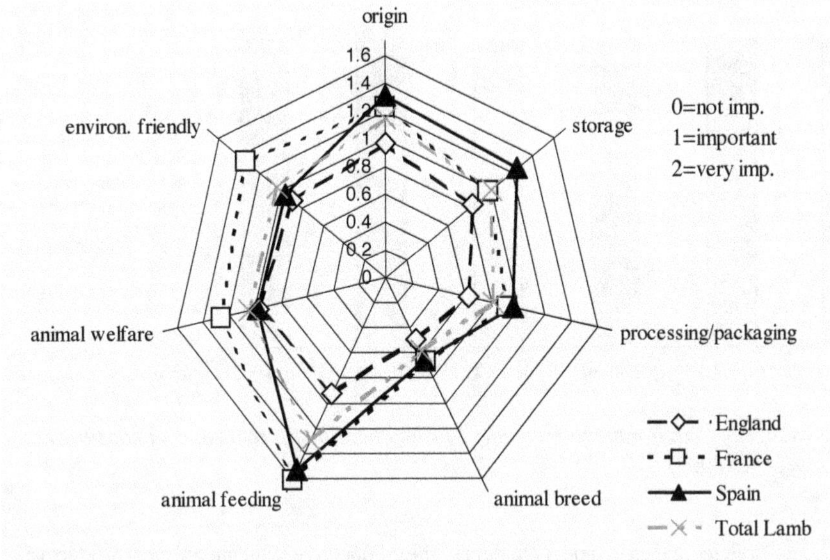

Figure 4. Importance of extrinsic attributes for lamb consumers per region.
Source: Marketing red meat in the European Union: extending the options FA-S2-98-909, final report.

animal welfare were also relatively important (75.9% of consumers considered environmentally friendly production as 'important' or 'very important'. This Figure was 76.7% in relation to animal welfare). Other attributes were considered less important.

In Aragón (N=300), the origin or region of production and the animal feeding regime were also considered to be the most important extrinsic attributes to define quality in lamb. Regarding the characteristics of the production system, the percentage of consumers that consider environment-friendly production as an important attribute in obtaining high quality lamb was similar to the average, although lower than other attributes like type of storage.

Moreover, when analysing the relationships between the variables that better explained the heterogeneity of consumers trough multivariate analytical methods, for both the global (Bernués *et al.*, 2003a) and the Spanish sample (Olaizola *et al.*, 2001) the most important factor identified was an "*ethical factor*", corresponding to consumer concerns on environmentally friendly production and animal welfare.

When establishing groups or segments of consumers according to the importance that extrinsic quality attributes had for them in Aragón (Olaizola *et al.*, 2001), 5 groups were identified (Figure 5).

In Group 3 (19% of the sample), 82% of respondents considered environmental friendly production as a very important attribute to define quality. In Groups 2 and 5 (37% and 14% of the sample, respectively) this attribute was mostly considered important. Only consumers of Group 4 (16% of the sample) did not attach any importance to the environment in relation to lamb quality.

Advantages of "typical" quality lamb

The demand for quality meat products, those that deliver specific quality attributes, seems to be increasing across the EU. The promotion of quality is regarded as a major driver of new EU

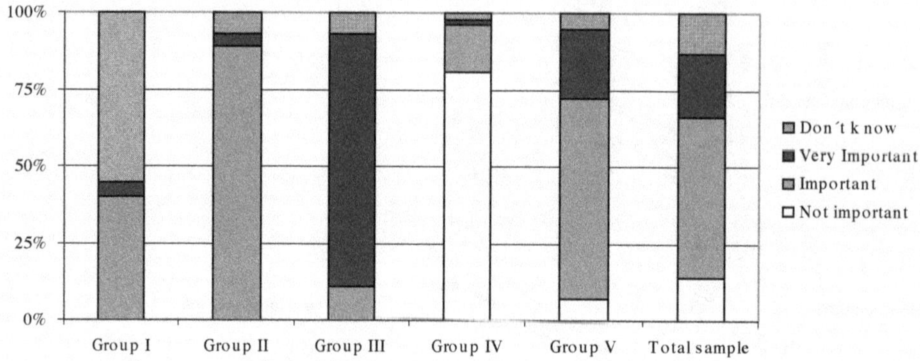

Figure 5. Importance of "environmental friendly" production in different types of consumers. Source: Olaizola et al. *(2001).*

agro-environmental policies and an essential element of the competitive strategies of meat producers and companies. These strategies aim at: differentiate meat products from those of competitors by enlarging product attractiveness or assuring the consumer a certain level of product quality; add value to the food product; have better access quality-driven segments of the market; increase consumer loyalty; and restore consumer confidence in meat products, that are suffering from a tarnished image.

When considering consumer attitudes towards specific quality lamb products (public or private collective brands and Protected Geographic Indications) in the same sample of consumers as described in the previous section, Bernués *et al.* (2003b) found that, together with the satisfaction derived from these products and the region of origin, one of the main factors positively influencing the purchase of this type of meats was the perception of higher guarantees delivered.

Similarly, the main factors that respondents felt could affect an increase in consumption of these products were a reduction in price, the availability of more information and better product guarantees. The lack of reliable information/guarantees delivered has been often highlighted as one of the main reasons that is causing the decrease in the consumption of red meat products in Europe (Corcoran *et al.*, 2000).

Many differences were observed between consumers of lamb quality products and those who did not (Table 1). Consumers of quality lamb tended to attach greater importance to extrinsic attributes such as environmental friendly production (the only attribute highly significant in all regions) and animal welfare. Consumers increasingly hold ethical concerns in relation to the impact of intensive rearing methods (Harrington, 1994; Wandel & Bugge, 1996) and therefore these attributes can become indicators of ethical demands in some groups of consumers. However, as Wandel and Bugge (1996) pointed out, the expressed concerns of consumers in relation to environmental and animal welfare issues do not mean that behaviour has changed accordingly.

When consumers were questioned on the sources of information they trusted more to assess the quality of these products, label and brand, as might be expected, formed important informational cues in all regions. As described in other studies (Nayga, 1999; Bernués *et al.*, 2003c), consumers who were more concerned with quality (safety, nutrition, health) also demanded more information and tended to rely on and use labels more.

In terms of labelling requirements, the name of the brand was a distinctive item between consumers of quality and standard lamb in all regions except in Spain. A possible explanation for this is that most meat is sold unbranded in this country. Origin of meat was important for consumers who had bought quality lamb as opposed to those who had not. A similar phenomenon could be observed for information related to system of production, meat traceability and system of quality control. Therefore, labelling of quality meat products seems to be essential to transmit information that target better concerns of consumers in terms of origin, characteristics of the production system and the traceability and quality control systems implemented.

Linking producers and consumers: consumer-led product development

The red meat industry, in particular the beef and lamb industry, is experiencing a long-running crisis, accentuated by recent sanitary scandals (Latouche *et al.*, 1998). Lack of consumer-oriented communication from the industry and consumer mistrust has often been given as one of the main problems of the meat sector (Northen, 2000).

From a marketing perspective, the aim of meat producers/suppliers should be to understand consumers' tangible and intangible demands with respect to meat quality and then to translate these into intrinsic (product) or extrinsic (production process) characteristics to respond to these demands.

Delivering the quality attributes demanded by the consumer, together with impartial and reliable information (cues), constitutes key actions to enable many meat industries to stay in business or to expand (Corcoran *et al.*, 2000). Within this context, new product development is a major success factor in competitive meat markets (Grunert and Valli, 2001), where the product is still mostly

Table 1. Chi-square analysis between buyers of quality lamb products and extrinsic quality characteristics; sources of information on quality and labelling preferences[1].

	England	France	Spain
Extrinsic quality	Environ. friendly**	Origin/region prod.***	Environ. friendly**
Attributes	Processing/pack.*	Environ. friendly*** Animal welfare** Animal breed**	Animal welfare** Storage**
Quality cues	Direct assessment** Label/brand** Price**	Label/brand*	Label/brand**
Labelling preferences	Brand name** Origin of meat*** Nutritional info.* Consumed by** Cooking inst.** Name of cut** Traceability/Q control**	Brand name** Origin of meat*** Nutritional info.** Maturation time* Name of cut** Production system*** Traceability/Q control**	Origin of meat** Consumed by* Name of cut** Production system**

[1]Text in table means positive relationship between consumers that have tried specific quality lamb and the importance give by the consumers to the variable.
Significances are: * $0.1 > P \geq 0.05$; ** $0.05 > P > 0.001$; *** $P \leq 0.001$. Source: Bernués *et al.* (2003b).

unbranded. Consumer-led product development should incorporate the emerging credence quality attributes that are important for an increasing number of consumers.

Figure 6 represents a conceptual model of industry supply, consumer perception and demand of food quality. The basic structure of the model fits with the 3 stages in the process of quality evaluation proposed by Steenkamp (1990): cue acquisition and categorisation; quality attribute belief formation; and integration of quality attribute beliefs into overall quality evaluation. The relationships between product characteristics (technical specifications), informational cues (cost cues, intrinsic, extrinsic) and attributes (quality judgments) are made explicit. Purchasing and consumption stages are separated and expected quality is differentiated from experience and credence quality. Meal preparation is also recognised as a very important factor for experience quality. The supply of quality by the industry is specifically represented, emphasizing the different stages of the chain and the implications for the intrinsic, extrinsic and cost characteristics of the product. Finally, the overall perceived quality, together with the dynamic and increasingly diverse personal and environmental factors, determine the purchasing motives, that are linked with credence and expected quality.

From this process the industry can work to translate purchase motivations (that integrate previous quality experience, values and concerns of the consumer, usage goals, etc. and also influences from the socio-economic environment) into consumer-led commercial strategies such as development of new products/attributes, further segmentation of the markets, etc.

Extensive livestock systems can have comparative advantages over intensified/industrialized production systems it they are able to demonstrate and deliver quality attributes that are related to the system of production (animal welfare, natural/traditional way of production, animal feeding assurance, protection of the environment in rural areas, etc). Translated to the case of the producers located in the SGNP (Figure 7), the strategy could consist in the exploitation of the extrinsic quality attribute "conservation of natural resoruces and landscape" of the lamb produced in the Park.

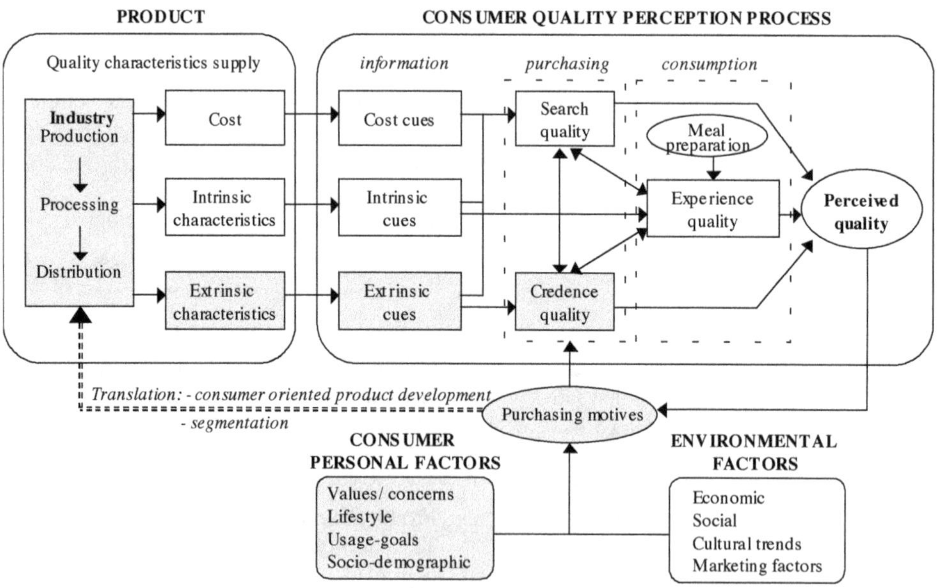

Figure 6. Supply, perception and demand of food quality.
Source: Bernués et al. *(2003a).*

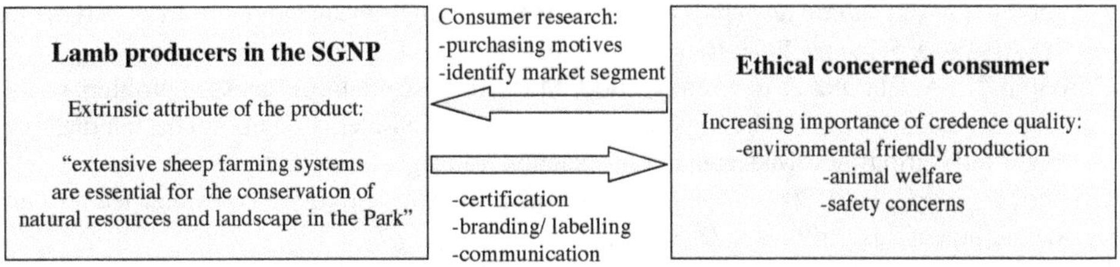

Figure 7. The process of consumer-led product development for lamb products in SGNP.

A number of steps need to be fulfilled for this strategy to be successful. First, it is necessary to perform marketing research to identify segments of consumers with preference criteria for extrinsic attributes related to their ethical concerns, specifically the protection of the environment. When analysing the purchasing motives of these consumers, another advantage is the fact that extrinsic attributes are very often related to consumers' credence quality in terms of meat safety, nutrition and health (Becker, 2000; Northen, 2000, Bernués *et al.*, 2003a).

As mentioned above, lack of consumer-oriented information is one of the main reasons for consumer mistrust of the red meat industry. Therefore, another stage would be the communication of the specificities of the quality lamb produced in the Park. For this communication to be effective and trusted by consumers, two conditions become essential.

It is necessary to differentiate the product trough the creation of a collective brand and/or label. This brand or label needs promotion in order to create product awareness. Meat labels and advertising need to go together to produce a consistent product image.

Last but not least, signalling of cues that inform on attributes of a production system in the label also requires independent and credible certification to reassure consumers of the quality specifications offered by the product.

References

Becker, T., 2000. Consumer perception of fresh meat quality: a framework for analysis. British Food Journal 102 (3), 158-176.

Bernués, A., A. Olaizola & K. Corcoran, 2003a. Extrinsic attributes of red meat as indicators of quality in Europe: an application for market segmentation. Food Qual. Prefer. 14(4): 265-276.

Bernués, A., A. Olaizola & K. Corcoran, 2003b. Consumer attitudes towards specific beef and lamb quality products in different European regions. In: Product quality based on local resources and its potential contribution to improved sustainability, 6th International Livestock Farming Systems Symposium EAAP, Benevento, Italy, 26-29 August 2003.

Bernués, A., A. Olaizola & K. Corcoran, 2003c. Labelling information demanded by European consumers and relationships with purchasing motives, quality and safety of meat. Meat Sci.65: 1095-1106.

Bernués A., J.L. Riedel, M.A. Asensio, M. Blanco, A. Sanz, R. Revilla & I. Casasús, 2005. An integrated approach to study the role of grazing farming systems in the conservation of rangelands in a protected natural park (Sierra de Guara, Spain). Liv. Prod. Sci. (in press)

Casasús, I., A. Bernués, A. Sanz, J.L. Riedel & R. Revilla, 2003. Utilization of Mediterranean forest pastures by suckler cows: animal performance and impact on vegetation dynamics. In: Animal production and natural resources utilisation in the Mediterranean mountain areas. XII symp. HSAP-FAO. 5-7 June 2003, Ioannina, Greece, pp. 65.

Corcoran, K., A. Bernués & R. Baines, 2000. Marketing Scottish beef and the problem of the changing consumer. In: 10th annual World Food and Agribusiness Congress of the International Food and Agribusiness Management Association. Chicago, June 2000.

Grunert, K.G. & C. Valli, 2001. Designer-made meat and dairy products: consumer-led product development. Liv. Prod. Sci. 72, 83-98.

Harrington, G., 1994. Consumer demands: major problems facing industry in a consumer-driven society. Meat Sci. 36, 5-18.

Kramer, K., T.A. Groen & S.E. van Wieren, 2003. The interacting effects of ungulates and fire on forest dynamics: an analysis using the model FORSPACE. Forest Ecol. Manag. 181, 205-222.

Latouche, K., P. Rainelli & D. Vermersch, 1998. Food safety issues and the BSE scare: some lessons from the French case. Food Policy 23(5), 347-356.

Monserrat, P., 2001. Importancia gestora y social del pastoralismo. Arch. Zootec. 3, 491-499.

Nayga, R.M., 1999. On consumers' perception about the reliability of nutrient content claims on food labels. Journal of International Food and Agribusiness Marketing 11(1), 43-55.

Northen, J. R., 2000. Quality attributes and quality cues: effective communication in the UK meat supply chain. British Food Journal 102(3), 230-245.

Olaizola A., A. Bernués, E. Manrique & M.T. Maza, 2001. Quality cues for lamb: identification of consumer profiles in Aragón, Spain. In: The food consumer in the early 21st century, 71st EAAE Seminar, Zaragoza, Spain, 19-20 April 2001.

Riedel, J.L., 2004. Interaciones entre el ganado y la vegetación de los pastos del Parque Natural de la Sierra y Cañones de Guara: implicaciones para la gestión de este espacio natural protegido. Thesis Master of Science CIHEAM, Zaragoza, Spain, pp. 183.

Riedel J. L., I. Casasús, A. Sanz, M. Blanco, R. Revilla & A. Bernués, 2004. Extensive livestock systems as tools for environmental management: impact of grazing on the vegetation of a protected mountain area. In: Abstracts of the Silvopastoralism and Sustainable Management International Congress, Universidad de Santiago de Compostela, Spain, p.148.

Steenkamp, J.-B.E.M., 1990. Conceptual model of the quality perception process. J. Bus. Res. 21, 309-333.

Thomson, P.B. & A. Nardone, 1999. Sustainable livestock production: methodological and ethical challenges. Liv. Prod. Sci. 61, 111-119.

Wandel, M. & A. Bugge, 1996. Environmental concern in consumer evaluation of food quality. Food Qual. Prefer. 8(1), 19-26.

Goat milk cheese: A mean of development of the Northern Moroccan provinces

A. Boukallouch

Association Nationale Ovine et Caprine, 457, Avenue Hassan II, Résidence Mariam Appartement B10, Agdal, Rabat, Morocco

Introduction

The history of the National Association for Sheep and Goat Breeders (ANOC) goes back to 1955, the year the Sheep Union for the Improvement and Utilization of Selected Races was created. It is necessary however to point out that, 1967 remains a critical phase in its history. The need for production and utilization of effective highly selected sheep felt among Moroccan breeders and the idea to create a professional association became a reality. Benefiting from the existence of a legal text regulating the creation and functioning of associations, a group of breeders had met and discussed the opportunity to create a professional association. In May 11, 1967, the Association of Pure and Selected Sheep Race Breeders in Morocco (AEROPESAM) was created.

The first presentation of the Moroccan breeders in the presence of the National Commission for Selection and Marking (CNSM) was launched in 1976. This was the event that astonished more than one among French breeders. This fact allowed few Moroccan breeders to become a member of the board of directors of the AEROPESAM. This was also the opportunity to decolonize the ovine Moroccan breeders. In 1978, the development of a draft sheep plan on the basis of 1975 version constitutes the cornerstone to set the policy of sheep raising development. All these ingredients helped prepare the creation of ANOC in July 7, 1980.

In 1999, in the frame of a new strategy for goat production improvement in the north of Morocco, the Direction Provinciale of Agriculture had trusted the management of this cooperative to the National Association for Sheep and Goat Breeders (ANOC). Then, a new concept of management permitted the collective valorization of goat milk by a network of 50 small breeders. By this management process, "Ajbane Chefchaouen" is the first and creative experience using the livestock product valorization for the social development and poverty eradication in rural areas.

The National Association for Sheep and Goat Breeder, ANOC

Status, objectives and goals of the association

The National Association for Sheep and Goat Breeders (ANOC) is a non profit association, governed by the terms and conditions of Dahir n° 1-58 of November 15, 1958 regulating the right and privileges of associations. ANOC was created in 1980 after restructuring of the Association of Pure and Selected Ovine Race Breeders in Morocco (AEROPESAM), created in May 11, 1967 in the terms of the same Dahir.

The creation of ANOC is the result of a private initiative of the founding members following the necessity and professional interest (sheep selection), that is production and selection of parents (males and females) of high productive.

These activities are greatly valued by the members of the association; and the government judges them important because regrouping breeders around a real dynamic of development in the principal sheep raising regions. It is in this context that ANOC obtained its recognition as a non profit association by the government in 1988 (decree n° 2.88.189 of March 30, 1988). It should be mentioned that ANOC is the unique Moroccan professional associations in agricultural sector that had benefited this recognition.

The ultimate objective of ANOC is to improve the breeder's income by the growth of the sheep and goat meat production in an economically profitable system, developing the breeder's profession, especially in hostile environments. This objective is part of the general terms and orientations of the government policy in livestock which aim to improving the sector of red meat.

Organization of the association

The organizational structure of ANOC is made of entities. At top of the pyramid, the General Assembly, the Board of Directors and the Office of the Board of Directors are the organs of direction of the association. In order to discuss issues and prepare propositions for the Board of Directors, several commissions are involved: technical, legal, financial, strategic and human resources. At the bottom of the pyramid, groupings of breeders represent the basic organs of the association. Every grouping is structured in Administrative Assembly, elects a President and Members of the Bureau. The President of the Group sits with 4 to 5 members elected by the Administrative Assembly of the grouping on the General Assembly Meetings. In the framework of regionalization launched lately by the association, a Regional Committee for the Follow-up, a structure of consultative character, has been established in four regions: Atlantic Coast, Middle Atlas, Oriental and South Center.

Mains activities of the organization

Initially the main activity of the association was about the selection of parents. Indeed, the association has deployed substantial efforts in the development of local rustic race (more than 50% of its activities). One of the most salient results was to demonstrate that local races possess important genetic and production potential of meat.

The association made remarkable progress since its creation in 1980. Its technical activities from the outset, limited to selection and breeding, have progressively been diversified to encompass all domains of small ruminant livestock and include animal health, training of extension agents, and supply of products and inputs to the breeders and organize professional workshops such as fairs. The technical actions, programmed in common between the breeders and the technical staff, are carried out in the field and taking into account local knowledge of the members. Technical interventions are often supported by pedagogic actions. These consist of training aiming to improve technician expertise and advance the skills of the members.

Another field of involvement is the organizational actions. The structure of ANOC encourages the breeders to actively commit in the general organization of the association and groupings (planning of activities, management, etc.). The outcome of this commitment lies in the organization of input supply, which started recently.

Direct and indirect recipients of the organization

The members, "direct recipients" of ANOC actions, are currently 4 200, organized in 50 groupings, distributed among 29 provinces in all cradle zones of races. The number of members has increased 19 folds between 1984 and 2005.

Regarding the livestock, the association supervises currently more than one million of sheep and goat heads of which 600 000 females, among all local and imported races, and 70 683 goats of which 41 952 ewes. The number controlled has increased 15 folds during 1984-2005.

The association also participates actively in the development of livestock by diffusing selected parents among breeders and extending these actions to non members' breeders "indirect recipients" (about 3 000 people).

Resources of the association

Human resources

ANOC employs currently 80 people in both administrative and technical personal, of which 21 are employees of the Ministry of the Agriculture who have obtained a leave of absence. The others have been directly recruited by the organization. More than 90% of the staff works directly in the field to serve and help the member breeders.

The Human Resources of ANOC increased significantly between 1984 and 2005. This can be explained by the necessity to spread out ANOC actions. ANOC had always addressed truthfully the needs of its members, which had evolved quantitatively and qualitatively during the same period. During the same lapse of time, ANOC tried not to increase the number of staff at headquarters, avoiding this way, ineffective growth of its central structure.

Financial resources

The total budget of ANOC in the last years approximates 8 millions of Dirham. It should be mentioned that the financial resources of the association achieved a substantial jump between 1984 and 2003, since they have increased roughly by 10 folds. Besides, the parts of ANOC own resources completed a sizeable increase which enables the organization to cover about 60% of the total budget (70% in 2003).

It is should also be pointed out that, since 1988, the association obtained the recognition by the government of its non profit character; the ANOC receives public financing through the Ministry of Agriculture in the framework of the contracts program mechanism. Agreed upon by both parties, these contracts programs assign to ANOC some actions of the public service for the Ministry of Agriculture.

Goat milk cheese

In 1992, the Direction of Animal Husbandry created a goat milk cheese production unit in the city of Chefchaouen, north of Morocco. The management of this plant was entrusted to the Bureau of "Ajbane Chefchaouen" cooperative with the support of the Direction Provinciale of Agriculture until 1998.

In 1999, in the frame of a new strategy for goat production improvement in the north of Morocco, the Direction Provinciale of Agriculture had trusted the management of this cooperative to the National

Association for Sheep and Goat Breeders (ANOC). Then, a new concept of management permitted the collective valorization of goat milk by a network of 50 small breeders. By this management process, "Ajbane Chefchaouen" is the first and creative experience using the livestock product valorization for the social development and poverty eradication in rural areas.

The main objectives of the strategy of ANOC are:

- The goat livestock improvement in the north by good coaching of the farmers and their motivation by milk collecting at a remunerative price of 6 Dirhams per litre (1 Euro = 11 Dirhams).
- The cooperative was also converted to a training centre of goat milk cheese making.

The average quantity of milk collected and processed increased form 72 litre per day in 1995 to a maximum of 750 litres per day in 2004. Otherwise, this dairy cooperative, specialized in fresh cheese making, valorizes also the milk produced by 1250 Alpine breed goats of the Bellouta station.

How to meet the customer requirements

The production and consumption of milk and its derivates meet the quality requirements. In order to satisfy consumers, ANOC applies strict rules in the milk production and transformation, assures pure goat products in the "Ajbane Chafchaouen" cooperative. Rules to be applied are identified and concern all the milk and cheese production chain. Continued efforts were made to meet the quality requirements.

Milk production, processing and marketing begin with the farmer and end with the consumer passing through milk collect, cheese factory and traders. Quality control of milk and its derivative follows the reverse route.

Quality control

The main actor is the consumer because he is at the end of chain of milk and derivative products especially cheese. The consumer evaluates three levels of the quality including sensory, dietetic and hygienic aspects. The sensory quality concerns taste, flavor and color. The dietetic aspect is related to the fat and protein contents. The hygienic quality includes the wholesomeness criteria.

After the consumer, comes the trader who looks for the ease of handling and storing products. The trader has also an important role in the preservation of quality and the sanitation requirements regarding the expiration date and storage conditions.

The cheese factory is the principal chain link in cheese production. The plant should meet the requirements of consumers and traders and the quality of raw milk and its transformation. It should also call for good yield products, processing property and supply consistency.

All these requirements must be taken into account when receiving milk. In this step, the milk must be of good quality, neither watered nor contaminated by chemical compounds or residues. The bacterial loading must be also lower.

Production: crucial phase

The farmer should comply with the regulations of the cheese factory. Cheese making requires good quality milk which in turn is related to animal and breeding practices. The farmer should control the quality of feeding and housing, the genetic selection and health status. The technical staff of ANOC participates in these activities by the extension of new methods of management, feeding, breeding, hygiene and health.

Regarding farming practices, the farmers should apply recommendations in the treatment of mastitis and respect the harvest time and avoid colostrum and cow's milk blending, watering, skimming and contamination. The farmer should implement the sanitation and hygiene procedure during milking and milk conservation.

Cheese making

Fresh or mild cheese making is well established. The goat milk is collected, pasteurized and acidized by using specific ferment. Then start the following operations: renneting, hoping, draining, stripping, salting and storage at + 4 °C. The fresh cheese obtained is a local product of good quality related to the quality of local goat milk. Cheese yield is 25% for dry matter and fat content is between 40 and 45%. It's marketed with the union label "Ajbane Chefchaouen".

Butter and buttermilk are also produced by the factory in order to meet the consumer's demand. The third of collected goat milk is processed to produce butter and butter milk.

Quality of fresh cheese

Microbiological quality

The microbiological quality follow-up is performed by routine tests every two months including:

- Low faecal flora contamination: total coliform count: less than 10 germs/g and faecal coliform counts: less than 1 germ/g. Results comply with those of the European Union standards of good practices.
- *Staphylococcus aureus*: less than 10 germs/g

Good sanitation during cheese making is very important whereas contamination could not occur only after pasteurization. Equipment and soil are being cleaned up and disinfected daily. High compliance of hygiene is performed by the staff of the cheese factory but not by all the farmers.

Sensory Quality

A sensorial study has demonstrated that "Ajbane Chefchaouen" presents the following particularities: firmness, low acidity, and acceptable bitterness. Another sensorial study conducted by Chababi and Salhi confirmed that among many sold cheeses "Ajbane Chefchaouen" remains the most appreciated by consumers.

Marketing

The approach of ANOC is to improve goat milk cheese making and also to organize its marketing in order to increase the income of the cheese factory. Presently, the first goal was achieved, especially the improvement of farmer's skills and the performance of animal production. However, no action was undertaken regarding the marketing aspect. The implementation of a marketing infrastructure will certainly promote these products towards prospective consumers. The customers are concentrated around the cheese factory. Sales are distributed as follow:

- *In-situ*: direct sale to tourists and travellers; (13%).
- Traders SUC AS groceries, supermarkets, etc.; (77%).
- The city of Chefchaouen; (36%).

- Other cities (51%) including Rabat (14.33%), Tetouan (11.40%), Casablanca and Tangier (5.73% each), Meknes and Fes (4.58% each), Kenitra and Sidi Kacem (2.29%).

The marketing of "Ajbane Chefchaouen" is characterized by irregular fluctuations due to irregular goat milk supply (higher and lower milk production periods). Spoilage of non sold products was considerably reduced from 19 tons in 2000 to 1.35 tons in 2004. This improvement could be explained by the extension of market assets (2 new clients).

The major trading rivals of "Ajbane Chefchaouen" are: first, goat milk cheese and cow's milk cheese produced in the north by other factories (Douiet, Colainord, scolait, etc.), second, "Jben beldi" (farm cheese) produced by rural women in this area and third, goat milk cheese produced by other small factories (Jben Zemzem and Rinco at F'nideq, Ajbane Zaer from Benslimane, Boufakrane, etc.)

Market research survey

A market research survey on "Ajbane Chefchaouen" conducted by researchers from the National School of Agriculture in Meknes demonstrated that this cheese is well appreciated by consumers. As a local product, "Ajbane Chefchaouen" has a great potential for development but this is hindered by two constraints: the price and the quality. According to traders, the trade name "Ajbane chefchaouen" carries the symbol for this local product in the region. Consumers are more familiarized with the name "Jeben" (Cheese in Arabic) which facilitate cheese identification and remainder of this product. In addition "Chefchaouen" is famous city in the north of Morocco which is well known for goat breeding and farm cheese home making.

The most relevant critics formulated by consumers are the high cost of this cheese (51% of responses) and the irregular availability of this product (25.5% of responses) which is observed even in the region of Chefchaouen. This irregular fluctuation of the availability of cheese induced some trader resistance or withholding to sell "Ajbane Chefchaouen" and also some consumer's resistance. Therefore in such conditions, milk products are substituted fraudulently. Additionally there seem to be a lack of promotional campaigns, non labelling to indicate fat content and unavailability of other weight portions: 500, 250, 150 and 80 g.

"Ajbane Chefchaouen" has built customer loyalty in different cities. The main reasons of this interest are due to the quality of cheese, its good packaging and its dietetic value. However, except Chefchaouen area where "Ajbane Chefchaouen" is well known by all the traders, only 30% of traders know this product. Therefore, an important promotional campaign is necessary.

The distribution network of "Ajbane Chefchaouen" permitted to build customer loyalty and to extend activities to many other urban centres. The product possesses intrinsic values being a local product, which was able to conquer supermarkets.

However, ANOC should improve its production and management strategy in order to ensure regular availability of products in the market and to develop a prospective process to meet customer requirements. Such strategy of brand differentiation based on an official label of quality will strengthen the image of this product and improve its marketing.

The market research survey reported also that there is a high demand for "Ajbane Chefchaouen" in the north as well as in other cities in Morocco. The establishment of efficient distribution circuit will be determinant for the sale volume.

Further developments

During 2004, the partnership between ANOC and the French Cheese Center (Centre Fromager de Carmejane, France), "Ajbane Chefchaouen" helped to diversify its production especially during high milk production by making ripened cheese, pressed paste, yogurt and herb aromatic cheese. These new products are well appreciated by the consumers.

In 2005, the Chefchaouen MEDA Project (Funded by EU) supported ANOC to establish a marketing strategy in collaboration with the School of Commerce (ESCAE). This study targeted the set up of a trading strategy for ANOC product valorization. Additionally, ANOC is establishing a label quality for fresh cheese produced by "Ajbane Chefchaouen".

Based on the "Ajbane Chefchaouen" experience, ANOC created three cheese factories in the northern part of Morocco (Zemzem, Asmir and Bellota). Such results encourage ANOC to include this strategy in rural development.

Conclusion

Advantages of the cheese factory:

- Local development and extension tool for cheese production in the north.
- Technology and hygiene control in the cheese factory.
- Good quality of the fresh cheese well appreciated by the consumers.
- More than 50% of the production is consumed in Chefchaouen region

Drawbacks of the cheese factory:

- The cheese factory operates intermittently (between 28 and 46% of its capacity).
- Lack of diversified products (patties of 240 g).
- Irregular fluctuation of milk supply (high and low milk production periods).

Herbage mass production and utilization in mountain pastures of northern Greece

A. Karalazos[1], B. Skapetas[1], D. Nitas[2] & I. Hatziminaoglou[1]

[1]Department of Animal Production, Faculty of Agriculture, Aristotle University, 541 24 Thessaloniki, Greece
[2]Department of Animal Production, Faculty of Agriculture, T.E.I., 541 01 Thessaloniki, Greece

Summary

This study was carried out in mountain regions of northern Greece. The reported study investigates the relationship between animals and pastures in an effort to improve the management of grazing sheep. During the period of June to October, five sampling zones with different altitudes were examined in the study to record botanical composition and herbage yield under grazing conditions using an experimental flock of 300 sheep.

The results of the experiment have shown that in the botanical composition of herbage mass of pasture the graminaceous species are dominant among the legume and other different species. This is explained by the preference of sheep to forbs and especially to legume species and secondly with the climate of the zone that is favourable for the development of graminaceous species (low temperatures). The average yield of herbage was 1 910 kg DM/ha. During the experimental period, a continuous increase of content of the herbage mass in dry matter (DM), crude fibre (CF), neutral detergent fibre (NDF), acid detergent fibre (ADF), cellulose (C), hemicelluloses (HC), and lignin (L) and a decrease in crude protein (CP) from June to October was observed. The average utilization percentage of herbage mass was 86%. This was higher than that of 50 %, which is considered as normal for the herbaceous vegetation. Declines in herbage yield and nutritive value occurred as a result of seasonal changes, which were affected by temperature and precipitation. The grassland was overgrazed and had the overgrazing continued the future productivity of the pasture would have decreased and the botanical composition would have been altered. This situation exists more or less in all regions of the Mediterranean basin and has unfavourable consequences on the application of extensive production systems of small ruminants and also on the grassland ecosystem.

Keywords: herbage mass, pasture, utilisation, overgrazing.

Introduction

Herbage production of grasslands depends on different factors such as the climate, the stocking rate and the climax stage (Zervas *et al.*, 1993, Tsiouvaras *et al.*, 1996). The variations of herbage production from year to year are correlated with the air temperature and rainfall, which constitute the regulator factors of photosynthesis. It has been observed that the precipitation of the April-July period influences the maximal herbage production (Espigares & Peco, 1995). During the summer period the quality of herbage mass of the grasslands is decreasing rapidly because of the increase of CF percentage (A.R.C., 1985). Under these conditions the animals are underfed because the intake of energy and also all the other necessary nutrients are less than the requirements for maintenance and growth or milk production (Walsh & Birrel, 1987; Boyazoglu & Flamant, 1991). Also, the content of CP, as well as any other nutrient, in the DM of herbage biomass is correlated with grazing (Heady, 1975). During the summer months the content of CP in the herbage biomass is 5-6%, sufficient only to cover the maintenance requirements of grazing animals. The lactation requirements in this case could be covered only by supplemental nutrition (Cook, 1972; Gunn *et al.*, 1992; Avondo *et al.*, 1995).

In the mountainous grasslands (>1 200 m) the vegetation activity starts in the spring and if the soil's moisture from winter's and also from summer's precipitation is satisfactory could continue during the summer period and the first months of autumn. The maximization of herbage production occurs as a rule at the end of July (Papanastasis, 1982). Generally, in the marginal semiarid and arid areas plant biomass matures very quickly, and the digestible protein content decreases rapidly (Boyazoglu & Flamant, 1991).

The objective of this study was to estimate the herbage production, the floristic and chemical composition of the herbage mass as well as the grazing capacity in mountain pastures. This information constitutes an important step on the application of rational grazing systems that would allow new management techniques associated with animal production in mountain areas of northern Greece.

Material and methods

The experimental work of the present study was carried out in mountain pastures of northern Greece. Five sampling zones of different altitude (1 400-1 800 m a.s.l.) were identified as ecologically and physiognomically homologues. In each zone, one grazing exclusion cage (5x5 m) was placed. Meteorological data were recorded at an altitude of 1 250 m.

Estimation of botanical composition and herbage production of the grassland were determined in the five vegetative stages (Table 1), under grazing conditions using the experimental flock of 300 sheep of the Animal Breeding and Training Centre of Vlasti. Measurements for the estimation of botanical composition were taken place inside of each cage. In each cage and each vegetative stage 10 measurements were carried out using the ten pins point frame (Brown 1963, Joint Committee, 1962). Sampling for the calculation of the herbage production of grassland was carried out at the same time under protective conditions from grazing (A) and also under grazing conditions (B), according the method of True Weight Measuring (Biswell and Liakos, 1977). In each sampling zone, two homologues to the floristic composition and herbage production surfaces (5x5 m) were used to select samples of herbage mass. On each vegetative stage, four samples of herbage mass were taken inside (A) and outside (B) of each cage using a metallic square frame 0.5 x 0.5 m (Biswell and Liakos, 1977). After any sampling period two new sampling surfaces for each zone were selected to move cages. In this way, the measurements of grassland's herbage production (kg/ha) were calculated (Nitas, 2000). Separation of dead material from growing green biomass was performed for each sampling period. The chemical analyses of samples of herbage mass were performed at the laboratory of Animal Nutrition, Faculty of Agriculture, Aristotle University of Thessaloniki, according the methods AOAC (1999) and Goering & Van Soest (1970).

All data were analyzed statistically using ANOVA, SAS General Linear Model procedure (SAS, 1998).

Table 1. Herbage production calculation.

Months			Progressive herbage production	Progressive grazed herbage production
June	A_1	B_1	A_1	A_1-B_1
July	A_2	B_2	A_2-B_1	A_2–B_2
August	A_3	B_3	A_3–B_2	A_3–B_3
September	A_4	B_4	A_4–B_3	A_4–B_4
October	A_5	B_5	A_5-B_4	A_5-B_5

Total herbage production (a)

$a = A_1+(A_2-B_1)+(A_3-B_2)+(A_4-B_3)+(A_5-B_4)$

Total grazed herbage production (b)

$b = (A_1-B_1)+(A_2-B_2)+(A_3-B_3)+(A_4-B_4)+(A_5-B_5)$

Legend:

A_1	=	from bursting of new vegetation till ear stage
B_1	=	non grazed till ear stage
A_2-B_1	=	from ear stage to flowering stage
A_3-B_2	=	from flowering stage to seed set stage
A_4-B_3	=	from seed set stage to seed maturity stage
A_5-B_4	=	from seed maturity stage to plant dehiscence stage
A_1-B_1	=	grazed till ear stage
A_2-B_2	=	grazed from ear stage to flowering stage
A_3-B_3	=	grazed from flowering stage to seed set stage
A_3-B_3		grazed from seed set stage to seed maturity stage
A_5-B_5	=	grazed from seed maturity stage to plant dehiscence stage

Utilization percentage of herbage mass = b/a x 100

Results and discussion

The estimation of botanical composition for three successive years suggests that the graminaceous species dominate the grassland compared to legumes and other (miscellaneous) species. This observation is explained by the greater resistance of graminaceous to low winter temperatures in the region of study (Nitas *et al.*, 1992). These species have a greater adaptability and a higher potential for regrowth, especially under grazing conditions by sheep, compared to legumes (Biswell and Liakos 1977). Sheep show a greater preference for the forbs, especially legumes (Papanikolaou *et al.*, 1993).

The stocking rate also influences in a great scale the botanical composition of the grassland. The intensive grazing (3 sheep/ha/year) decreases plant cover, increases the growth of annual species compared to perennial species leading in this way to degradation of grassland (Tsiouvaras *et al.*, 1996). In this study, botanical composition varied between years for graminaceous species ranging from 63 to 71% of total plant species. Legume species varied from 2.5 to 9.5% with significant differences observed between the vegetative stages. The legumes percentage decreased rapidly from July to October. The percentage of other herbaceous species fluctuated from 26 to 28% with significant differences between the years.

Herbage production till the end of July was satisfactory for the region as a result of satisfactory precipitation (Table 2). During the period August to September, the herbage production decreased to very low levels due to the intensive xyrothermic conditions of this period. The mean aerial temperature in this period was high, whereas the levels of precipitation were low (Figure 1).

Table 2. Progressive herbage mass yield (kg DM/ha)[1].

Vegetation stage	Year 1st	2nd	3rd	Mean
June	691^{a}	539^{ab}	739^{a}	657
July	1 185^{a}	851^{a}	363^{b}	799
August	104^{b}	241^{b}	328^{b}	224
September	46^{b}	169^{b}	73^{b}	96
October	129^{b}	318^{b}	195^{b}	214
Standard error.	11.5	7.6	6.1	8.7

[1]Mean of 5 cages.

[a,b,c] Within columns, means not sharing a common superscript differ significantly ($P<0.05$).

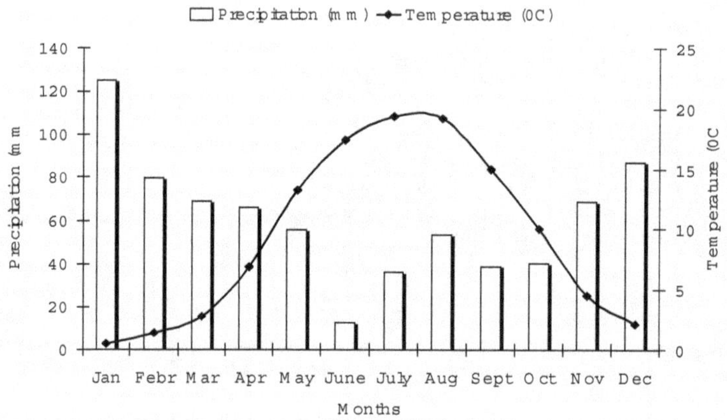

Figure 1. Precipitation (mm) and air temperature (°C), mean of 3 years.

The mean herbage production of grassland of the region for three consecutive years was 1 910 kg DM/ha (Table 3). The fluctuations from one year to other may be explained by the corresponding variations of precipitation during the period April-August for each year (Table 4). An important index for grassland productivity is the precipitation during the period April to August (Papanastasis, 1982).

The consumed herbage mass was found to be at very high level (1 650 kg DM/ha, mean of three years), thus the average utilization of herbage as a percentage of herbage production was 86.4%. The very high percentage of utilization of herbage mass production by grazing sheep shows that the grassland was overgrazed (the proper utilization percentage should be 50%) and if this phenomenon would be continued for the coming years the productivity of grassland would be decreased and the botanical composition would be changed (increase of undesirable herbaceous species). The proper stocking rate (1 sheep/ha/year), on the contrary, would increase the herbage production keeping at the same time the equilibrium of mountainous ecosystem.

The chemical composition of herbage mass is shown in table 5. The content of CF and lignin at the first vegetative stages (ear and flowering stage) was in low levels. The percentage of these two

Table 2. Progressive herbage mass yield (kg DM/ha)[1].

Vegetation stage	Year 1st	2nd	3rd	Mean
June	691[a]	539[ab]	739[a]	657
July	1 185[a]	851[a]	363[b]	799
August	104[b]	241[b]	328[b]	224
September	46[b]	169[b]	73[b]	96
October	129[b]	318[b]	195[b]	214
Standard error.	11.5	7.6	6.1	8.7

[1]Mean of 5 cages.

[a,b,c] Within columns, means not sharing a common superscript differ significantly ($P<0.05$).

Table 3. Total yield, consumed herbage mass (kg DM/ha) and utilization percentage of herbage mass during grazing[1].

	Year 1st	2nd	3rd	Mean
Total yield	1 910	2 080	1 740	1 910
Consumed herbage mass	1 700	1 690	1 550	1 650
Utilization percentage	89	81.2	89	86.4

[1]Mean of 3 years.

Table 4. Pasture's grazing capacity[1] per month.

Period	Yield kg DM/ha	Grazing capacity Ewe/ha/month	Ha/ewe/month
June	657	7.3	0.14
July	799	8.9	0.11
August	224	2.5	0.40
September	96	1.1	0.91
October	214	2.4	0.42

[1]Mean of 5 cages and 3 years.

nutrients, however, increased significantly during the seed set stage and seed maturity stage and this tendency continued till the end of October. The same evolution was observed for the NDF, ADF, cellulose and hemicelluloses. This tendency could be explained by the increase of herbage cell walls against the cell contents (Frames, 1992). These changes lead to a decrease of nutritive value of herbage mass, mostly during the latest vegetative stages. The decrease of nutritive value became more serious by the simultaneous decrease of CP percentage (from 9-11% at the first stages reduced to 5–6% at the latest). The fluctuations of CF and CP percentages among the different vegetative stages are in a good agreement with the results of other authors (Papanastasis, 1982; Walsh and Birrel, 1987). The increase of CF and the decrease of CP in the content of herbage mass DM contributes to animal underfeeding. In this case, sheep could not meet their energy and protein requirements even for maintenance (Walsh and Birrel, 1987).

Table 5. Chemical composition (% DM) of herbage mass according the grazing period[1].

Vegetative stage nutrients	June	July	August	September	October	S.E.
Dry Matter (DM)	23.5^{a}	30.9^{b}	56.6^{c}	59.8^{c}	68.5^{d}	1.8
Organic Matter (OM)	93.3^{a}	93.6^{a}	93.9^{a}	94.9^{b}	93.9^{a}	0.5
Crude Protein (CP)	10.6^{a}	8.8^{b}	6.6^{c}	5.4^{c}	6.0^{c}	0.6
Ether Extract (EE)	1.3^{a}	1.4^{a}	1.5^{ab}	1.6^{b}	1.6^{b}	0.05
Crude Fiber (CF)	24.9^{a}	29.1^{ab}	32.7^{bc}	35.6^{c}	36.7^{c}	2.1
Neutral Detergent Fiber (NDF)	35.4^{a}	41.6^{b}	51.3^{c}	65.2^{d}	66.9^{d}	1.8
Acid Detergent Fiber (ADF)	24.7^{a}	29.4^{b}	35.8^{c}	41.1^{d}	43.8^{e}	0.8
Cellulose (C)	19.5^{a}	23.9^{b}	29.5^{c}	32.1^{d}	34.3^{d}	0.6
Hemicelluloses (H)	10.7^{a}	12.3^{b}	15.8^{c}	23.8^{d}	24.0^{d}	0.5
Lignin (L)	5.2^{a}	5.5^{ab}	6.3^{bc}	9.0^{d}	9.5^{d}	0.4

[1] Mean of 5 cages and 3 years.

$^{a, b, c, d, e}$ Within lines, means not sharing a common superscript differ significantly ($P<0.05$).

Table 6. Chemical composition (% DM) of herbage mass according to the altitude[1].

Cage nutrients	1st cage	2nd cage	3rd cage	4th cage	5th cage	S.E.
Dry Matter (DM)	49.1^{a}	48.0^{ab}	47.5^{b}	45.6^{bc}	43.1^{c}	0.34
Organic Matter (OM)	93.4^{a}	93.8^{a}	94.1^{a}	93.7^{a}	93.5^{a}	0.03
Crude Protein (CP)	7.5^{a}	7.3^{a}	7.9^{ab}	7.9^{ab}	8.5^{b}	0.16
Ether Extract (EE)	1.6^{a}	1.5^{a}	1.5^{a}	1.5^{a}	1.5^{a}	0.04
Crude Fiber (CF)	32.6^{a}	32.1^{a}	30.7^{b}	29.3^{b}	28.4^{c}	0.32
Lignin (L)	6.6^{a}	6.1^{a}	5.9^{a}	6.3^{a}	6.4^{a}	0.11

[1] Mean of 3 years.

$^{a, b, c}$ Within rows, means not sharing a common superscript differ significantly ($P<0.05$).

The factor "cage" (altitude) didn't influence substantially the chemical composition of herbage mass (Table 6). A decrease of DM and CF of the herbage mass, which was observed in cages 4 and 5 (greater altitude) could be explained by the fact that the drying up of herbaceous vegetation on the greater altitudes is relatively slower.

In conclusion the results of this study suggest that in mountainous grassland the grasses are the dominant herbaceous species among the legumes and other floristic groups. The maximization of herbage production takes place at the middle of July. During the period August–September the productivity decreases rapidly because of high temperatures and the lack of soil moisture (xyrothermic climate). During October takes place a limited regrowth of vegetation but the autumn frosts interrupt this tendency. The total herbage production could be characterized as mediocre (1 910 kg DM/ha), but the grassland's productivity has a strong seasonal character and moreover the quality of herbage production decreases significantly during the period August to September. This situation makes necessary the flock removing to a new grassland region and the supplemental nutrition. The very high percentage of utilization of herbage mass (86.4%) shows an intensive overgrazing of the grassland. This is no rare phenomenon for all the mountainous pastures in Greece. The overgrazing combined with the rapid decrease of the quality of herbage mass (increase of CF, NDF, ADF and L, also decrease of CP) suggest that mountainous grasslands do not meet the nutritional requirements of grazing sheep during summer and autumn. The strong seasonal character of production of the natural vegetation (quality and quantity) can be considered as a main constraint on the evolution and development of small ruminant production systems in mountain regions of northern Greece.

References

AOAC, 1999. International official methods of analysis (16th ed). Assoc. Offic. Anal. Chem., Arlington, V.A., USA.

A.R.C. (Agricultural Research Council), 1985. The nutrient requirements of ruminant livestock. Commonwealth Agricultural Bureaux, Slough.

Avondo, M., G., Licitra, M., Bognánno, A.N., Keshtkaran, D., Marletta, & G., D'Urso, 1995. Effects of the type and level of supplementation on grazing behaviour of lactating ewes in a Mediterranean natural pasture. Livest. Prod. Sci. 44: 237-244.

Biswell, H. & L. Liakos, 1977. Range management. Thessaloniki, p. 539 (in Greek).

Boyazoglu, J. & J.C. Flamant, 1991. The actual state and the future of animal production in the Mediterranean rangelands, 4th International Rangeland Congress, Montpellier, France, pp. 1017-1025.

Brown, D., 1963. Methods of surveying and measuring vegetation. Comm. Bu. Past. Field Crops Bull. 42.

Cook, C.W., 1972. Symposium on nutrition of forages and pastures: collecting forage samples representative of ingested material of grazing animals for nutritional studies. J. Anim. Sci., 23: 265-270.

Espigares, I. & B. Peco, 1995. Mediterranean annual pasture dynamies: impact of autumn drought. J. Ecology, 83: 135-142.

Frames, J., 1992. Improved grassland management. Farming Press, U.K.

Goering, H.K. & P.J. Van Soest, 1970. Forage fibre analysis. Agricultural Handbook no. 379. U.S.D.A., Washington, D.C.

Gunn, R.G., J.A. Milne, A.J. Senior, & A.M. Sibbald, 1992. The effect of feeding supplements in the autumn on the reproductive performance of grazing ewes. 1. Feeding fixed amounts of supplements before and during mating. Anim. Prod., 54: 243-248.

Heady, H,F, 1975. Rangeland management. McGraw Hill Co. N.Y.

Joint Committee, 1962. Basic problems and techniques in research. Nat. Acad. Sci. - Nat. Res. Council. Publ. no. 890.

Nitas, D., 2000. Fodder plants and pasturing grounds. Technological Educational institute of Thessaloniki. Thessaloniki, p. 114 (in Greek).

Nitas, D., K. Papanikolau, & A. Karalazos, 1992. The alpine rangelands of mount Olympus. Animal Science Review, 16: 33-51 (in Greek).

Papanastasis, Â., 1982. Production of natural grasslands in relation to air temperature and precipitation in northern Greece. Forest Research Institute, Thessaloniki, Greece. p. 128 (in Greek).

Papanikolau, Ê., A. Karalazos, D. Dotas, & D. Nitas, 1993. Grazing capacity and herbage mass quality fluctuation and goats and sheep grazing behaviour on grassland that came from burned shrubland. Animal Science Review, 12: 63-64 (in Greek).

SAS, 1998. SAS/STAT User's Guide. (Release 7,0). SAS Inst., Inc. Cary, NC.

Tsiouvaras, C.N., Z. Koukoura, A. Ainalis, & P. Platis, 1996. Effect of grazing intensity on the productivity of a semiarid grassland in Macedonia, Greece. In: Zervas, N. P., Hatziminaoglou, I. (Eds), Proc. International Symposium on the optimal exploitation of marginal Mediterranean areas by extensive ruminant production systems, June 18-20, 1994, Thessaloniki, Greece, EAAP Publication, no.83, pp. 376-379

Walsh, G.L. & H.A. Birrel, 1987. Seasonal variations in the chemical composition and nutritive value of five pasture species in south-western Victoria. Aust. J. Exp. Agric., 27: 807-816.

Zervas, Í., I. Hatziminaoglou, A. Georgoudis, Á. Karalazos, A. Psihoudhaqis, K. Papanikolau, D. Dhimos, G. Gavrilis & K. Konsultu, 1993. Present situation and proposals for primary animal production and rangelands improvement (Farm and economic improvement study of Trikala villages Grizanio, Ahladohori, Diassello, Agrelia and Liopraso). Aristotle University of Thessaloniki (in Greek).

Natural resources feeding free grazing bovines in Peneda mountain

J. Côrte-Real Santos

DRAEDM, Rua do Pinhó 80, 4800-475 São Torcato, Portugal

Summary

The free grazing livestock farming systems based on natural resources are those that fulfilling welfare conditions can offer wealthy and safe products with traditional quality (Hovi *et al.*, 2003). In the present work it we tried to make the synthesis and the interconnection between several studies carried out within 2 different projects. It was intend to show the relationships between bovines and plants as the basis of a distinguished meat product. It is raised up some questions about product valorisation and try to question this farming system sustainability. It was studied floristic diversity with square point adapted, ingestive behaviour and intake of the free grazing bovines. It was evaluated the evolution trend of the farming sustainability. In the present work is pointed out some questions related with PDO certification system in cattle meat from the Northwest of Portugal. As a conclusion, it can be said that the main bottle-neck to the sustainability of this system, is the farmers ageing because within 20 years there will be no one in Peneda's mountain villages, unless policies preventing this issue are taken into place.

Keywords: farming system, natural resources, free grazing bovines, Peneda's mountain.

Introduction

The free grazing livestock farming systems based on natural resources are those that fulfilling welfare conditions can offer wealthy and safe products with traditional quality (Hovi *et al.*, 2003). Although, it is necessary that these systems are economically viable, socially acceptable, technically correct, environmentally non-degrading and should conserve land, water, plant and animal genetic resources (Vitalis, 2001). The best way to assure enough income to the farmer-families is to improve the market value of the animals' products. Different products characteristics lead to different quality concepts, where the price to consumer, is still one of the most important desiderates (Knap *et al.*, 2003). The PDO certification system is not an enough warranty of the product quality, because in the EU, the origin labelled products concept varies from country to country (Casabianca, 2002).

In Peneda's mountain at a 500 to 600 m high there are small villages, from 50 to 160 persons, where private land, between 1.6 to 2.2 hectares/farm in average (Pires, 2000), is used to produce hay (54%) maize corn (14%) and home grown vegetables (10%). Between 800 and 1 000 m high there are summer houses ("brandas") where farmers go when weather gets hotter (between February and November). In "brandas", natural grass hay is produced (56% of total arable land) and is harvested in June in order to feed animals. Between 1 000 to 1 400 m high, where land is communal "*baldio*" natural herbaceous and shrubs are used as animal feed. It's in this environment that 3 000 cow-calves are free grazing from February to November (10% of which are free grazing all year around), 3 000 horses are free grazing all the year around and 1 000 sheep and goats graze with shepherd, all year around.

The present work tries to make the synthesis and the interconnection between several studies carried out within 2 different projects. It was intended to show the relationships between bovines and plants as the basis of a distinguished meat product. We also try to rise up some questions about product valorisation and try to question this farming system sustainability. In PAMAF0036 project "Sustainable agriculture development: study methodologies and definition of criteria for intervention in mountain areas" (1998-2000), it was carried out the farming system characterisation and also the floristic diversity in communal land from Peneda's mountain. We have also studied ingestive behaviour with the direct observation method and intake with Balent and Gibon method. Afterwards we have complemented the bovines feeding preferences study with the micro histological faeces analysis method under the PARLE project "Improvement of quality products and resources in the farming livestock system of the northern mountains", (2001-2004). Still within PARLE project, we have assessed the chemical composition of the herbaceous plants and from the main four shrubs preferred by the bovines.

Material and methods

Studied area and farming system

Peneda´s mountain is located in the Northwest of Portugal and belongs to the National Park of Peneda-Gerês. It has a rainy climate (in average more than 2000 mm) with a big thermal amplitude (#T>20°C), very steep slopes (about 20% of the territory has slopes greater than 30%). The cow-calves free-graze on communal land, natural vegetation from February till December (warm period) between 1000 and 1400 m high. From December till February (cold period) the animals sleep in the sheds and in the morning they graze in private natural pastures while in the afternoon they graze on communal land near the villages. During the warm period, farmers have to look for their cattle in communal land (6 hours to walk more than 8 km surpassing differences of 500 m in high), in order to evaluate if there are still all the animals that they own, but also to check if cows have gave birth.

The cultivable land is divided into very small patches of ground (2 100 m^2 in average) (Pires, 2000) where ploughing with a pair of well trained cows can take about 60 hours per hectare (Santos, 2000). The natural permanent pastures hay is harvested in May and June (tasks such as harvesting, piling up, transporting to the tractor and to the barn and to stack up, varies from 3.3 to 38.7 hours/person/are). The Peneda´s cows population average live weight is about 343±51.3 (n=47) and for males is 372±61.7 (n=19) (Santos, 2000). The commercial product from Peneda's farms is the calf which is weaned at 7 months old with about 125 kg LW. The average daily gain is about 0.544 kg from 3 to 7 months of age. These calves suckle in their mothers and also graze with the adults in common land. They produce a 70 to 80 kg carcass (Cacho, 1999). Only 34% of these bovines belong to two local breeds (Barrosão and Cachena) while 66% are crosses among between these tow breeds (Santos, 2004).

Floristic diversity

To study floristic diversity of communal land (baldio) it was used square point adapted method (Canfield, 1941), which consists in 39 interceptions along a 20 m stretched rope, in six different places along the mountain (Barros & Santos, 1999). We have repeated the observation twice in each place, except in "Branda de Real" where we have repeated four times.

To estimate dry matter (DM) production of the herbaceous plants we have used the cage (1 m^2) method with successive samples in Spring and Autumn in 2004. The chemical composition analyses of the plants were done in an official laboratory (AGROS, feedstuff laboratory). It was analysed DM, crude protein, crude fibre, NDF, ADF and ADL. We have calculated the digestibility of organic matter (DOM) with Van Soest equation.

Ingestive behaviour and intake

In order to know the feeding strategy of the free grazing cow-calves we have studied ingestive behaviour and intake. In this mountain it is not possible to use measuring equipment on the animals like Ethosys proposed by Decuq *et al.* (1996), due to topography, animal behaviour and type of vegetation. This problem was experienced on previous work when 8 out of 12 radio tracking collars were lost due to the bovines scratching their necks against rocks and trees. The intake of free-grazing suckler cows', was evaluated by Balent and Gibon (1986) method (BGm), which proposes the following equation,

$$QI=\Sigma_i nBPm_i \times Q_i \times D_i \qquad (1)$$

where:
QI represents the daily intake per animal (kg DM);
$nBPm_i$ represents the bite rate during feeding (bites/min);
Q_i represents bite size (g DM/bite) and
D_i represents the grazing time (min/day).

The number of bites was counted every one-minute period of time, after comparing the accuracy of three different periods; one, five and ten minutes. It was done five repetitions per animal and per type of vegetation (grass and 5 different shrubs) in order to obtain $nBPm_i$. To obtain bite size we have collected two samples S_0 and S_1 from the same bite, 10 samples per plant. The S_0 was collected in the same place where the animal was grazing. The S_1 was collected within a circle of 1 meter of diameter, from the point where the S_0 was collected. It was tried to collect the same amount of plant that it was seen done by the animal.

The grazing time was measured with focal-animal sampling method (Altmann, 1974), by direct observation. The animal observed in each period, was an adult female (more than 3 years), non-pregnant, of about 350 kg LW, which was the herd's leader. An adaptation period of 30 min to 1 h was found necessary in order to let the animals get used to the observer. Since it was considered of great importance to know bovine's activity during the night-period, we have carried out the observation during 24 hour-periods, by teams of 6 persons in 2 hours turns, instead of the fixed intermittent periods of 5 minutes each hour proposed by Altmann (1974). To minimise the animals' distress during the night-period, flashlights were turned on, one hour before sunset.

In order to study feeding preferences and animal behaviour we have recorded the following animal activities: "grazing" means the time that the animal spends in feeding or chewing even if the animal is in motion; "walking" means the time that the animal spends in movement without chewing; "other behaviours" means all the behaviours not directly related with feeding, like urinating, defecating, licking, scratching, interaction with other animals; "rumination" means the time spend by the animal since the first regurgitation until the last swallowing; "resting" means the time when the animal is motionless, usually laid down, without chewing and without being alert; "alert" means that the animal is on watch. We have carried out three observations of 24 hour-periods in spring, two in summer and autumn and only one in winter in the years of 1998 and 1999.

During rumination activity, the following parameters were studied: number and time of the day when rumination periods occurred during 24 hour-period of observation (PR); time spent in each rumination period (DPR); number of regurgitations (NR) in each rumination period; number of chewing's per regurgitation (NMR); and time spent per regurgitation (TR). The number of mastication's per regurgitation counting was carried out by turn with its registration in a paper. We used JMP version 3.2.2 form SAS institute Inc (1989) as statistical software.

As we are aware of the weaknesses of the BGm, namely on bite weight (Mayes & Dove, 2000) we have chosen to use the micro-histological faeces analysis (Sparks & Malecheck, 1968; and Holecheck, 1982) in order to validate the results. Still, in our opinion, the bite weight method is the most feasible one, because it tries to copy exactly what the animal is doing when grazing. For details about micro histological faeces analysis method, see Santos & Ferreira (2003) work.

Farming sustainability

As it was of concern, it was carried out a study in order to know this livestock farming system sustainability trend, (Santos, 2003). The objective of this work was to evaluate the role of farmer's management practices in the evolution of livestock farming system sustainability. After the typology (A- farms with no bovines; B-farms with 1-4 bovines; C- farms with 5-9 bovines; D- farms with 10-17 bovines and E-farms with equal or more 18 bovines) built with Bertin's methodology (Simões, 2000) it was chosen a sample of six farms within the types, as follows: one farm from B and D and two farms from C and E types. After OECD indicators methodology and Université Libre de Bruxelles indicators proposals, it was built the following indicators according to MESMIS indicator concept.

Results and discussion

Floristic diversity

Figure 1 presents the diversity of the feeding resource that exists in communal land from Peneda's mountain, where 75 species were identified and classified in 11 families.

There is a great floristic diversity in feeding resources that varies with places. There are places (Garganta do Eiró) where shrubs are the main soil occupation but there are other places (Branda de

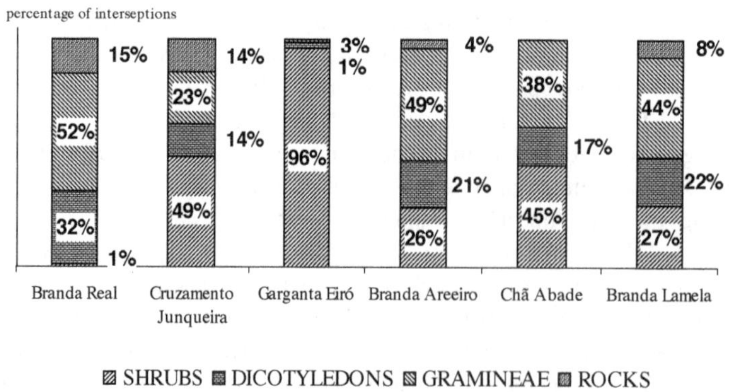

Figure 1. Feeding resources diversity in communal land.

Real) where herbaceous plants are the main soil occupation. In all other places there is a certain equality in soil occupation by shrubs and herbaceous. The floristic diversity varies with the high, soil fertility, water content of the soil and also grazing pressure (Melo *et al.*, 2000). In spring the average production per hectare in DM of the herbaceous plants in communal land was 1 951 kg while in autumn was 2 144 kg. At 500 to 600 m high (private land in villages) we have obtained DM productions per hectare varying between 2 207 kg (low soil water content), 5 214 kg (medium soil water content) and 4 482 kg (high soil water content), in average Melo *et al.* (2000).

In figure 2 to 5 it is shown the chemical composition for the herbaceous plants and for the four shrubs families studied. Pacheco & Xavier (1999) found minor CP values (9.8 to 14.4 % DM) for herbaceous plants in a mountain next to Peneda´s. For the shrubs families they found CP similar values. Lachaux *et al.* (1987) have also found similar values to CP for *Eryca* and *Cytisus* families. DOM (digestibility of organic matter) is higher in herbaceous (53% in spring and 43% in autumn) than shrubs (31% in spring and 26% in autumn) but it varies more with season of the year in herbaceous than in shrubs. From the 4 shrubs families studied only *Cytisus* show variation between spring (43%) and autumn (31%). All the other three families show a great stability (the greater difference was of 5%) along the year.

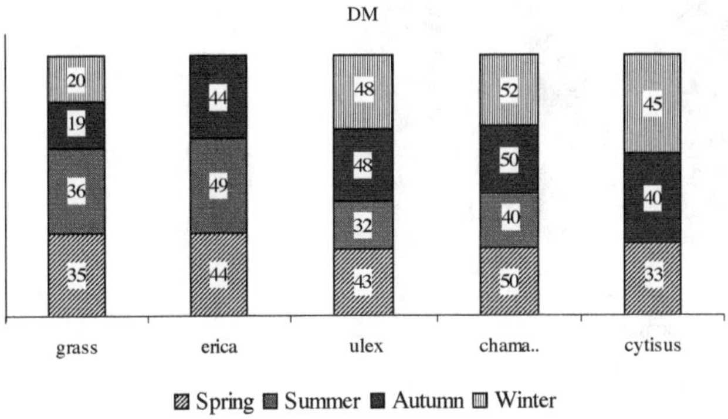

Figure 2. DM(%), along the year.

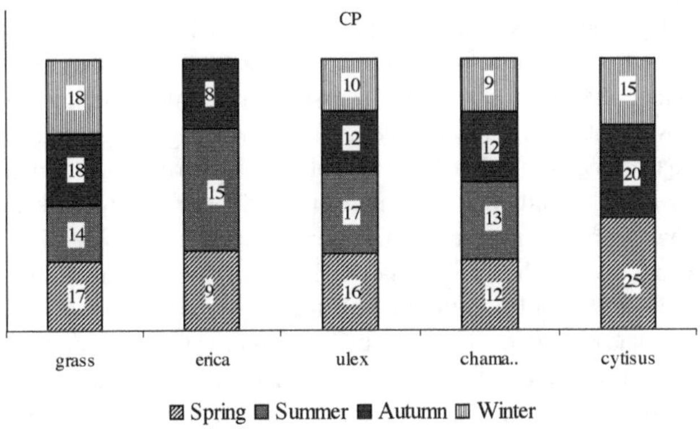

Figure 3. CP as % of DM, along the year.

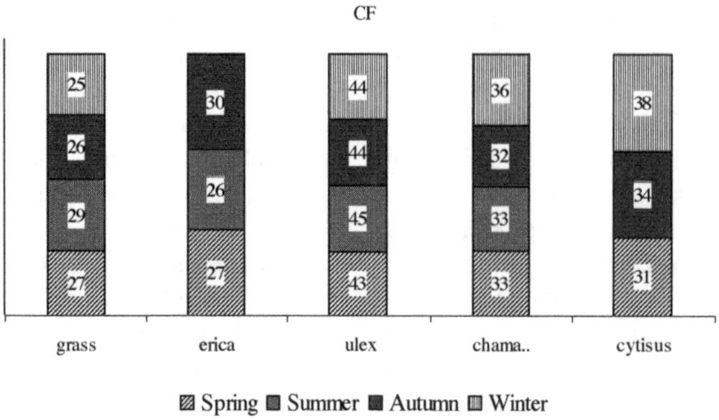

Figure 4. CF as % of DM, along the year.

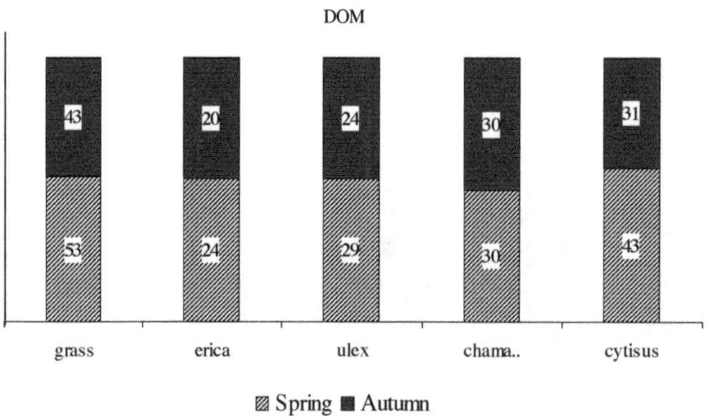

Figure 5. DOM by Van Soest equation.

Ingestive behaviour and intake

The ingestive behaviour, in 24 hour period of observation, divided into AM (ant-meridium) and PM (post-meridium), it's shown in figures 6 to 9. One can see that behaviour patterns are similar in either AM and PM sectors. The general ingestive behaviour pattern in hot weather (spring and summer) is very similar to the one obtained by Ayantude *et al* (1999) in Nigéria in bovines grazing during dry season. Both these patterns seem to demonstrate that cattle are able to adapt to different climates and rough conditions of feeding availability, although maintaining its feeding efficiency (Tolkamp, 1999). Mandaluniz *et al* (2000) have carried out a very similar work in Bask Country. Despite the daylight has less hours than in Peneda and that they have registered less behaviour activities (only 3) the ingestive behaviour pattern of Peneda's cattle is different from the one find out in Bask Country because grazing activity (50 to 75%) and resting activity (20 to 40%) was much higher in Mandaluniz *et al* work. There is no apparent reason for the rumination activity is higher in hot season (25 to 31%) than in cold season (21%) because the diet DM and CF are equal in both seasons (40% DM-hot *vs* 41%-cold; CF= 33% DM-hot *vs* 34% DM-cold). In winter we have not registered

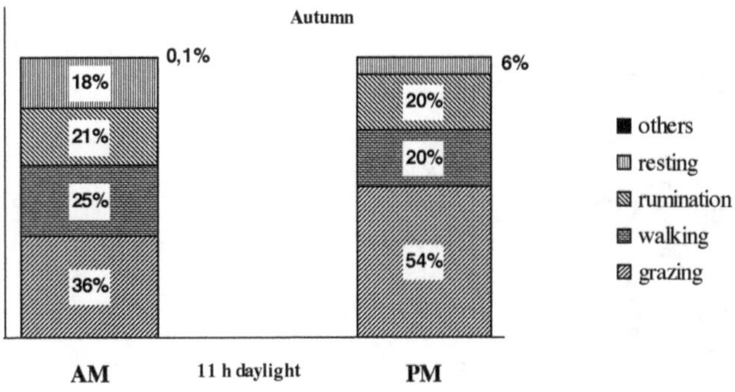

Figure 6. Ingestive behaviour in autumn.

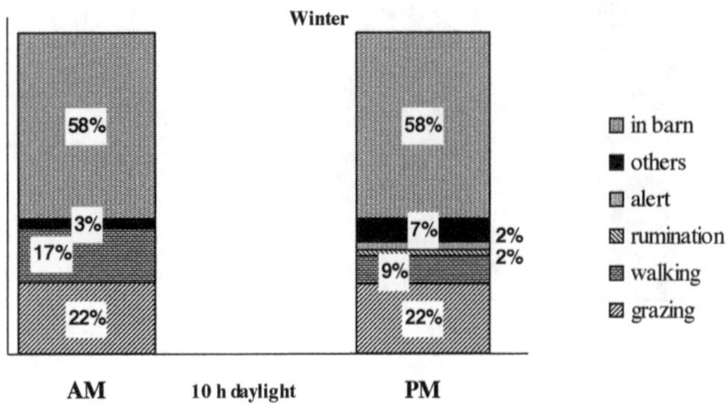

Figure 7. Ingestive behaviour in winter.

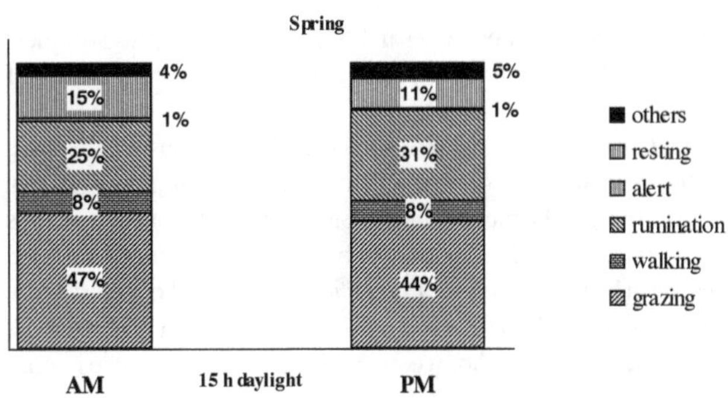

Figure 8. Ingestive behaviour in spring.

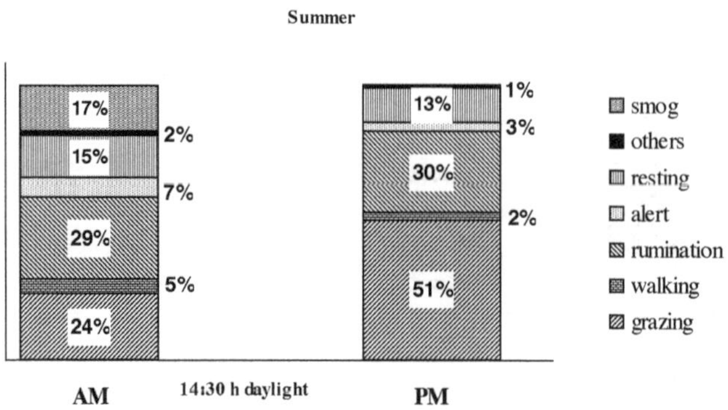

Figure 9. Ingestive behaviour in summer.

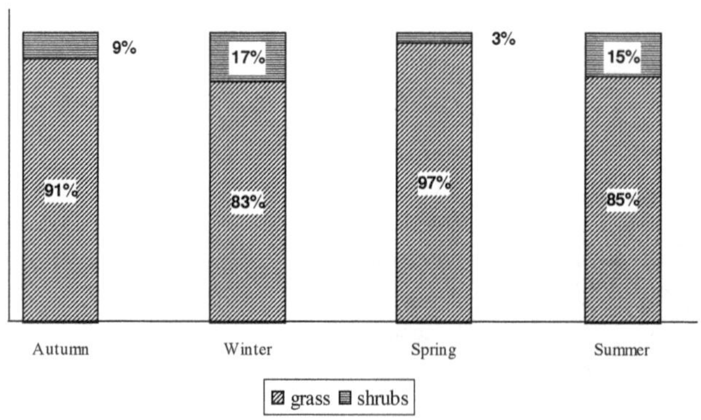

Figure 10. Feeding preferences.

resting activity because the animals are locked in their sheds for about 15 h (17 PM to 8 AM). All the subsequent feeding strategy is regulated because the animals "know" that they have little time to feed themselves.

On the other hand, the resting activity has its higher values in hot season (spring plus summer) because bovines, between 11 AM and 17 PM, look for protection against heat in the shadows of shrubs and little woods.

The feeding preferences are shown in figures 10 and 11. As expected, herbaceous are the main feeding option by bovines all along the year, also stated by Mandaluniz *et al* (2000). Shrubs have a very important role (Mandaluniz *et al.*, 2000) in the diet, mainly in Summer, due to that herbaceous suffer more with the high temperatures (30 to 40°C) than the shrubs. Shrub preferences have the highest values in Winter, because they are in better physiological conditions and have a better resistance to snow and frost than herbaceous plants. *Erica* family is the most chosen by bovines. *Chamasespartium* family is mainly preferred during hot weather while *Cytisus* its chose in cold season.

We present in figure 12 diet composition of the free grazing bovines obtained with micro histological faeces analysis (MHFA) method (Santos & Ferreira, 2003). The results obtained with this method are very similar to those obtained with the direct observation method in what concerns the importance of shrubs in the diet and also the animals general preference by the *Erica* family plants. *Cytisus* family had higher values in hot season, with the MHFA method than with direct observation method. Although, as these are residual data (< 1% of the total frequencies) we can say that the MHFA method has confirmed the results obtained with the direct observation method in what concerns bovines feeding preferences, which was also stated by Holecheck *et al* (1982) and Gordon (1994).

Table 1 presents the free grazing bovines' intake data with the BGm (1986). It was found (*t* test) a season of the year (autumn, winter, spring, summer) effect ($P<0.001$) and also a type of vegetation (shrubs, herbaceous) effect ($P<0.001$), in $nBPm_i$. It was also found a type of vegetation effect ($P<0.001$) in Q_i. There was no operator (the person who collected the sample) effect nor in $nBPm_i$ neither in Q_i. The intake values we had with BGm for an adult cow with 350 kg LW are much higher than those proposed by the main feeding systems such as the UF French system (5 to 6 kg DM/day), the American NRC system (7.5 kg DM/day) or the British ARC system (8 to 9.4 kg DM/day). We have found good correlations between daily intake by BGm and rumination time (r=0.70), despite

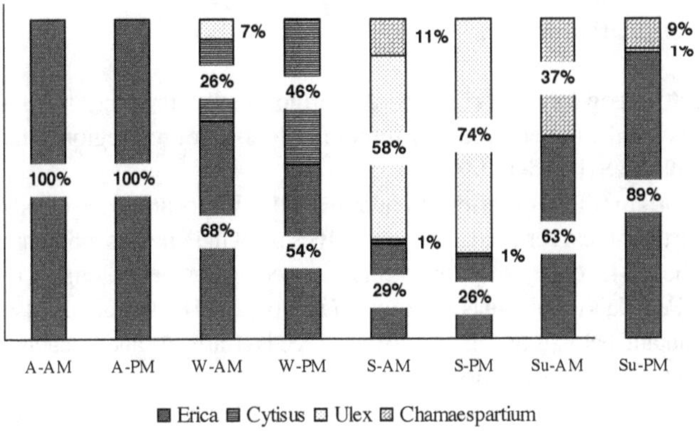

Figure 11. Shrub preferences.

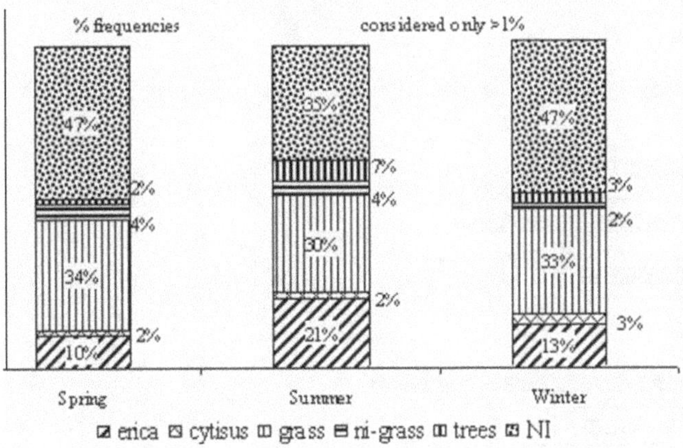

Figure 12. Diet composition.
Legend: ni-grass=non-identified grass. NI=non-identified fragments.

the few number of observations (n=8). There is also a good correlation (r=0.93) between number of regurgitations and rumination time in each rumination period. We found a negative but very weak correlation (r= - 0.06) between rumination time and regurgitation time.

Farming sustainability

The sustainability of a livestock farming system it is of difficult objective assessment and contains a very wide and different range of concepts (Wit *el al*, 2000). It is presented in figure 13 and tables 2 to 5 the indicators that were built (Santos, 2003).

The population ageing, carries along with it, the diminishing of the human power capacity which is the main bottle-neck to the sustainability of this system. The farmers have been able to adapt to the social and economic evolution shifting land use from permanent crops to permanent pastures, adapting the use of the territory to their available human power capacity (Santos, 2003).

Some questions about PDO certification system

The EU regulation 2081/92 has settled down the rules related to the origin labelled products in order to promote the value of products for the development of remote or less-favoured regions and improving farming incomes (Barjolle & Sylvander, 1999).

The Protected Designation of Origin (PDO) certification rules from the two local bovine breeds only permits pure animals to have their meat certified. The cow-calves meat in Peneda's mountain has a potential of production of about 548 tonnes while the meat produced by bovines belonging to the two local breeds that exists in Peneda's mountain is 168 tonnes (Santos, 2004). Only about 34% of the bovines from Peneda's mountain, belongs two these two local breeds (Barrosão and Cachena) while 66% are from crosses from those breeds (Santos, 2000).

We have carried out a sensorial evaluation with 21 consumers panel where we found out that panellists could not relate the meat that they eat with the animal's breed that produced the meat (Santos, 2004). The question we wish to leave is: why should the PDO certification system from cow-calves meat in Northwest of Portugal, only allow pure breed animals to produce meat to be protected by labelled origin laws?

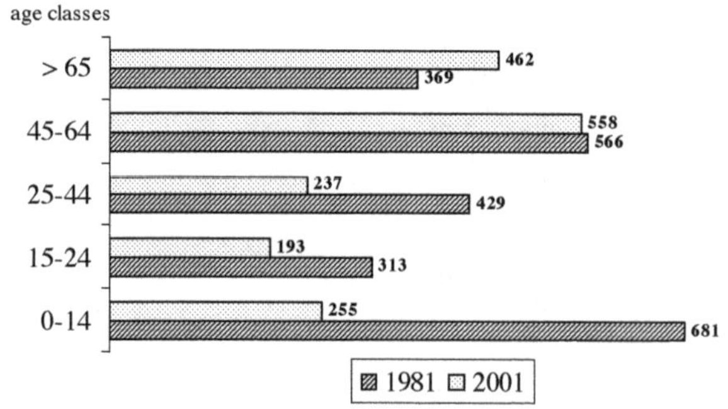

Figure 13. Population age pyramid (INE 1981, 2001).

Table 1. Daily intake of an adult grazing cow with 350 kg live weight (mean±S.D.).

Season	$nBPm_i$	Q_i	D_i	QI
Autumn – grass	55.2±12.42	0.37±0.073	592	12.1
Autumn – shrubs	37.0±13.87	1.29±0.583	56	2.7
Winter – grass	53.6±11.23	0.21±0.107	261	2.9
Winter – shrubs	27.2±16.19	2.40±1.233	53	3.5
Spring – grass	57.3±9.96	0.32±0.216	644	11.8
Spring – shrubs	40.2±13.35	1.04±0.643	16	0.7
Summer – grass	54.1±5.27	0.36±0.183	461	9.0
Summer – shrubs	33.3±4.50	0.96±0.222	82	2.6

Legend: $nBPm_i$ – number of prehension bites per minute. Q_i – bite weight (g). D_i – grazing time (min). QI – daily intake (kg DM).

Table 2. Indicators related with animals and farms.

Year	cpu65	cpa65	nTB	nF
1989			1 531	353
1994	9.0*	5.7*	1 336*	
1997	10**	2.9**		
1999	14.6***	6.5***	1 819	306

Source: *= Santos, 2000. ** Simões, 2000. *** data only from Gavieira parish. INE (RGA, 1989 and 1999).
Legend: person = chief of the family-farm. cpu65=n^{er} of cows per person under 65 years old; Cpa65 = n^{er} of cows per person above 65 years old; nTB = n^{er} of total bovines in Peneda's mountain; nF = n^{er} of farms in Peneda's mountain.

Table 3. Indicators related with animal production.

Year	SP	CDR	GP
1999	50%	4%	264
2000	38%	20%	220
2001	47%	37%	210

Legend: SP = bovines system productivity; CDR = calves death-rate until 12 months of age; GP = growing period in days. Source: Data collected within R & D PARLE Project.

Table 4. Indicators related with land.

Year	hc vs af	al/pc
1989		85.6%
1997	85%	
1999		5.5%
2001	61%	

Source: al/pc: INE (RGA 1989 and 1999). hc *vs* af: data collected within R & D PARLE Project.
Legend: hc vs af = ratio in DM between animal herbaceous intake and herbaceous availability on fallow; al/pc = ratio between hectares of arable land and permanent crops.

Table 5. Number of bovines belonging to Cachena breed.

Year	No. of young animals	No. of adult animals
1998	98	343
1999	225	526
2001		515

Source: 1998 1999, 2001-Programa nacional de melhoramento animal.

Conclusions

The livestock farming system in Peneda's mountain can assure the welfare conditions to the animals, it respects biodiversity and environmental now a days demands, and offer a product of distinguished quality. We can state that is needed more research work in order to improve shrubs as bovines feed. It is also desirable to implement territory management techniques in order to augment the calf weaning weight, without major system changing's. In order to obtain a better animal product it will be of great help to establish a selection scheme. To improve meat product, it is necessary to develop a certification system for a label that links the product to the territory and its farming system, instead of the actual PDO certification. It is of great importance, to alert the politicians who have the power to decide, for the actual problem of the farmers ageing, because within twenty years, there will be no one in Peneda's mountain villages.

Acknowledgements

We acknowledge the students of Professional School of Agriculture Conde S.Bento in Santo Tirso and our colleagues from DRAEDM the help to collect data in the 24 hour-period of observation. We are also very grateful to our scientific coordinator – D[er] Laura Larcher Graça from PARLE project, for her efficient and effectiveness coordination and support in doing the work and writing this paper. We also wish to thanks to our colleague Filipe Pacheco for his article review.

References

Altmann, J.,1974. Observational study of behaviour: sampling methods. Behaviour, 49: 227-265.

Ayantunde, A.A., Fernandez-Rivera, A., Hiernaux, P., van Keulen H., Udo H.M.J., 1998. Forage intake and feeding behaviour of day and/or night grazing cattle in sahelian rangelands. In: Regulation of feed intake, D.van der Heide, E.A. Huisman, E. Kanis, J.W.M. Osse and M.W.A. Verstegen (Eds). CABI Publishing.

Balent, G, Gibon, A., 1986. Mesure de l'ingestion des ovins et des bovines au pâturage hors domaine experimental dans les pyrenees centrales. Cahiers de la Recherche–Dévelopment, no. 9-10: 84-91.

Barjolle, D. & Sylvander, B., 1999. Protected designations of origin and protected geographical indications in Europe: regulation or policy?. Final report of PDO and PGI products: market, supply chains and institutions. Concerted Action DOLPHINS, contract QLK5-2000-0593.

Barros, M.M.N.M. & Santos, J.C-R., 1999. Criação de bovinos na Serra da Peneda. Racionalidade do sistema alimentar. Revista "Pastagens e Forragens", vol. 20: 39-54.

Cacho, B.M.F.S., 1999. Propostas de intervenção e desenvolviemnto ao nível da criação animal para zonas de montanha do Alto Minho - contribuição para o seu estudo. Engineer degree thesis in Escola Superior Agrária de Ponte de Lima, pp. 82.

Canfield, R.H., 1941.Application of the line interception method in sampling range vegetation. Journal of Forestry 39: 388-394.

Casabianca, F., 2002. Synthesis WP1. In: WP1 Origin labelled products: Definitions, characteristics and legal protection, final report. Concerted Action DOLPHINS, contract QLK5-2000-0593.

Decuq, F., Micol, D., Dubroeucq, H., 1996. Utilisation du système d'enregistrement automatique du comportemen alimentaire «Ethosys» sur des troupeaux de bovins et de cheveaux. Renc. Rech. Ruminants. 3: 74.

Gordon, I. J., 1994. Animal-based techniques for grazing ecology research. Small ruminant research, 16: 203-214.

Holecheck, J.L., 1982. Sample preparation techniques for michohistological analysis. Journal of range Management, 35: 267-268.

Holecheck, J.L., Vavra, M. & Pieper, R.D., 1982. Botanical composition determination of range herbivores diets: a review. Journal of range management, 35(3): 309-314.

Hovi, M., Sundrum, A. & Thamsborg, S.M., 2003. Animal health and welfare in organic livestock production in Europe: current state and future challenges. Livest. Prod. Sci. 80: 41-53.

Knap, P.W., Klont, R.E. & Diestre, A., 2003. Animal breeding and product quality. In: Proceedings of the 54th Annual meeting of EAAP. Paper G6.1, Abstract 711.

Lachaux, M., Meuret, M., Simiane, M. de., 1987. Composition chimique des vegetaux ligneux pautres en region mediterraneenne française: problemes posees par l'interpretation des analyses. Fourrages – Acquis actuels, pastoraux et zootechniques, sur le paturage en foret, 231-267

Mandaluniz, N., Aldezabal, A. & Ortegui, L., 2000. Mountain grazing system of beef cattle ina natural park of the Basque Country, preliminary data. Livestock farming systems. Integrating animal science advances into the search for sustainability. EAAP Publication no. 97: 179-182.

Mayes, R.W., Dove, H., 2000. Measurement of dietary nutrient intake in free-ranging mammalian herbivores. Nutrition Research Reviews, 13: 107-138.

Melo, M., Fernandes, F., Amaro, R., Xavier, D., Santos, J.C-R., 2000. Avaliação da produção e qualidade das pastagens na serra da Peneda. Proceedings da III Reunião Ibérica de Pastagens e Forragens, 67-72.

Pacheco, L.F. & Xavier, D., 2000. Prásticas de pastoreio. Imaginar, observar e aprender, no. 15 da colecção Estudos. Edição da DRAEDM, Braga.

Pires, C.B. 2000. Estrutura das explorações agrícolas de Cabreiro, Gavieira e Sistelo. Cadernos da Montanha I. Laura Graça & Henrique Santos editors. I: 48-57.

Santos, J. C-R. 2000. Sistema de criação dos bovines na serra da Peneda. In Cadernos da Montanha I. Graça, L. L. & Santos, H.M.R. dos (Editors), DRAEDM, Braga, Portugal, 106-119.

Santos, J. C-R., 2003. Peneda's mountain sustainable development and its cattle management practices. Proceedings of the 6th International Livestock Farming System Symposium, 169-174.

Santos, J. C-R., 2004. Será a carne de bovino das serras da Peneda e do Soajo de qualidade organoléptica diferente, em função da raça do animal que a produz, do ponto de vista do consumidor? In: Cadernos da Montanha III. Carvalho & Miranda Jorge (Editors), ARDAL, Arcos Valdevez, Portugal, 59-68.

Santos, J. C-R. & Ferreira, A.C., 2003. Botanical diet composition of free grazing bovines: micro histological faeces analysis. Revista Portuguesa de Zootecnia, Ano X-no. 2: 85-94.

Simões, S., 2000. In Cadernos da Montanha I. Graça, L. L. & Santos, H.M.R. dos (Editors), DRAEDM, Braga, Portugal, 58-69.

Sparks, R.D. & Malecheck, J.C., 1968. Estimating percentage dry weight in diets using a microscopic technique, Journal of Range Management, 21:264-265.

Tolkamp, B.J., 1998. Limitations in the use of constraints for intake predictions. In: Regulation of feed intake, D.van der heide, E.A. Huisman, E.Kanis, J.W.M. Osse & M.W.A. Verstegen (Editors). CABI Publishing.

Vitalis, V., 2001. Sustainable development: measuring what?. OECD 6th Round table on sustainable development. Background paper.

Wit, de J., Oldenbroek, J.K., Keulen, van H. & Zwart, D., 2000. Criteria for a sustainable livestock production: a proposal for implementation. Agriculture, Ecosystems & Environment, 53: 219-229.

Characterising and improving the traditional goat production system on the highlands of eastern Mediterranean region of Turkey

O. Gürsoy

University of Cukurova Faculty of Agriculture Department of Animal Science, 01330 Adana, Turkey

Summary

The traditional extensive nature of goat production on the Anti-Taurus Mountains of the Eastern Mediterranean Region of Turkey is characterized as resource poor, primitive, resistant to change and are far from meeting the economical and social needs of the people engaged. The production levels of the goats indicate quite wide variation and may permit many areas of intervention for the betterment of livelihood of the producers.

The local goats of the small scaled vertically integrated enterprises make use of the shrubs and herbaceous vegetation almost all year round. The milk produced is converted to typical traditional products highly demanded with reasonably good prices.

Trials and research indicated that the productivity of the goats and the profitability could be increased substantially in terms of reproductive efficiency and milk production of the does, survival rate, growth and, feedlot performance of kids. The overall farm income for a small scale farm of having 40 was seen to generate 60 % more income following the up to date management techniques.

Keywords: kilis goats, improvement interventions, goat milk, kid fattening.

Introduction

Over the ages and still, small ruminant production has been playing a vital role in sustaining the livelihood of the rural families primarily in the developing countries, more specifically in the resource poor areas of these countries (Gürsoy, 2005; Boyazoglu and Morand-Fehr, 2001; Haenlein, 2001; Rubino *et al.*, 1999).

However, food security and famine concerns of governments in the second half of the last century diverted support regimes to field crops, poultry and dairy cattle production on the expense of small ruminant production, the major agricultural activity in the resource poor mountainous and arid regions of Turkey and many other countries. As a result resource rich areas absorbed all the benefits of the support regimes and policies while off balancing the socio-economical make up leading to the shrinkage of small ruminant production and the products produced compared to the poultry and dairy cattle products (Gürsoy, 2004).

Turkey has well distinguished nine ecological regions. One of them is the Mediterranean region, possessing narrow coastal strips below the steep Anti-Taurus mountains of Hatay province, bordering Syria and the famous Taurus Mountains extending from the east to the west along the Mediterranean sea coast of Turkey. Being poor in arable land, the mountains are covered with the typical Mediterranean shrubs (maqui) which constitute the body of the Mediterranean vegetative cover. Livestock production provides on the average 65 % of the total family income and there are many families totally relying on goat production for their livelihood (Erkan *et al.*, 2001). The highlands are not very suitable to sheep and cattle production as these species require graze land or pastures.

Anti-Taurus Mountains dominate the total land area of Hatay Province. Yayladagý county of Hatay province lies over the highlands of Anti-Taurus Mountains locally named Amanos Mountains. It is the county where goat production dominates followed by tobacco cultivation, olive, cattle and sheep. Mean arable land possession of very low quality class is less than 1 ha per family (enterprise). Wheat and chickpea are the most cultivated crops diverted to household consumption to guarantee food safety.

In the year 2000, the government limited the tobacco plantation by restricting the purchase of tobacco crop by Turkish Monopoly (official tobacco buyer) to 200 kg/year/family. This policy reduced the incomes of the people tremendously. University of Cukurova designed a project to characterise the goat production system as possible alternative to compensate the income loss via possible productivity increments in 2002. This paper deals mainly with the characterisation of the traditional goat production system of the area and puts forth the possible ways of increasing the productivity of the enterprises as well as the economical welfare of the respective farmers. Therefore, the objective of this paper to characterize the actual goat husbandry system, enterprises and the environment limiting production as well as the socio-economic welfare of the people engaged.

Goat husbandry on the Anti-Taurus mountains

The results of the survey on 103 goat farmers of the Yayladagi County of Hatay Province indicated that the traditional husbandry remained almost unchanged over the ages. With the exception of vaccination against epidemic diseases the annual management cycle is believed to remain unchanged.

Management system

Goat production is well characterised as extensive system. There is no semi-intensive or intensive goat production. Flock sizes vary between 10-250 heads. In the survey covering 25 villages the mean flock composition was found to be 34 does, 36 kids, 11 yearling females, 6 yearling males and 1 buck (Acuz, 2005). In another study with one flock from each of five villages, the flock compositions at weaning were found to be similar (Table 1). The number of weaned kids seem to be low due to the fact that the producers give away one of the twins so that the doe can safely nurse the remaining kid. They also have the habit of replacing their buck in every two years generally from their own flock. The main trait for selection is the milk yield of the dam.

Mating takes place during late August till the end of October. The bucks are kept within the flock all year round. Hence it is quite common to have few out of season kidding. Kidding takes place during January-March. The kids suckle their dams twice a day, namely in the morning prior to and on their return late in the afternoon for a period of two months. They continue to suckle once a day until weaning at 3-4 months of age. Weaning takes place during April-May. The weaned kids are herded by children or women in the vicinity of the villages.

The goats are herded by men mainly on the shrubs which have lost their forest properties. The villages such as Denizgoren, Candýr, Yeditepe etc., keep their flocks around 10 months on Keldag, a steep mountain covered with very unique vegetation *ferula communis* from April to July as well as typical shrubs of the Mediterranean. During the milking season the milk is processed to cheese (feta cheese) and is transported by mules or horses to Yeditepe where the wholesalers or middlemen collect and market the products in Yayladagi, Samandagi and Antakya.

Table 1. Flock composition at weaning time.

Villages	Mature goats	Kids	Yearling females	Yearling males	Bucks	Total
Candir	70	50	17	3	0	140
Karakose	30	30	18	2	1	81
Kulac	30	48	8	1	0	87
Saksak	41	55	30	4	1	131
Saksak	71	65	5	2	2	145
Mean	48	50	16	2	1	117

Source: Acuz, 2005.

The majority of villages use the nearby rangelands, covered mostly by the shrubs and very little grasses and legumes, process their milk to feta cheese, salty yoghurt, sürk, jara cheese, butter depending on the stage of the lactation. They use a unique system, called 'nobat', 'lobat' or 'odünclesme' for processing their milk in an efficient way. About 5-6 families share the days of the week and on each day the milk is measured and accumulated for a family to process it in a bigger volume. It also economizes the time spent for processing. Hence instead of processing everyday small quantities, the family processes every 5-6 days.

It is women's job to milk the goats and process the milk. Men usually take the responsibility of marketing the product and enjoying the spending of the money earned.

Goats are dried out in September and October. No hand feeding is offered all year round even during kidding, milking or mating for flushing purposes. Some straw, bran and cut shrub branches are offered on the heavy snowy and rainy days preventing grazing.

The male kids are grazed on the same ranges as separate flocks the first year and are castrated and grazed another year before being sold at an average weight of 40-45 kg. The people in the area prefer goat meat to sheep and cattle. There are no intensive or semi intensive kid fattening in Yayladagi with the exception of farmers supplementing barley to castrated yearlings especially prior to Kurban Bayram (religious sacrifice feast). The yearling females are used as replacements and the surpluss are sold in the animal market before mating.

The rangelands of Yayladagi

Yayladagi rangelands are of typical Mediterranean vegetation, maqui which consists mainly Kermes oak (*Quercus coccifera),* Dikensiz Pinar (*Rhamnus alaternus*), dephnia (*Laurus nobilis*), *Juniperus oxycedrus, Pistacia terebentis, Myrtus communis, Melia azedarach, Daphne sericea, Junuperus excelsa, ferula communis* ve *Cystus salviaefolius*. (Gürsoy *et al.*, 2003). The evaluation of the rangelands in four seasons starting spring(every three months) indicated that the biomas/day of the fresh forage plants in Saksak village were 249.5 kg/day, 201.4 kg/day, 39.7 kg/day and 169.3 kg/day. In Candir village the same values were 267.5 kg/day, 486.2 kg/day, 71.6 kg/day and 267.3 kg/day (Gürsoy *et al.*, 2003).

Technical characterisation of the enterprises

Goat production in the area is practiced is traditional way. Characterising the enterprises' present reproductive performance, milk and meat production potentials could give clues for the magnitude of possible improvements.

Reproduction

The survey right after the kidding season in the 23 villages indicated that twinning rate and the kids/kiddin doe were 22.6% and 123.3 % respectively. The survival rate until weaning was 90.4% (Table 2).

In another study covering three villages the reproductive performances of goats in the area revealed fairly satisfactory reproductive performances in the flocks studied. The kidding rate was found to be 90 %. The rate of infertility, abortion and still births were in expected normal ranges. The mean twinning rates under the extensive system were similar in the villages, more specifically slightly less than 40 % in the three villages.

Keskin *et al.* (1997) reported very similar twinning in the Baglama village (42.9 %). The goats may be considered as moderately prolific.

Lactation

The mean lactation length for Saksak, Kulac, Candir and Karakose villages were 235, 280, 242 and 245 days respectively. The overall mean being 245 days seem to be acceptable for the extensive system. The situation of Karakose is almost one month shorter and in Kulac the farmers milk their goats until the end of the mating season (Table 3).

The mean lactation yields and milked yields were found to be quite variable in the four villages. Kulac village excells with mean milk yield of 346 kg under the extensive system without any

Table 2. Reproductive performance of goats in Yayladagi.

	Saksak		Kulac		Karakose		Mean	
Reproduction traits	n	%	n	%	n	%	n	%
Goats mated	50	100.0	32	100.0	48	100.0	130	100.0
Infertile does	3	6.0	1	3.1	2	4.2	6	4.6
Conception rate	47	94.0	31	96.9	46	95.8	124	95.3
Abortion rate	5	10.6	2	6.5	3	6.5	10	8.1
Does kidded	42	84.0	29	90.6	43	89.5	114	87.8
Still birth	-	-	-	-	2	4.7	2	1.8
Single kidding	24	57.1	14	48.3	30	69.8	68	59.7
Twin kidding	18	42.9	13	44.8	11	25.6	42	37.6
Triplet kidding	-	-	2	6.9	-	-	2	1.8
Kids born	60	-	46	-	52	-	158	-
Kids/doe kidding	-	142.9	-	158.6	-	120.9	-	138.6
Kids/doe mated	-	120.0	-	143.8	-	108.3	-	121.5
Surv. rate at weaning.	59	98.3	46	100.0	49	94.2	154	97.5
Surv. rate at 6th month	59	98.3	46	100.0	48	92.3	153	96.8
Surv. rate at 9th month	59	98.3	46	100.0	48	92.3	153	96.8

Source: Acuz, 2005.

Table 3. Lactation milk yield in the villages of Yayladagi.

Village	n	Lactation yield $\bar{X} \pm S\bar{x}$	%V	Min-Max	Milked yield $\bar{X} \pm S\bar{x}$	%V	Min-Max
Saksak	21	187.0±3.3[b]	17.9	108.3-255.5	123.7±2.0[b]	16.4	77.1-167.9
Kulac	28	346.1±8.6[a]	24.8	156.7-516.2	210.2±5.1[a]	24.5	111.6-317.7
Candir	21	177.5±6.8[b]	38.0	77.5-344.3	85.0±2.9[c]	34.5	41.1-135.8
Karakose	29	141.8±4.6[c]	32.1	69.5-249.0	82.3±2.7[c]	33.1	40.4-140.8
Genel	99	216.8±10.3	47.8	69.5-516.2	127.8±6.4	47.8	40.4-317.7

Source: Acuz, 2005.

[a,b,c] Different superscript within a column indicate significant difference(P<0.01).

supplementation. It may be due to higher blood level of Damascus breed as this village is bordering Syria. Both Karakose and Candir villages had inferior management as compared to the other two villages. They also had very low milked yields due to prolonged suckling periods widely believed to induce better growth of the kids.

It was very striking to find goats producing as much as 516 kg of milk as lactation yield and the range in the flocks of Yayladagi was 69.5-516.2 kg. This was extremely important for the future breeding strategies. Previous studies by Ozcan (1977) and Sengonca *et al.* (2003) reported 91 and 81 kg of milk for the Kil goats which supported the argument that the goats of Yayladagi were different from the primitive Kil (Hair) goats found extensively in the mountainous areas of Turkey. Sonmez *et al.* (1974) reported 238 kg of milk for Kilis goats of all ages. Mavrogenis and Papachiristoforou (2000) reported 318 g for the Damascus goats in Cyprus which is quite similar to the Kulac village. Ozuyanik (2004) estimated the lactation milk yield of Damascus goats under intensive management conditions in the Turkish Republic of Northern Cyprus significantly higher than the figures found in Yayladagi (453.8 kg).

Growth

Table 4 gives the growth performance of the 22 female and 18 male kids under the extensive conditions was studied in Saksak Village. The kids were weaned at a mean age of 75 days. It was observed that the kids failed to achieve marketable weight of 40-45 kg within one year. This is considered highly important because it both affects the income of the family and also the carcass quality when slaughtered at almost 2 years of age.

In the traditional system the mean daily gains were found to be low. The kids were not supplemented at all. Daily gains were seen to be high during the suckling period because the does were not milked. After weaning the daily gains dropped in a steady trend due to the nutritional status of the shrubs available. These figures indicate that the kids are underfed and had there been any hand feeding the kids would have performed better and would be causing less damage to the shrubs that they graze on.

Productivity increasing interventions

Like most goat producing areas of Turkey (Tuncel & Bayindir, 1983), the goat production system on the Anti-Taurus Mountain villages of Hatay Province was found to be extremely unproductive.

Table 4. Growth performance of the kids in the villages of Yayladagi (kg).

	Live weights		Daily Gains	
	Singles (n=17)	Twins (n=23)	Singles (n=17)	Twins (n=23)
Age	$\overline{X} \pm S\overline{x}$	$\overline{X} \pm S\overline{x}$	$\overline{X} \pm S\overline{x}$	$\overline{X} \pm S\overline{x}$
Birth	4.0±0.56	3.8±0.61	-	-
Weaning[1]	14.4±1.95	11.8±1.79**	141.6±3.5	107.0±1.6
105th day	16.5±2.22	14.2±2.07**	71.0±4.6	81.0±2.9
135th day	19.0±2.70	16.6±2.48**	82.6±2.5	77.3±2.6
165th day	19.9±3.27	18.5±2.54	29.8±4.3	66.2±3.0
195th day	22.0±3.81	20.8±2.96	56.9±7.0	73.8±4.5*
225th day	23.2±4.06	22.1±2.95	55.1±4.7	44.5±4.9
255th day	24.4±4.40	23.1±3.07	39.5±3.2	33.2±2.5
285th day	25.6±4.77	24.1±3.27	38.6±2.2	33.5±1.7
Birth-285 days		-	75.8±1.6	71.0±1.1*

Source: Acuz, 2005.
[1]Weaning age: 75 days;
** *P*<0.01; * *P*<0.05.

However close study of the enterprises revealed the fact that it was possible to increase their production potential with simple interventions such as internal and external parasite control, vaccination, simple improvements of the available housing, using genetically better bucks, flushing, supplementation with a well balanced ration during late pregnancy, post kidding, milking, for growth and feedlot performance of the male kids (Hatziminaoglou *et al.*, 1995). The results were extremely satisfactory.

Reproductive efficiency

Two main interventions were made in order to wean more kids per doe. The first was to prohibit giving away one of the twins. This had been a traditional practice arising primarily from the inadequate feeding of the newly kidded doe producing insufficient amount of milk for two kids (Table 5). Another reason might be to honour the relatives and neighbours by presenting them a kid or two for religious sacrifice during the Islamic Feast of Kurban Bayram. Hence the does were supplemented by half a kilo of well balanced concentrate feed 4-6 weeks before kidding and 0.75 kg of the same feed after kidding for 45 days in addition to grazing.

The second intervention was a simple flushing application. The does were provided varying amounts of concentrate feed starting two weeks before mating and during mating. The twinning rate in one producer, possessing 40 does, increased from 33 % to 45 % gaining 6 kids from flushing and 18 kids from keeping all the twins. Similarly in the other producer's flock of 39, twinning was increased from 31% to 51% gaining 8 kids from flushing and 20 kids from keeping all the twins.

Lactation

Milk constitutes very important share of the annual income of the goat producer. However there was no significant effort to increase the milk production. The previous survey showed that supplementation of concentrate feed would have significant benefits: increase milk production, decrease the grazing

Table 5. Reproductive response to flushing

Reproductive criteria A+B	Farmer A	Farmer B	Combined
Does mated	40	40	80
Does kidding	40	39	79
Single kidding	22	19	41
Twin kidding	18	20	38
Kids/doe mated	1.45	1.48	1.46
Kids/doe kidding	1.45	1.51	1.48

pressure on the rangeland and improve the conditions of the does for the mating season. Table 6 gives the effect of supplementation. Labour was not taken into consideration as the whole flock was grazed together. It was seen clearly that supplementation of 0.5 and 1.0 kg of well balanced concentrate mix increased marketable milk by 40 and 75 % respectively. This alone generated appreciable amount of income because the area was very specialized in converting milk into cheese and salty yoghurt which were highly demanded with reasonably high prices.

Survival rate, growth and feedlot performance

Survival rate of the kids were improved tremendously due to vaccination, parasite control, supplementation, vitamin and mineral administration. The survival rates between birth and weaning increased from 90% to 98 %.

In an on farm trial in Saksak village 25 male kids of a producer were randomly divided into two groups post weaning (approximately 90 days of age) and one group (n=13) was supplemented with a balanced ration of 17 % CP and 2.6 Mcal for 180 days. The other group was treated in the traditional system, more specifically grazed the available shrubs and rangeland without supplementation. Supplemented kids reached a mean weight of 43 kg while the control group remained at 22 kg. This trial opened the way to studying the growth performance and carcass traits of these kids as a graduate thesis. Sentut (2005) offered supplementation of concentrate mix and other health improving interventions as well as intensive kid feedlot. The results are presented in Table 7 which indicates that the kids could be marketed within the same year reaching around 40 kg.

Table 6. Supplementation of goats during lactation.

	Concentrate supplementation (kg/d)		
Costs and returns	0.0	0.5	1.0
Marketable milk yield (kg)	125.7	175.8	220.6
Feed consumed (kg)	-	90.0	180.0
Feed cost	-	25.2	50.4
Gross returns (YTL)	104.0	145.5	187.5
Net returns (USD)	76.8	88.8	101.2

1 USD= 1 355 YTL.

Table 7. Growth and feedlot performance of kids.

Feedlot criteria	Management systems		
	Traditional	Semi-intensive	Intensive
Initial weight	16.3	17.8	19.2
Days on feed	170	170	170
Initial age	138	138	138
Final weight	21.3	37.5	42.1
ADG	29.5	115.8	134.8
Feed cons.	-	4.7	5.2[1]

Source: Sentut, 2005.
[1]Includes concentrate + roughage.

Phenotypic variation and possible breeding efforts

The great variation in milk production both within the flocks and between the villages led the way to changing the bucks of the three producers: two in Saksak and one in Kulac. The bucks were selected from high producing Damascus does from Reyhanlý County which is closer to Syria. The producers sold their bucks and the difference was compensated from the Research Project. Their progeny are being milked this year and substantial increases are expected.

Economical prospects

This Project was initiated in 2002 with a donation from a private company as well as Cukurova University Research Fund. It was estimated that in the year 2003 the gross returns for two producers increased by 60 % in terms of extra kids raised (male and female), male kid sales after supplementation or feedlot, milk product (cheese and salty yoghurt) sales. In the year 2004 their incomes increased over 100 % as compared to their incomes in the year 2002. Furthermore, they doubled the sizes of their flocks. Erkan *et al.* (2001) reported that the farm incomes on the Taurus Mountains Project were increased by 65% due to technology transfer to the farms involved with livestock production as the major enterprise.

Conclusions and recommendations

Similar to most Mediterranean, goat production on the highlands of the Anti-Taurus Mountains is practiced in a very traditional way. The farmers are resource poor and far from keeping up pace with the general development of the total economy (Gürsoy, 2004; Zervas *et al.*, 1983). They are in remote areas and are not enjoying the benefits of the central economy. There is practically no solid government service. They are not organized for cooperative activities for purchasing cheap inputs and marketing their valuable products which could easily be converted to organic production.

Small ruminant sector is not subsidized at all (Gürsoy, 2004). Furthermore the government is banning goat production in many places in Turkey with extreme penalties. These policies of exterminating goats in Turkey have been put into action without sound evaluations and assessments by the authorities. It is widely believed that 'goats destroy forests and must be reduced in number'(Tuncel & Bayindir, 1983) and must be replaced immediately by either sheep (Awassi or

Chios) or exotic goats like Saanen or their crosses. Unfortunately the decision makers are unaware of the fact that the intended breeds of sheep and goats may not even survive under the harsh conditions of the highlands.

Besides, most of the grazing areas are far from being classified as forests. They have been subjected to heavy erosion over thousands of years. They have been and are still being subjected to destruction by man for fuel needs and other purposes. Had there been an unbiased research on the sources of damage to the forests as well as shrubs the result could be that the damage caused by man for fuel and wood (commercial) as well as opening up farmlands via fire may triple the damage caused by goats.

Hence as long as these people live in and around forests they can not be separated from their traditional ways of goat production (Tuncal & Bayindir, 1983). They would keep preferentially eating goat meat, goat milk products. Exterminating the goats while keeping the people in and around the forests or shrub lands will not decrease the damage caused by man.

Intensification of management systems along with the application of up to date breeding technologies and rotational grazing may decrease the pressure of the goats on the forests while providing them with substantial incomes to elevate their economical and social welfare.

As organic farming and organic products gain increasing popularity in the rich nations of the world, one would expect better support regimes for the traditional products produced from goat and sheep milk and meat. For instance there are over hundred different varieties of cheese, butter, yoghurt, ice-cream produced as typical delicacies of sheep and goat milk in Turkey as well as in Europe (Zervas *et al.*, 1983, Rubino *et al.*, 1999, Gürsoy, 2005).

It would be a tragedy to lose all the cultural heritage accumulated for thousands of years by forcing the goat producers from these less favoured remote areas to urban areas. Urgent policies must be put in action for preserving and promoting these unique products and keep those farmers in their lands happily and with higher standards of living in line with the urban areas.

Acknowledgements

The author greatly acknowledges the generous supports of Mr. Erol Makzume and Mr. Semih Baki of Antoine Makzume International Transport and Trading Limited Company, Turkey and University of Cukurova Research Fund. Special thanks are due to my graduate students Servet Acuz, Tugba Sentut and Isil Bilgic for their efforts in Yayladagi County.

References

Acuz. S., 2005. Hatay ili Yayladagi ilcesinde yetistirilen keci irkinin karakterizasyonu, büyüme ve süt verimlerinde varyasyonun belirlenmesi [Characterization of the goat breed in the Yayladagi county of Hatay province and determination of variation in growth and milk production]. Yayimlanmamis Yüksek Lisans Tezi-C.Ü. Fen Bilimleri Enstitüsü. [Unpublished M.Sc. Thesis submitted to the University of Cukurova - Institute of Science], Adana, Turkey, pp. 77.

Boyazoglu, J. P., Morand-Fehr, P. 2001. Mediterranean dairy sheep and goat products and their quality. A critical review. Small Rumin. Res, 40: 1-11.

Erkan,. O., Gül, A., Yilmaz, I., Sengül, H. 2001. Impact of the Project. In: Sustainable development of small-scale farmers of the Taurus Mountains of Turkey, O. Erkan, S.P.S. Beniwal, J. Ryan & M. Bounejmate (Editors). ICARDA. Integrated Natural Resource Management Technical Report Series, Aleppo, Syria: 86-97.

Gürsoy, O. 2004. Economics and profitability of sheep and goat production in Turkey under new support regimes and market conditions. 55th Annual Meeting of the European Association for Animal Production. 5-9 September, Bled, Slovenia, pp. 22.

Gürsoy, O. 2005. Small ruminant breeds of Turkey. I: Characterization of the Small Ruminant Breeds in West Asia and North Africa. Vol. 1. West Asia. L. Iniguez (Ed.). International Center for Agricultural Research in the Dry Areas (ICARDA) Aleppo, Syria: 239-416.

Gürsoy, O., Kokten, K., Acuz, S., 2003. Hatay Ili Yayladagi Ilcesi Keciciliginin Gelistirilmesi ve Mera Karakterizasyonu Projesi-Mera Olcüm Raporlari (Yayimlanmamis). [Unpublished Data of the Project-Improvement of the Goat Production and Characterization of the Rangelands of Yayladagi County of Hatay Province].

Haenlein, F.W. 2001. Past, present, and future perspectives of small ruminant research. J. Dairy Sci. 84: 2097-2115.

Hatziminaoglou, J., Zervas, N.P., Boyazoglu, J. 1995. Goat production systems in the Mediterranean area: the case study of Greece. In: Goat Production Systems in the Mediterranean, A. El Aich, A., Landau, S., Bourbouze, A., Rubino, R. and Morand-Fer, P., (Editors). Wageningen Pers, Wageningen, Netherlands, EAAP Publ. 71: 82 - 109.

Keskin, M., Kaya, S., Ozcan, L., Bicer, O. 1996. Hatay Bolgesinde yetistirilen kecilerin bazi morfolojik ve fizyolojik ozellikleri üzerine bir arastirma [A study on certain morphological and physiological characteristics of goats rared in Hatay region], M.K.Ü. Ziraat Fakültesi Dergisi 1(1): 69-84.

Mavrogenis, A.P., Papachiristoforou, C., 2000. Genetic and phenotypic relationships between milk production and body weight in Chios sheep and Damascus goats. Livest. Prod. Sci. 67: 81-87.

Ozcan, L., 1977. Cukurova Üniversitesi Ziraat Fakültesinde yetistirilen Kilis ve Kil Kecilerinin islahinda Saanen ve G1 genotipinden yararlanma olanaklari [Possibilities of using Saanen and G1 genotypes in improving Kilis and Kil goats rared at the University of Cukurova Faculty of Agriculture], C.Ü.Z.F. Yayinlari:122, Bilimsel Inceleme ve Arastirma Tezleri No:19, Adana, Turkey, pp. 68.

Ozuyanik, 2004. KKTC kurak iklim kosullarinda yetistirilen Damascus kecilerinin adaptasyon mekanizmalari üzerine bir arastirma [A study on the adaptation mechanisms of the Damascus goats managed under the dry conditions of Turkish Republic of Northern Cyprus] Unpublished M.Sc. Thesis submitted to the University of Cukurova - Institute of Science, Adana, Turkey, pp. 50.

Rubino, R., Morand-Fer, P., Renieri, C., Peraza, C., F.M. Sarti. 1999. Typical products of the small ruminant sector and the factors affecting their quality. Small Rumin. Res. 34 : 289-302.

Sengonca, M., Taskin, T., Kosum, N., 2003. Saanen x Kil melezlerinin ve saf Kil kecilerinin kimi verim ozelliklerinin belirlenmesi üzerine es zamanli bir arastirma [A simultaneous study on determining certain production characteristics of pure Kil and Saanen x Kil crossbred goats]. Turk J. Vet. Anim. Sci. 27: 1319-1325.

Sentut, T. 2005. Hatay ili Yayladagi ilcesinde erkek oglaklarin besi gücü ve karkas ozellikleri. [Growth performance and carcass traits of mail kids at the Yayladagi county of Hatay province]. Unpublished M.Sc. Thesis submitted to the University of Cukurova - Institute of Science, Adana, Turkey, pp. 67.

Sonmez, R., Sengonca, M., Alpbaz, A. G. 1974. Ege Üniversitesi Ziraat Fakültesinde yetistirilen Kilis Kecilerinin verimleri üzerinde bir arastirma [A study on the production characteristics of Kilis goats rared at Ege University Faculty of Agriculture], Ege Üniversitesi Ziraat Fakültesi Yayinlari No:239, Bornova Ege Üniversitesi Matbaasi, Izmir, Turkey, pp. 20.

Tuncel, E. & Bayindir, S. 1983. Genetic improvement of goats in Turkey (Meat, milk, hair and skin), In: Proceedings: International Symposium on Production of Sheep and Goat in Mediterranean Area. 17-21 Oct., 1983, Ankara, Turkey: 33-45.

Zervas, N., Boyazoglu, J., Hatjiminaoglou, J. 1983. The potential of Mediterranean sheep breeds for milk and meat production: strategies of improvement. In Proceedings: International Symposium on Production of Sheep and Goat in Mediterranean Area. 17-21 October 1983, Ankara, Turkey, 1-23.

Session 3. Possibilities for improving traditional systems

Main Papers

Animal genetic resources and sustainable development in the Mediterranean region

L. Telo da Gama

Estação Zootécnica Nacional - INIAP, 2005-048 Vale de Santarém, Portugal

Summary

The richness of the Mediterranean area in Animal Genetic Resources places this region in a key position regarding overall biodiversity, with particular responsibilities in their management. Native breeds of domestic animals, especially in this region, are used mostly in low-input low-output systems, developed over the years, which fit well with the challenges placed by sustainable development, including product quality, biodiversity, economic viability, respect of social and environmental constraints and acceptability by consumers (including ethics and animal welfare). Those breeds play, therefore, a key role in ecological and social equilibrium, and are the basis of local high-quality products. Nevertheless, many of the native breeds have already been lost or are endangered of extinction, and steps must be taken to invert that trend, of which the development of sustainable breeding programs, aiming at improving their lifetime overall efficiency (including both production and non-production traits), could improve their competitiveness relative to other genotypes. In addition, the valuation of their products would give further support to these native breeds, and to the systems and human communities to which they are associated.

Keywords: native breeds, sustainability, production systems, conservation, animal breeding.

Introduction

In addition to being the center where domestication took place for several domestic animal species (sheep, goats, cattle) some 100 centuries ago, the Mediterranean region is one of the areas of the world considered to be a hotspot of biodiversity (Cincotta *et al.*, 2000). Native animal breeds in the Mediterranean region have played a major role in utilizing resources available under extensive production systems and marginal areas, thus contributing for environmental and socio-economic stability, and they have been an integral part of local culture, tradition and gastronomy. According to the DAD-IS database (FAO, 2005), the 21 Mediterranean countries have a total of 334 cattle breeds, 149 goat breeds, 162 horse breeds, 179 pig breeds and 408 sheep breeds. Relative to the total number of livestock breeds in the world, Mediterranean breeds represent about 16% for cattle, 18% for goats, 16% for horses, 18% for pigs and 21% for sheep, which is, undoubtedly, a tremendous contribution of this region to the diversity of Animal Genetic Resources (AnGR) in the world.

The intensification of agriculture, which took place mostly during the second half of the 20th century, resulted in the extinction or endangerment of several livestock breeds worldwide. In Europe, for example, 18 % of the breeds existing in the beginning of that century have already disappeared and nearly 40 % are currently in an endangered status. Notwithstanding the fact that Europe has more than 2 500 recognized livestock breeds (Rege & Gibson, 2003), many of the local breeds have a minimal economic impact, such that, depending on the species considered, 80% of the production is provided by 19% to 34% of the breeds (Martyniuk & Planchenault, 1998), and nearly 60% of the cattle in the European Union are considered to derive from the Holstein Friesian breed (Drucker *et al.*, 2001).

In recent years, emphasis on the need to implement sustainable development programs, of which native breeds are an important component, has caused a re-evaluation of the strategies followed in the past, and the disappearance of native breeds at an alarming rate has prompted the development of national and international programs aiming at the characterization, conservation and sustainable utilization of AnGR.

Sustainable development

Even though several definitions of sustainable development have been proposed, the Brundtland Report considers it to be "the development that meets the needs of the present generation without compromising the ability of future generations to meet their own needs" (WCED-UN, 1987). More generally, according to the FAO definition, sustainable development is concerned with:

- management and conservation of the natural resources base, including land, water, plant and animal genetic resources;
- orientation of technological and institutional change in such a manner as to ensure the attainment and continued satisfaction of human needs for present and future generations;
- agricultural practices which are environmentally non-degradable, technically appropriate, economically viable and socially acceptable.

When considered in the context of sustainable agriculture, animal production/breeding objectives and practices must face several challenges, which have been the subject of research and discussion in recent years (Gibson & Witlton, 1998, Gibon *et al.*, 1999, Olesen *et al.*, 2000, Liinamo & Neeteson, 2001, Rege & Gibson, 2003). Even though several perspectives can be taken (with more emphasis on issues of environmental, social, economical, etc., nature), the overall challenges can be translated into a few major objectives and goals of animal production and breeding systems under sustainable development, as briefly outlined in Table 1.

Interpreted in a broad sense, quality is one of the major issues about which consumers are particularly aware and concerned with nowadays, in particular in what regards the safety of food products, with several bitter examples over the last few years (BSE, Foot and Mouth Disease

Table 1. Objectives and goals of animal production and breeding systems under sustainable development (adapted from Liinamo & Neeteson, 2001).

Objectives	Goals
Quality	Safety and healthiness for consumers
	Product quality
Acceptability by consumers	Ethics
	Animal welfare
Diversity	Biodiversity
	Adaptation to different environmental constraints
Environmental constraints	Pollution
	Land use
Economic viability	Short run
	Long run
Social constraints	Culture and tradition
	Rural development

outbreaks, dioxins, etc.) making this issue of tremendous importance. In addition, the inherent quality of a food product, in the sense of being the "sum of properties and characteristics of a product which confer on it the ability to satisfy explicit and implicit requirements" (Sorensen & Jakobsen, 2005) has been progressively more important in recent years for a wider segment of the population. The properties and features associated with quality include both real and perceived attributes of the product (Edwards, 2005), with the first ones encompassing objective characteristics (physical, chemical, organoleptic, etc.) while the second group relates to how the consumer perceives the product and the mode of production associated with it (animal welfare, environmental impact, cultural value, etc.).

In this broad definition of quality, it is not only the end-product itself, but the way it is obtained, which are of importance for consumer acceptability. This issue is of particular importance in developed countries, where consumers are progressively more concerned with practices and questions underlying the production systems, such as ethics of production, animal welfare, environmental issues, organic production, minimum use of drugs, etc. This perception by consumers is shaping current breeding and production practices, and will probably be a major driving force in the future of animal production.

The Convention on Biological Diversity, approved in 1992 in the Rio Summit, emphasized the conservation of biological diversity as a key component of sustainable development, and recognized AnGR, both domestic and wild, as an integral part of overall biodiversity. Domestic animal breeds, resulting from a long process of domestication and different selection practices, evolved in specific environments to which they gradually adapted, fitting to different constraints (environment, market, social, etc.), and developing specific attributes. Among these, hardiness (a broad expression which includes high longevity, ability to withstand periods of scarcity, disease resistance, etc.) is a trait of major importance, especially in the Mediterranean region, where native breeds tend to be used mostly in marginal areas, in which the ability to survive and produce is frequently the major breeding objective.

Undoubtedly, one of the major deterrents to the intensified livestock production systems currently used in many countries is the detrimental impact on the environment, and several countries have placed restrictions on animal production due to excessive effluent output, especially nitrogen, from intensive farming. On the other hand, sustainable use of grazing lands has been promoted as a viable way to satisfy human needs, while maintaining or enhancing the quality of the environment and conserving natural resources (Boyazoglu, 1998).

To subsist over time, a sustainable system must be and remain economically viable over the short- and long-run, because otherwise it will be abandoned and the system will no longer be sustainable. Thompson & Nardone (1999) discuss this issue, both from an economic and ethical standpoint, comparing two approaches, i.e., resource sufficiency and functional integrity. In any case, a critical component of sustainability is the input/output ratio generated by the system (with both inputs and outputs including factors other than just financial considerations) and, in many instances, the need to maintain the overall ecosystem, based on collective interests, implies the support of good practices, to compensate for the lower profitability (financial or otherwise) which the system often attains.

Several social constraints may frame sustainable development, and social sustainability is an issue by itself. In a broad perspective, agricultural practices and livestock production/uses are often part of local traditions and cultural values, and constitute an essential component of rural development, maintaining in use lands which otherwise would be left abandoned or used for non-sustainable purposes. Thus, society must look at sustainable agricultural systems as a fundamental cornerstone of ecological and social equilibrium, and be willing to support its maintenance, or the price to be paid afterwards will probably be too high.

The situation recently experienced in Portugal may illustrate how different factors interact to maintain the viability of a sustainable system. The Central region of Portugal is characterized by mountainous areas, often forested, which have traditionally been associated with the production of small ruminants. Due to urban migration, the human population in this region decreased dramatically in the second half of the 20th century, and animal production was progressively abandoned, such that the population of goats decreased by 22% between 1990 and 2000. The summer of 2003 was extremely hot in Portugal, and fires in that period consumed nearly 400 000 ha of forest and agricultural areas (i.e., almost 5% of the surface area of the country), with the vast majority of those fires occurring in the Central region of Portugal. It is difficult not to think that this situation could have been prevented, or at least minimized, if small ruminants were still used in controlling shrubs and keeping forest areas clean, while producing meat and milk; nevertheless, their reduction resulted in the abandonment of those areas, with a highly negative social and environmental impact, and consequences which affected society as a whole.

Importance of animal genetic resources

Most likely, the major feature of agriculture and animal production in the Mediterranean area is diversity - of climates, orography, agriculture, peoples, traditions, etc. As a result of the very distinct conditions, local breeds were developed and have gone through a long process of adaptation to specific environments, making this region one of the richest in the world in what concerns animal biodiversity. These breeds represent, therefore, an immense direct contribution to the overall diversity of AnGR, and also an indirect effect, through their strong influence on local breeds developed in other parts of the world (e.g., diffusion of Merino sheep worldwide, development of Criollo breeds of all livestock species in South America, etc.).

Native breeds in the Mediterranean are also an important part of local culture and tradition, and the large assortment of high quality local specialties reflects this specificity. For example, the variety of sheep cheeses or the very large number of pork transformed products, which are unique to this part of the world (not to mention other well known non-animal products, such as wine, olive oil, bread, etc.), are a demonstration of the importance that agriculture and local traditions have had in the Mediterranean region throughout the years. One very important point is that many of these quality-products are associated with specific breeds and production systems, and if these breeds are abandoned or extinct, so will the products with which they are associated.

Another important role of AnGR is in supporting the livelihood of local communities, often dependent upon animal production in marginal areas for their maintenance. In the harsh conditions of these areas, only local breeds are able to survive and produce, the alternative being abandonment and desertification. Hence, the contribution of native AnGR in rural development and human fixation must be acknowledged, with important consequences in ecological and social equilibrium.

Integration of local breeds and livestock production systems

The experience observed in high-input, high-output systems, which resulted from the intensification of agriculture and increasing demand for food, has resulted in relatively uniform production systems throughout the world, based on a very narrow genetic pool. The examples of poultry, swine and dairy, where very few breeds are still commercially used and the number breeding companies intervening in the market is progressively more reduced, gives an indication of what would have happened with the other species if the trend towards intensification had continued. Intensive production systems, and the species to which they are associated, certainly have made a major contribution to food security for humans, contributing with up to 75% of the protein of animal origin in human diets in developed countries (Boyazoglu, 1998). Nevertheless, the repercussions of these systems (biodiversity, animal welfare, pollution, product quality, etc.) must also be taken into account, and the use of a multidisciplinary "farming systems approach" seems justified (Gibon *et al.*, 1999).

On the other hand, livestock species which are kept in low-input systems tend to be used in extensive systems and marginal areas, with diversified environmental constraints, which often represent a serious limiting factor to their production efficiency. In general, extensive systems of animal production are characterized by a limited stocking rate, little use of advanced technology, low productivity per animal and per hectare, reduced use of fossil energy and feeding mainly based on natural grazing and use of agricultural by-products (Boyazoglu, 1998). Native breeds selected over time under those specific constraints are usually better adapted, and the only ones able to take advantage of the scarce and irregular resources which available in those conditions. As a result, an equilibrium has been reached between the environment and native genetic resources, which fits well within the framework of sustainable development. Nevertheless, this is a fragile equilibrium and some of the elements of sustainability may be below a critical threshold in these systems (e.g., overgrazing, insufficient profitability, social constraints, etc.), and steps may need to be taken (e.g., organized breeding programs, social support/funding, etc.) to overcome these limitations.

Overall, it becomes clear that, in nearly all items associated with sustainable development in Mediterranean systems, native breeds of livestock have played a major role over the years, and they are expected to be an integral part and one of its major pillars in the future.

Challenges for native breeds

Several reasons may be identified as having caused the situation of endangerment in which a very large number of native breeds is currently in. Among those, probably the most important one is intensification of agriculture, which took place after World War II. This intensification was extremely important in providing the means of responding to the need to feed a growing human population in the world, but changed traditional farming practices completely, and threatened the maintenance of their foundations, including AnGR.

Competitiveness

One of the major limiting factors hindering the development of native AnGR is their low level of production, which threatens the profitability of their use and, therefore, their maintenance and survival. As a result, native breeds may not be competitive in some production systems, and more so when intensification takes place. Assuming that native breeds are crucial components of sustainable development, increasing their competitiveness is an important step, which may be accomplished through the development of adequate breeding programs, aimed at selection for lifetime production efficiency and adaptation, within the constraints imposed by sustainable breeding programs (Olesen *et al.*, 2000).

On the other hand, native breeds have a clear advantage in their ability to utilize marginal lands, and convert their scarce resources into high quality products. This is clearly a competitive advantage, but the price differential that the consumer is willing to pay for the attributes of the product (both real and perceived) must be compensatory for the system to be economically viable. In this area, certification programs (DOPs, etc.) are important support mechanisms, which will certainly make a major contribution towards the competitiveness and survival of local breeds .

Overall, this is an expression of genotype-environment interaction, such that the competitiveness of native breeds will depend upon the priority given to the implementation and maintenance of sustainable systems where those breeds excel, especially through their capabilities of adaptation and generation of quality products.

Conservation strategies

Given the decline in census observed in the majority of native breeds kept in marginal areas, steps have been taken over the last few years to develop conservation programs aimed conservation of native AnGR. Among these efforts, the role of the FAO should be underlined, through its "Global Programme for the Management of Animal Genetic Resources", which is now yielding results expressed in country reports on the status of AnGR, followed by an overall balance to be made in the near future and development of regional strategies, under the leadership of the FAO. In addition, policy measures taken by the European Union as well as local governments, have had a considerable impact in slowing down the census decline which has consistently been observed in several native breeds. Other measures, such as the development of *ex situ* conservation programs, are important additional measures, which complement, but do not replace, *in situ* conservation (Oldenbroek, 1999).

The vast majority of conservation programs currently implemented in native livestock breeds are aimed at minimizing the risk of extinction of any given breed, through the use of both *in situ* and *ex situ* conservation measures. Among these, the use of germplasm banks and financial support to farmers keeping endangered breeds have been major steps of the conservation programs. Nevertheless, it can be argued that financial constraints could lead to a situation where priorities need to be established in terms of the breeds to be conserved, to maintain as much genetic diversity for the future as possible. In this context, alternatives have been proposed to establish conservation priorities based on Weitzman approach (Thaon d'Arnoldi *et al.*, 1998; Simianer, 2003).

One of the criticisms often presented to conservation strategies (Gama & Delgado, 2000; Roosen *et al.*, 2003) is that a major effort is often made towards avoiding breed extinction, while the conservation of within-breed genetic diversity has frequently been ignored. Nevertheless, genetic variation within breeds has been found to represent the majority of the overall genetic variation in a given livestock species (Hammond & Leitch, 1996), and must therefore not be ignored. The situation is particularly critical in livestock species with small numbers, especially when breeding practices lead to high rates of inbreeding and, consequently, to a high loss of genetic variation. For example, the Alentejana breed of cattle in Portugal has a number of females around 8 000 but, due to the influence of some herds and the excessive use of some bulls, retrospective calculations indicate that the effective population size is only 25 (Gama & Carolino, 2002), i.e., about one-half of what has been recommended by the FAO as the minimum tolerable. Other examples could be found, where it becomes clear the need to carefully manage within-breed diversity, to prevent further losses of genetic variation, and thus maintain the ability of those breeds to adjust to possible unforeseen environmental changes in the future.

Sustainable animal breeding

Selection applied in most organized breeding programs in domestic animal species has been quite successful in genetic improvement of production efficiency (Smith, 1984), and it is expected that introducing new technologies in selection programs could improve the rates of genetic improvement currently attained. Nevertheless, undesirable correlated changes in traits not in the breeding objective have often been observed, namely in behavioural, physiological and immunological traits (see review by Rauw *et al.*, 1998), such that animal welfare and health status may be affected. These results are a strong argument in favour of a more comprehensive approach to animal breeding programs, based on a better understanding of the physiological mechanisms underlying overall lifetime production efficiency, and where the breeding goals are defined to include both production and adaptation traits.

The need for redefinition of breeding programs has been recognized in most species, and more so when sustainable development is a major issue and a long term perspective is considered (Gibson & Wilton, 1998). The implementation of sustainable breeding programs, designed to "produce animals that fit better in the environment for the production of good quality products for present and future generations in an acceptable manner" should, therefore, consider factors such as genetic diversity, use of natural resources, animal health and welfare, food quality and safety, and drawbacks of the technologies to be used (Liinamo & Neeteson van Nieuwenhoven, 2002). One possible approach to combine production and non-production factors is to define in the breeding objectives the value of both market (i.e., cost/benefit of products) and non-market (animal welfare, ecological impact, etc.) genetic gains, and use selection index theory to establish the breeding goals, such that animal breeding programs become concerned not only with production efficiency, but also with their impact on genetic diversity, environment and society (Olesen *et al.*, 2000). This approach has been used in the framework of the SEFABAR network (Liinamo & Neeteson-van Nieuwenhoven, 2002), with a critical review of sustainability issues and alternative breeding programs for different domestic species

Quality products and AnGR: two historical examples of sustainable development

Throughout the years, and until the 20th century, wool was a highly valued product, and fine wool produced by Merino sheep was exceedingly appreciated for clothing. Up until the 18th century, Merinos existed only in the Iberian Peninsula, and export of live Merino sheep to other countries was forbidden by a body called the "Honoured Council of the Mesta", which was created in 1273 to protect sheep and wool production and harmonize sheep transhumance, a practice which was very common at that time. In the circumstances of that time, the Merino was therefore a highly prized genetic resource, valued by the quality of its major product, and integrated in a production system with a tremendous social and economic impact. These systems were highly sustainable and, apparently, quite profitable, but of course this monopoly could not be kept forever, and permission to export outside the Peninsula was finally granted in the 18th century, and Merinos are now found around the world.

Another example, also with a long tradition but still in existence nowadays, is the production of Iberian pigs raised outdoors under silvopastoral systems (known as "montado" in Portugal, and "*dehesa*" in Spain). These animals are usually finished in Quercus forests (oak and cork trees), using acorns as the major source of energy, and then slaughtered at an age of about 15-18 months, with 150-160 kg. When compared with white breeds, Iberian pigs have a body composition with a higher proportion of monounsaturated fatty acids (Fallola *et al.*, 1998; Edwards, 2005) and a higher ability to digest fat and fibre in the diet (Freire *et al.*, 1998). Products processed from acorn-fed Iberian pigs (ham, sausages, etc.) have unique organoleptic properties, largely determined by the amount and quality of intramuscular fat, and are delicacies for which the consumer is willing to pay a premium price. This is, indeed, an excellent example in which several of the challenges posed by sustainable development have been fulfilled, where a local breed, raised in an environmentally-sound manner, produces a high-quality product and attains economic self-sustainment. Of course, this is still a fragile system, highly dependent on climate stability and, in particular, on minimization of health risks, since the breed and the corresponding production system were highly threatened by African Swine Fever in the mid-20th century.

Future outlook

The need to establish conservation and breeding programs for AnGR, as key components of sustainable development, has received more attention in the last few years from different levels of society. In opposition to the intensification trend seen in the past, in recent years there was an increased social recognition and understanding of rural values and traditions, and a better appreciation of the need for sustainable development programs (Boyazoglu & Hatziminaoglou, 2002), and policies followed by the European Union have changed towards the extensification of production systems, as a result of a re-evaluation of the models of agricultural development followed in the past and consumer's pressure regarding the quality and safety of food products. Currently, additional emphasis is being placed on the need to reconsider the strategies used for AnGR management, and it is generally recognized that efforts must be made towards better conservation and sustainable utilization programs.

We are now in a situation where society as a whole is becoming more aware and concerned about issues such as sustainable development, biodiversity, environmental protection, etc., and these issues have become a top priority for policy-makers. It is, therefore, a turning point, with completely different challenges and demands ahead of us, but where the background and surrounding conditions are more favourable than before for the development of native AnGR.

References

Boyazoglu, J. 1998. Livestock farming as a factor of environmental, social and economic stability with special reference to research. Livest. Prod. Sci. 57: 1-14.

Boyazoglu, J. & Y. Hatziminaoglou. 2002. Livestock genetic resources and production systems: a Mediterranean overview. Arch. Latinoam. Prod. Anim. 10(1): 62-67.

Cincotta, R.P., J. Wisnewski, & R. Engelman. 2000. Human population in the biodiversity hotspots. Nature 404: 990-992.

Drucker, A.G., V. Gomez & S. Anderson. 2001. The economic valuation of farm animal genetic resources: a survey of available methods. Ecological Economics 36: 1-18.

Edwards, S.A. 2005. Product quality associated with outdoor pig production. Livest. Prod. Sci. 94: 5-14.

Fallola, A., C. Sanabria, E. Sabio, M.C. Vidal-Aragon & F. Esparrago. 1998. The extensive pig quality. Proc. International Symposium on basis of the quality of typical Mediterranean animal products. EAAP Publication No. 90: 320-338.

FAO. 2005. Domestic Animal Diversity Information System / DAD-IS (http://dad.fao.org/en/home.htm).

Freire, J.P.B., J. Peiniau, L.F. Cunha, J.A.A. Almeida & A. Aumaitre. 1998. Comparative effects of dietary fat and fibre in Alentejano and Large White piglets: digestibility, digestive enzymes and metabolic data. Livest. Prod. Sci. 53: 37-47.

Gama, L.T. & J.V. Delgado. 2000. Assessing the risk status of a breed. 5th Global Conference on Conservation of Domestic Animal Genetic Resources. Brasilia.

Gama L.T. & N. Carolino. 2002. Demographic analysis of the Alentejana breed of cattle. Proc. 6th World Congress on Genetics Applied to Livestock Production. 29: 395.

Gibon, A., A.R. Sibbald, J.C. Flamant, P. Lhoste, R. Revilla, R. Rubino & J.T. Sorensen. 1999. Livestock farming systems research in Europe and its potential contribution for managing towards sustainability in livestock farming. Livest. Prod. Sci. 61: 121-137.
Gibson, J.P. & J.W. Wilton. 1998. Defining multiple-trait objectives for sustainable genetic improvement. J. Anim. Sci. 76: 2303-2307.
Hammond, K. & H. Leitch. 1996. The FAO Global Programme for the Management of Animal Genetic Resources. In: Biotechnology's role in the genetic improvement of farm animals. Miller, R., V. Pursel & H. Norman (Editors). American Society of Animal Science: 24-42.
Liinamo, A.E. & A.M. Neeteson. 2001. Sustainable breeding for farm animals: overview of ongoing research and business efforts in Europe. Proceedings of the 52nd Annual Meeting of the EAAP, Budapest.
Liinamo, A.E. & A.M. Neeteson van Nieuwenhoven. 2002. Inventory and options for sustainable farm animal breeding and reproduction. SEFABAR first annual report. (www.sefabar.org).
Martyniuk, E. & D. Planchenault. 1998. Animal genetic resources and sustainable development in Europe. Proc. 6th World Congress on Genetics Applied to Livestock Production. 28: 35-42.
Oldenbroek, J.K. (Editor). 1999. Genebanks and the conservation of farm animal genetic resources. DLO Institute for Animal Science and Health. The Netherlands.
Olesen, I., A.F. Groen & B. Gjerde. 2000. Definition of breeding goals for sustainable production systems. J.Anim. Sci. 78: 570-582.
Rauw, W.M., E. Kanis, E.N. Noordhuizen-Stassen & F.J. Grommers. 1998. Undesirable side effects of selection for high production efficiency in farm animals: a review. Livest. Prod. Sci. 56: 15-33.
Rege, J.E.O. & J.P. Gibson. 2003. Animal genetic resources and economic development: issues in relation to economic valuation. Ecological Economics 45: 319-330.
Roosen, J., A. Fadlaoui & M. Bertaglia. 2003. Economic Evaluation and Biodiversity Conservation of Animal Genetic Resources. Paper FE 0304 - Christian-Albrechts-Universität zu Kiel (www.food-econ.uni-kiel.de/Workingpaper/wopap.html).
Simianer, H., S.B. Marti, J. Gibson, O. Hanotte & J.E.O. Rege. 2003. An approach to the optimal allocation of conservation funds to minimize loss of genetic diversity between livestock breeds. Ecological Economics 45: 377-392.
Smith, C. 1984. Rates of genetic change in farm livestock. Res. Dev. Agric. 1: 79-85.
Sorensen, J.T. & K. Jakobsen. 2005. Product quality and livestock systems. Livest. Prod. Sci. 94: 1.
Thaon d'Arnoldi, C., J.L. Foulley & L. Ollivier. 1998. An overview of the Weitzman approach to diversity. Genet. Sel. Evol. 30: 149-161.
Thompson, P.B. & A. Nardone. 1999. Sustainable livestock production: methodological and ethical challenges. Livest. Prod. Sci. 61: 111-119.
WCED-UN. 1987. Our Common Future – the Brundtland Report. World Commission on Environment and Development (www.are.admin.ch/imperia/md/content/are/nachhaltigeentwicklung/brundtland_bericht.pdf).

Genetic characterization of local genetic resources and its use for sustainable management

A. Georgoudis[1]*, Ch. Ligda*[2] *& J. Al Ôarayreh*[1]

[1]*Aristotle University of Thessaloniki - Faculty of Agriculture, Dept. of Animal Production, Thessaloniki, Greece*
[2]*National Agricultural Research Foundation, 57 001, Thessaloniki, Greece*

Summary

In the Mediterranean area during the period of domestication, a large quantity of genetic variability accumulated leading to a wide range of breeds used and management practices adopted. Sheep and goat breeding is one of the most widespread agricultural sectors of Greece, and the most important livestock sector. Sheep are kept under the following production systems: the extensive system with transhumance, the extensive or semi-intensive system without transhumance, and the intensive system. Data from 10 sheep breeds of Greece were analysed in order to assess their level of sustainability. The variables considered to have the higher impact on the sustainability of sheep farming were: the trends of the flocks size, the age of the farmer, the existence of successor or not, the years of education of the farmer, the existence of non agricultural income, the percentage of subsidies in the total revenues, the quality of the product, the resistance to diseases, the adaptation to environment, the production level, the mortality of the lambs until weaning and the percentage of farm produced forages used. Information on the genetic characterisation of the breeds was used in order to set the priorities for the sustainable conservation of the breeds.

Keywords: genetic resources, sustainable management, conservation.

Introduction

Livestock production often has a profound impact on the regions within is situated, economically, socially and environmentally. In many rural areas livestock production has historically been the mainstay of economic and social activity, supporting large numbers of rural communities. Livestock production has also been responsible, over centuries, for creating the landscapes and habitats valued by so many. This is still the case in the regions around the Mediterranean basin, although recent years have seen severe rural depopulation and the intensification of livestock production in fertile lowland areas well served by major infrastructure. In these areas livestock production is increasingly associated with deleterious effects on environmental quality (Boyazoglu, 1998, 1999) .

In the Mediterranean area during the period of domestication, a large quantity of genetic variability accumulated. Traditional populations, breeds and selected lines are now in different stages of the evolution process conducted by breeders. Beginning with the industrial development, the economic and social situation has greatly changed, the demand for animal protein is increasing with an intensification and specialization of both the plant and animal sectors, thanks to the scientific and technical achievements which permit the increase in animal production through the use of genetically improved material, more productive and technically better managed (Nardone and Gibon, 1998).

The existing great variations in land, climate and cultures in the northern Mediterranean area where they dominate for millennia lead to many variations in breeds and production systems in the situation and prospects of the sheep and goats genetic resources. There is a vast range of breeds used and a wide range of management practices adopted. The breeds used, the product and the management system are all designed to suit local human social needs as well as the farmed and natural environment. This diversity of production systems is in marked contrast to many other sectors, for example the dairy industry where the breed of cow, the management systems and the feeding systems are becoming remarkably uniform (Morand-Fehr and Boyazoglu, 1999).

Production systems are always in an interaction with their economic and social environment. Such interaction is more pronounced in disadvantaged areas that experience rapid decline and important adverse changes in their economic and social organisation. A major challenge is, then, to characterise the problem and suggest policy options in the development of productive sectors that would reverse the economic decline and will generate a process of sustainable development in the disadvantaged areas (Boyazoglu, 1998, 1999).

Dairy sheep are traditionally bred in Mediterranean areas on a wide range of production systems. Each region has its own local breed well-adapted to that specific environment and to cope with the seasonal availability of pastures. Breeding environments and production systems may greatly differ, from a breed to another but also within breed, on landscape, soil, climate, rainfall distribution, vegetation seasonality and grazing availability. Usual husbandry systems include a suckling period of at least one month, and have a dual purpose, with income from lamb's meat and ewe's milk, that is generally processed into high-quality cheese. Very often there is a complete association between a region of production, a breed and the brand name of a cheese, sometimes qualified as a PDO (Ligda *et al.*, 2002).

Sheep and goat dairy products of the Mediterranean regions face a great future, since they are perceived as quality products and they are also products without surpluses. Furthermore, considering that in these regions sheep and goats generally belong to extensive production systems and since they use marginal spaces and resources, they keep geographical zones populated and maintain their productivity, which otherwise would be abandoned if production of this type did not exist.

Nevertheless, the development of small ruminants' production faces strong economic constraints limiting capacity to provide higher levels of income and employment to the population. There are five types of such constraints comprising factors related to: 1) labour and employment; 2) management; 3) grazing resources; 4) relations with forests and the environment; and 5) available infrastructure. The existing ties, however, between livestock development and economic well-being of the population in the disadvantaged areas, needs to be understood in order to derive strategic and policy recommendations (Morand-Fehr and Boyazoglu, 1999).

With sustainable development and multi-functional agriculture, we are moving towards a new type of agreement between society and livestock farmers, in particular farmers in areas less-favored at agronomic level, but which often have a rich heritage and potential for tourism. Political measures are being taken, at European, national and regional level, which make it possible to propose means of support to favor the emergence or consolidation of projects to develop quality products, in the framework of the Rural Development plan. But like any local development initiative, those relating to quality sectors have to be able to rely on people, jobs and very different skills, all motivated by a same project and sharing a certain number of values in common.

Sheep production systems in Greece

Sheep and goat breeding is one of the most widespread agricultural sectors of Greece, and the most important livestock sector, according to their share in the animal production (about 43%) and in the total agricultural production (about 13%).

Small ruminants breeding are the most traditional forms of livestock-farming that have been sufficiently adapted to the Greek natural conditions and make use in high degree of the pasture lands, marginal and inadequate for culturelands, as well as those with seasonal remains (stubbles of cereals, etc.). A percentage of 78% of sheep and 91% of goats are raised under low input production system in the mountainous and marginal regions of the country.

The number of raised sheep in Greece amounts 9 244 000 animals, representing about 10.25% of the total sheep population bred in the EE. The medium size of sheep exploitations in our country amounts about 60 animals. The average size of the units with more than 10 animals is 84 sheep. A percentage of 80% of the total sheep population is kept in less favored areas of the country, and belongs to dual-purpose breeds (milk and meat). About 93% of the ewes are milked.

The changes in sheep numbers were not significant in the last 20-30 years. However, the number of farms has declined significantly. Thus, the flock/herd size has increased due to specialization and reorganization of the sector. Greece has three important regions of husbandry of sheep, the areas of Macedonia, Western Greece and Crete. In these three regions are found the 46% of the total raised population of sheep in our country.

Small ruminants are kept under the following production systems: the extensive system with transhumance, the extensive or semi-intensive system without transhumance, and the intensive system. There is lack of reliable and consistent data and it is often not possible to draw a clear line between the different types of the production systems. Many farmers practicing essentially extensive forms of agriculture have introduced modern, intensive practices in recent years. Sheep production systems are usually divided into the extensive system with transhumance (applied on about 85% of the total population), the extensive or semi-intensive system without transhumance (77% of the total population), and the intensive system (about 15% of the total population).

In most cases, intensive exploitation, with its high levels of specialization and integration of productive and well-organized process, gives better economic results. Intensification therefore, of production systems appears likely to be a continuing trend, and it is important that efforts are made to keep these developments consistent with efficient flock management and attention to animal health. The younger farmers tend to more intensive farming systems, often with high total invested farm capital for modern housing, new and more productive breeds and higher living standards for their families (Georgoudis and Ligda, 1999).

In the frame of the ECONOGENE EU project on the conservation of sheep and goat biodiversity and their importance for rural development in European rural marginal areas, molecular, geographical, socio-economics and husbandry data from 10 Greek sheep breeds were collected.

The questionnaire used included, among others, personal information of the farmer and the structure of his family, data on the husbandry system and characteristics of the breed kept, breeding and nutritional strategies, marketing strategies and total revenues as well as contribution of revenues from agricultural and other outside-farm activities and finally the role of subsidies in the total farm revenues (Ajmone-Marsan *et al.*, 2005).

Data for each breed were collected from 11 farms except for the Kefallinias breed, where data from only 9 farms were available. In total, for the 10 breeds collected, 108 farm questionnaires were analysed. The names of the breeds sampled and their breeding regions, demographic trends, effective population size (Ne) and rate of inbreeding (ÄF), are presented in table 1.

Table 1. Sheep breeds sampled and their breeding regions and demographic trends.

	Breeding region	Population size	Ne	ΔF	Population trend
Anogeiano	Anogeia, other regions in Crete island	4 500	838	0.06	Increase
Kalarritiko	Tzoumerka mountains – Plains of Thessaly (winter)	5 350	670	0.07	Stable
Karagouniko	Plains of Thessaly	180 000	23 200	0.002	Increase
Kefallinias	Kefallonia island	40 000	5 154	0.010	Stable
Kymi	Around Kymi (Euboia island)	559	117	0.43	Decreasing
Lesvos	In Lesvos and Limnos islands	180 000	23 200	0.002	Stable
Orino	Mountains of Epirus and West Macedonia	15 000	1 933	0.026	Decreasing
Pilioritiko	In Magnisia, Pilio mountains	2 462	494	0.1	Stable
Sfakia	In Crete island - Sfakia	60 000	7 733	0.006	Stable
Skopelos	Islands of Skopelos and Skiathos and in Magnisia	1 824	368	0.13	Stable

Table 2. Class frequencies and means of the 12 variables with high impact on the sustainability .

	Trends			Age of the farmer			Existence of successor			Years of education		
Breeds	Declining	Stable	Growing	<40 years	40-60 years	>60 years	No	Yes	< 60 years old	<6 years	6 … 12 years	> 12 years
Anogeiano	0	0	100	37	36	27	0.0	27.3	72.7	73	27	0
Kalarritiko	9.1	27.3	63.6	45	55		18.2	0.0	81.8	73	27	0
Karagouniko	18.2	36.4	45.5	18	36	46	9.1	45.5	45.5	64	36	0
Kefallinias	11.1	66.7	22.2	11	67	22	11.1	11.1	77.8	78	22	0
Kymi	18.2	63.6	18.2	18	18	64	27.3	27.3	45.5	73	27	0
Lesvos	9.1	45.5	45.5	18	82		0.0	0.0	100.0	64	18	18
Orino	27.3	63.6	9.1	9	64	27	18.2	18.2	63.6	82	18	0
Pilioritiko	9.1	27.3	63.6	18	55	27	18.2	9.1	72.7	90	10	0
Sfakia	9.1	27.3	63.6	46	45	9	9.1	0.0	90.9	45	55	0
Skopelos	36.4	36.4	27.3	9	55	36	36.4	0.0	63.6	90	10	0
Total	14.8	38.9	46.3	23	51	32	14.8	13.9	71.3	73	25	2

From the total variables collected, 12 variables considered to have high impact on the sustainability of sheep farming have been selected, according to the results of the Econogene project (Bertaglia, 2004; Fadlaoui *et al.*, 2004; Kliambas, 2005). These variables, which refer to animal population, production system and social economics parameters, were: the trend of the flocks size, the age of the farmer, the existence of successor or not, the years of education of the farmer, the existence of non agricultural income and the percentage of subsidies in the total farm revenues, the special characteristics of the breeds (adaptation to environment, resistance to diseases, production level and quality of the product), the mortality of the lambs until weaning and the percentage of farm produced forages of the total amount used.

The impact of these variables on the sustainability of the production system has been considered as plus or minus, according to the following rules: for the categorical variables, frequencies of the classes have been calculated for each breed and the total dataset. The positive or negative contribution of each variable has been derived from the comparison of the class considered to contribute more to the sustainability with the relevant class average of the total dataset. In the case of the continuous variables, the mean of each breed has been compared with the respective mean of the total. The algebraic sum of all variables has been considered to contribute positive or negative to the sustainability of the breed. The results are presented in tables 2, 3 and 4.

Breeds with minus or zero end-score are considered as not sustainable. This is the case for the Kymi, Skopelos, Pilioritiko and the Orino breeds, a result consisted with their existed characterisation as endangered (for the first three) or with decreasing population size (the last). However in order to prioritise the conservation strategies for these breeds, their genetic characteristics should be taken into consideration.

Table 3. Class frequencies and means of the 12 variables with high impact on the sustainability.

	Non-agricultural activities					Breed characteristics					
Breeds	no	yes	0-25%	25-50%	50-75%	Quality of the product	Disease resistance	Adaptation to bio-physical conditions	Production level	Farm produced forages	Mortality until weaning
Anogeiano	90.9	9.1	18.2	81.8	0.0	37.9	10.3	34.5	17.2	0.0	7.6
Kalarritiko	100.0	0.0	27.3	63.6	9.1	32.1	28.6	39.3	0.0	11.8	3.9
Karagouniko	72.7	27.3	81.8	18.2	0.0	14.3	28.6	35.7	21.4	54.5	5.9
Kefallinias	66.7	33.3	11.1	88.9	0.0	13.6	31.8	31.8	22.7	8.0	2.9
Kymi	54.5	45.5	18.2	81.8	0.0	12.5	37.5	41.7	8.3	37.3	4.4
Lesvos	9.1	90.9	45.5	54.5	0.0	32.1	7.1	35.7	25.0	0.0	3.5
Orino	54.5	45.5	90.9	9.1	0.0	30.0	33.3	36.7	0.0	35.7	5.3
Pilioritiko	81.8	18.2	36.4	63.6	0.0	34.5	34.5	31.0	0.0	9.1	2.9
Sfakia	54.5	45.5	27.3	63.6	9.1	52.4	0.0	33.3	14.3	5.0	7.2
Skopelos	81.8	18.2	27.3	72.7	0.0	26.7	30.0	30.0	13.3	4.5	2.3
Total	66.7	33.3	38.0	59.3	1.9	28.6	24.5	34.9	11.9	16.5	4.7

Table 4. Sustainability of the production systems of sheep breeds .

	Trends of flocks	Age of the farmer	Existence of successor	Years of education	Non agricultural income	Subsidies	Quality of the product	Disease resistance	Adaptation to bio-physical conditions	High productivity	Mortality	Farm produced forages	Total score
Anogeiano	+	+	+	+	+	-	+	+	+	+	-	-	+
Kalarritiko	+	+	-	+	+	-	+	+	+	-	+	-	+
Karagouniko	+	-	+	+	+	+	-	+	+	+	-	+	+
Kefallinias	+	+	+	-	+	-	-	+	-	+	+	-	+
Kymi	-	-		+	-	-	-	+	+	-	+	+	-
Lesvos	+	+	+	+	-	+	+	-	+	+	+	-	+
Orino	-	+	-	-	-	+	+	+	+	-	-	+	0
Pilioritiko	+	+	-	-	+	-	+	+	-	-	+	-	0
Sfakia	+	+	+	+	+	-	+	+	+	+	-	-	+
Skopelos	-	-	-	-	+	-	+	+	-	+	+	-	-

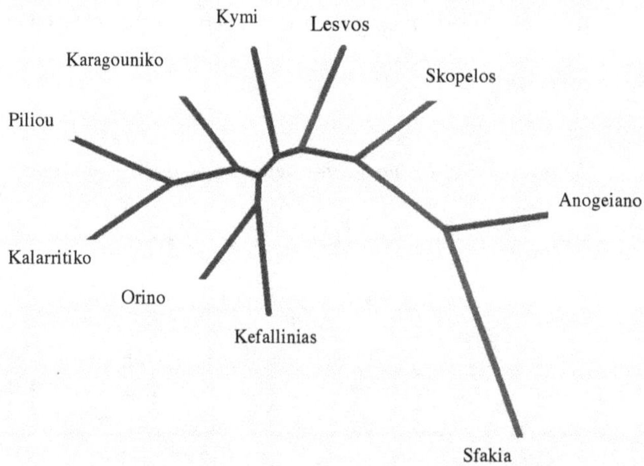

Figure 1. Phylogenetic tree based on microsatellite markers of Greek sheep breeds.

Genetic characterisation of sheep breeds in Greece

In order to obtain information on the genetic characterisation of the 10 sheep breeds studied, DNA from 31 samples from each breed has been analysed based on 31 microsatellite markers. Those markers were: *BM1329*, *BM1824*, *BM8125*, *DYMS1*, *HUJ616*, *ILSTS005*, *ILSTS011*, *ILSTS028*, *INRA63*, *MAF33*, *MAF65*, *MAF70*, *MAF209*, *MAF214*, *McM140*, *McM527*, *OarAE129*, *OarCP34*, *OarCP38*, *OarFCB20*, *OarFCB128*, *OarFCB193*, *OarFCB193*, *OarFCB304*, *OarHH47*, *OarJMP29*, *OarJMP58*, *OarVH72*, *SR-CRSP-1*, *SR-CRSP-5* and *SR-CRSP-9*

(Ajmone-Marsan *et al.*, 2005). Statistical analyses were performed using the statistical package FSTAT (Goudet, 2002). The following parameters contributing to the genetic diversity of each breed have been calculated: total number of alleles, mean number of alleles, observed heterozygosity and expected heterozygosity. According to the results obtained, the Skopelos breed showed the lowest mean number of alleles (6.4), while for the rest of the breeds the values ranged from 8.1 for Sfakia to 9.3 to Orino breed. Orino breed presented also the highest number of private alleles (12) and Skopelos the lowest (2). Genetic diversity ranged from 0.680 (Skopelos) to 0.761 (Karagouniko), while observed heterozygosity ranged from 0.625 (Skopelos) to 0.739 (Kefallinias).

The phylogenetic tree was obtained with the Phylip software (Felsenstein, 2004), using Reynolds' genetic distances after 1000 bootstraps. The tree (Figure 1), showed a clear differentiation of the Anogeiano and Sfakia breeds from the rest of the breeds (with a high bootstapping value 99%), a separation of the Skopelos breed (45% bootstraps) and a third group with the remaining breeds, with the differentiation of Kymi breed (lower than 40%).

A number of criteria, covering a wide range of different aspects of breed characterisation and breed value are used when prioritizing breeds for conservation. Ranking of breeds based on the degree of endangerment is the first step, while in a second step the breeds should be weighted according to the following criteria, genetic uniqueness, adaptation to a specific environment, traits of economic importance, unique traits, cultural or historical value, (Ruane, 1999). Based on the results regarding the sustainability of the production systems (Table 4), Skopelos and Kymi breeds are not sustainable. Furthermore, taking into consideration the demographic trends, the high level of inbreeding (Table 1) and the results of the genetic analysis, where Skopelos and Kymi showed a differentiation from the rest of the breeds, these two breeds should be set high in the priority list for conservation.

Measures should be undertaken to support breeders in order to drive the breeds towards self sustainability, as indicated by the results from the socio economic and husbandry data. Major drawbacks toward this goal will be, the age of the farmers and the absence of successors, the limited years of education of the farmers, the low percentage of farm produced forages and the continuing decreasing trend of the flocks. However the strong elements of the production system are the low mortality of the lambs until weaning and the specific characteristics of the breed, in particular the quality of the product for human consumption, the disease resistance and the high productivity. These are characteristics on which marketing strategies could be developed in order to increase the market values of the breeds and their products.

References

Ajmone-Marsan, P. & Econogene Consortium, 2005. Overview of ECONOGENE, a European project that integrates genetics, socio-economics and geo-statistics for the sustainable conservation of sheep and goat genetic resources. International Workshop on "The role of Biotechnology", Turin, Italy, 5-7 March 2005.

Bertaglia., M., 2004. Livestock biodiversity conservation: The case of sheep and goat breeds in European marginal areas. Doctorate dissertation, Kiel.

Boyazoglu, J., 1998. Livestock farming as a factor of environmental, social and economic stability with special reference to research. Livest. Prod. Sci., 57(1):1-14

Boyazoglu J., 1999. Livestock Production Systems and local Animal Genetic Resources with Special Reference to the Mediterranean Region. VII Congress of the Mediterranean Federeation for Ruminant Health & Production, Santarem, Portugal.

Fadlaoui, J., Roosen & P.V. Baret, 2005. Genetics and Economics: Prioritizing Breeds for Conservation. International Workshop. The role of Biotechnology for the characterisation and conservation of crop, forestry, animal and fishery genetic resources. 5-7 March, Turin, Italy

Felsenstein, J. 2004. Phylip v.3.6. Phylogeny Inference Package. University of Washington.

Georgoudis, A. & Ch. Ligda, 1999. Small Ruminant Breeding Programmes in Greece: A case study for the Workshop on Animal Breeding Strategies. ICAR Technical Series no. 3: 493-502.

Goudet, J., 2002. FSTAT, a program to estimate and test gene diversities and fixation indices (v.2.9.3).

Kliambas, G., 2005. Investigation of sheep production system with use of multivariate techniques. Master of Science Thesis, (in Greek).

Ligda, Ch., Th. Papadopoulos & A. Georgoudis, 2002. Use of autochthonous breeds for rural development and production of quality products. EAAP Publication no. 106: 207-209.

Morand-Fehr P. & J. Boyazoglu, 1999. Present state and future outlook of the small ruminant sector. Small Rum. Res., 34, 175-188.

Nardone, A. & A. Gibon, 1998. Livestock farming systems research and development issue. In Consejería de agricultura y pesca (Ed.). Technical and social systems approaches for sustainable rural development. Congresos y jornadas. 45. pp. 85.

Ruane, J., 1999. Selecting breeds for conservation. Genebanks and the conservation of Farm Animal Genetic Resources, J.K. Oldenbroek (Ed.), Lelystad, NL.

Local breeds and genetic improvement

G. Gandini[1], R. Rizzi[1], A. Bagnato[1], A. Montironi[1], F. Panzitta[1], F. Pizzi[2], A. Stella[3] & G. Pagnacco[1]

[1]Department VSA, University of Milan, Via Celoria 10, 20133 Milan, Italy
[2]IBBA- CNR, Milan, Italy
[3]FPTP-CERSA, Lodi, Italy

Summary

Genetic improvement is generally associated to large breeds and management of local small breeds is seen as maintenance of the current genetic profile, however, selection programmes may increase sustainability of local breeds. The paper discusses some issues associated to selection in local breeds. First, in general farm animal populations are under some mass selection: in these cases the commonly used formula of S. Wright to compute effective size is inappropriate and its use results in considerably lower estimates of inbreeding rates. Secondly, research has been developed in the last years to design breeding schemes which maximise genetic gain while limiting the rate of inbreeding. Although these investigations generally refer to intensively selected nuclei, they set a useful framework for selection in local breeds of small size, but its implementation needs further research. As an example it is shown that index selection adjusted for average relationship can produce appreciable genetic progress in a breed as small as 750 cows with acceptable inbreeding increments: compared to the Italian Friesian, expected response per year was 0.29, 0.27 and 0.29 respectively for milk kg, fat kg and protein kg. Thirdly, which traits should be considered? It is suggested that genetic improvement should maintain all characteristics contributing to low production costs, the environmental and cultural values of their farming systems and the links between breeds and their farming areas, that can add value to their products within the European certification systems of agricultural products.

Keywords: selection, inbreeding, local breeds, conservation, animal genetic resources.

Introduction

The dynamics of the decline of local breeds in Europe is poorly known: cultural, social and food demand changes, transformation of the food production chain, country regulations and technological modifications might affect in various ways different breeds (Gandini & Oldenbroek, 1999). In most decline situations it is likely that these factors result in a lack in economic profitability of the local breeds compared to other breeds or other economic activities. EC Regulations 1257/99 and 445/02 provide incentive payments to the farming of endangered local breeds in order to increase their economic profitability. Economic incentives can be effective to halt the decline of local breeds only in a short term scenario and, in many cases, in spite of the EU's support to farmers, rearing local breeds still remains unprofitable (Signorello & Pappalardo, 2003).

Therefore, the question to be answered is which opportunities do we have for self-sustaining local breeds? Among the different options available, genetic improvement has been poorly investigated and it is often considered not compatible with conservation principles. Selection is generally associated with the framework of large breeds and conservation is seen as maintenance of genetic variation and of the current genetic profile. In this paper issues associated to selection in local breeds are presented and discussed, considering some experiences developed in Italy.

Selection and local breeds

In general, local breeds do not benefit from modern breeding techniques. Selection programmes may increase the genetic ability for productivity and consequently the profitability of local breeds. Two major considerations have to be forwarded. Selection schemes should take into account the maintenance of genetic variation within the breed and risks associated with high inbreeding increments. Secondly, traits proposed for selection should be accurately evaluated with respect to genetic correlations with those traits determining the conservation value of the local breed.

Selection and inbreeding

In general, farm animal populations are under some mass selection. With selection, families have different probabilities to contribute to the following generation, both for non-inherited and inherited causes. In this case the effective population size (Ne) is smaller that the number (N) of breeding animals and consequently inbreeding rates (ΔF) are higher (N_e=1/(2ΔF), where ΔF is per generation). The commonly used formula to compute effective size, N_e=4MF/(M+F) (Wright, 1931), where M and F are respectively the number of reproductive males and females, can be extended to take into account variances of family size (e.g. Hill, 1979). These variances can be partially estimated from population age distribution (Ollivier and James, 2004), but these formulas cannot be applied when selective advantages are inherited. The inappropriate application of the Wright formula in local breeds undergoing mass selection leads to underestimation of rates of inbreeding and losses of genetic variation. Equations for estimating Ne in populations undergoing mass selection have been proposed (e.g. Santiago & Caballero, 1995; Nomura, 1996; Bijma *et al.*, 2000) although they require, besides M and F, other parameters such as heritability and selection intensity for their application. In table 1 effective population sizes are computed for populations of 10 males and a varying number of female parents, both with the Wright (1931) equation, which assumes no mass selection, and the Santiago and Caballero (1995) formula which takes into account mass selection. Heritability of 0.4 and a number of individuals scored per sex, family and generation ranging from 1.5 to 6 were used. Such parameters represent the selection intensities that might be observed in European local breeds. Table 1

Table 1. Effective population sizes for populations of 10 males and F female parents, assuming no selection (Wright, 1931) and mass selection (Santiago & Caballero, 1995) on a trait with heritability = 0.4.

			Mass selection			
M	F	No selection	n[1] = 6	n = 4	n = 2	n = 1.5
10	50	33.3	20.6	22.4	26.6	29.1
10	100	36.4	22.3	24.0	27.6	29.7
10	500	39.2	23.0	24.4	27.4	28.9
10	1 000	39.6	22.6	24.0	26.8	28.2
10	10 000	40.0	20.7	21.9	24.2	25.4

[1]n = number of individuals scored per sex, family and generation.

shows as the assumption of absence of selection in mass selected populations may result in considerably lower Ne estimates.

When BLUP selection is considered, inbreeding rates from 2.5 times (heritability = 0.4) to 4 times (heritability = 0.1) higher than with random selection have been observed (Wray and Thompson, 1990), and Ne estimation becomes more complex.

A theoretical framework has been developed in the last years to design breeding schemes which maximise genetic gain while limiting the rate of inbreeding. Besides increasing the number of selected sires and dams, rates of inbreeding can be reduced by using overestimated heritabilities in EBV estimation (Grundy & Hill, 1993), or by modifying the weight of within and between family components of EBVs (Verrier *et al.* 1993). Wray and Goddard (1994) and Brisbane and Gibson (1995) proposed the use of a selection index with penalties on the average relationship among selected sires and dams, considering that inbreeding is expected to increase at the same rate of average relationship. A method of selection with a constraint on inbreeding rate has been proposed for discrete (Meuwissen, 1997) and overlapping (Meuwissen, & Sonesson, 1998) generations.

These investigations generally refer to intensively selected nuclei. Nevertheless they set up a potentially useful framework for selection in local breeds of small size. The application of these methods to local breeds has been poorly investigated. Their implementation as a function of population size, percentage of the recorded populations, selection scheme (e.g. progeny tested or young sires), management of selection (central or peripherical decisions) requires further investigation.

As an example, Figure 1 and Table 1 show the results of a simulation study (Gandini *et al.*, 1998) of index selection in a small local dairy cattle breed of 723 cows and heifers. A lactation trait was stochastically simulated. EBVs were estimated yearly using a BLUP animal model. Dams were selected based on the most recent phenotypic performance. Sires were selected among male calves and used for one year. Selection of sires was performed on EBV – kRel, where Rel is the average relationship among selected sires and among current dams and k is a weighting factor, as suggested by Brisbane and Gibson (1995). With k = 0, truncation selection on EBV was simulated. Figure 1 reports selection response and rate of inbreeding per year for different selection schemes (n. of males = 8, 10, 12, 16; k = 1, 1/2, 1/3, 1/4, 1/5, 1/6, 1/7, 1/15, 0).

We can observe that, within a certain range of k values, by varying k selection response shows limited variation, while inbreeding rate per year changes consistently. This suggests opportunities for

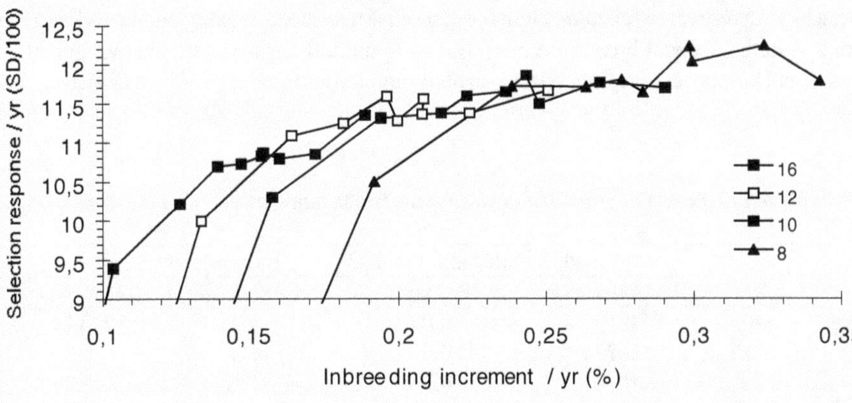

Figure 1. Selection response versus inbreeding increment for different selection schemes: n. of males = 8, 10, 12, 16; k = 1, 1/2, 1/3, 1/4, 1/5, 1/6, 1/7, 1/15, 0 (from left to right).

Table 2. Reduction of the number of males (M) at constant selection response and inbreeding.

M	k	Selection response	Inbreeding rate
8	1/2	11.7	0.24
10	1/7	11.5	0.25
12	0	11.7	0.25

selection on EBV – kRel in small populations. Moreover, by varying k, responses in terms of inbreeding with different numbers of males overlap. This allows, at a given selection response and inbreeding rate, to reduce the number of sires per year and the associated costs, as shown in table 2.

To compare the selection responses that can be obtained in the simulated small dairy cattle breed with those obtainable in the Italian Holstein Freisian, expected genetic responses in 5 characters were computed following selection index theory for the scheme with 8 sires and k = 1/ 2 (Table 2) (Gandini *et al.*, 1999). The realised selection intensity was 1.17, with an accuracy of 0.28 and a generation interval of 2.8 years. In table 3 responses are compared with those expected in the Italian Friesian for the ILQ selection index.

Expected responses per generation in the small local dairy cattle breed were 0.16, 0.15 and 0.16 of those expected in the Italian Holstein Friesian, respectively for milk kg, fat kg and protein kg. These proportions increased to 0.29, 0.27 and 0.29, respectively for milk kg, fat kg and protein kg, when expected gains per year were considered, because the shorter generation interval of the selection scheme of the local breed. Results in Table 3 show that selection of young sires on the EBV-kRel strategy can produce appreciable genetic progress in a breed as small as 750 cows with acceptable inbreeding increments.

Genetic improvement in local breeds: for which traits?

Which are the major differences, from our perspective, between local breeds and intensively selected breeds? i.) local breeds have not undergone the intensive selection programmes that in the last decades were used in the large and cosmopolitan breeds; ii.) local breeds, differently from the highly selected breeds, generally perform in low input-output production systems; iii.) local breeds often show better performances with respect to functional traits (e.g. disease resistance, fertility, longevity); iv.) the production systems of local breeds are often linked to natural agro-ecosystems; v.) genetics, environment and farming cultures are reflected in high variation and quality of animal products.

Table 3. Expected response in standard deviation units per generation and per year.

		Local breed: 8 males and k = 1/ 2		Italian Holstein Freisian	
Trait	R.E.[1]	Per generation	Per year	Per generation	Per year
Milk kg	-0.4	0.19	0.07	1.21	0.24
Fat kg	0.1	0.22	0.08	1.47	0.30
Protein kg	1	0.28	0.10	1.75	0.34
Fat %	0	0.04	0.02		
Protein %	0	0.08	0.04		

[1]Relative emphasis.

For some of the above characteristics local breeds well fit in the current European animal production scenario. In fact, quality and diversity of products, farming as management tool of agro-ecosystems of conservation value, a multifunctional character of agriculture and reduction of production costs are priority objectives of the Common Agricultural Policy (CAP).

With some generalisation, genetic improvement in local breeds should: i.) avoid to match the quantitative performances of the highly selected breeds; ii.) maintain all characteristics contributing to low production costs; iii.) maintain the environmental and cultural values of their farming systems. Therefore, traits proposed for selection should be accurately evaluated for their genetic correlations with those traits that determine the conservation value of the breed, in order to avoid its deterioration. These may include adaptation to a harsh environment or to low-input production systems or functional traits like longevity, fertility, meat and milk quality. In addition, genetic improvement in local breeds should iv.) maintain their links with the farming area, that can add value to their products within the current European certification systems of agricultural products PDO (Protected Designation of Origin), PGI (Protected Geographical Indication) and TSG (Traditional Speciality Guaranteed), or through specific commercial brands. To this respect it should be stressed that the diversity of breeds and production systems in the Mediterranean areas is associated with a noticeable variety of animal products, food preferences and cultural traditions. In many cases strong links between specific breeds and products were present at the half of last century which, following the industrialisation of production and food processing systems, become more and more subtle and in some cases disappeared. Maintenance and restoration of these breed-product links is advisable for several reasons. Their contribute to maintain the diversity of animal products in Europe advocated by CAP. Secondly, the relationship between breed and product can be helpful to diversify the breed products and to sell them at higher prices, i.e. to improve the economic profitability of the breed.

In conclusion, the introduction of selection programmes in local breeds, in particular in those with small population sizes, requires to consider two major aspects: i.) maintenance of breed genetic variation and inbreeding control, ii.) maintenance of breed genetic originality, i.e. maintenance of between breed variation.

References

Bijma P., van Arendonk, J.A.M. & Woolliams J.A., 2000. A general procedure for predicting rates of inbreeding in populations undergoing mass selection. Genetics 154: 1865-1877.

Brisbane, J.R. & Gibson J.P., 1995. Balancing selection response and inbreeding by including predicted stabilised genetic contributions in selection decisions. Genet. Sel. Evol. 27: 541-549.

Gandini, G., A. Bagnato & R. Rizzi, 1999. Expected genetic response from aggregate indexes in the Reggiana cattle. XIII Congresso ASPA, 152. Piacenza.

Gandini, G., De Filippi, P., Bagnato, A. & Pagnacco, G., 1998, Selection and inbreeding rate in a small endangered dairy cattle breed: the reggiana. Proceedings of the 6th World Congress on Genetics Applied to Livestock Production. Armidale, 28: 131-134.

Gandini G. & OldenbroekJ.K., 1999. Choosing the conservation strategy. In J.K Oldenbroek (Editor) Genebanks and the conservation of farm animal genetic resources. ID-DLO, Lelystad.

Grundy, B. & Hill, W.G., 1993. A method for reducing inbreeding with best linear unbiased prediction. Anim. Prod. 56: 427.

Hill W. G., 1979. A note on effective size with overlapping generations. Genetics 92: 317-322

Nomura T., 1996. Effective size of selected populations with overlapping generations. J. Anim. Breed. Genet. 113: 1-16

Meuwissen T.H.E., 1997. Maximising the response of selection with a predefined rate of inbreeding. J. Anim. Sci. 75: 934-940

Meuwissen T.H.E. & Sonesson A.K., 1998. Maximising the response of selection with a predefined rate of inbreeding – overlapping generations. J. Anim. Sci. 76: 2575-2583

Ollivier L. & James W., 2004. Predicting the annual effective size of livestock populations. Genet. Res. (in press)

Santiago E. & Caballero A., 1995. Effective size of populations under selection. Genetics 139: 1013-1030

Signorello G. & Pappalardo G., 2003. Domestic animal biodiversity conservation: a case study of rural development plans in the European Union. Ecol. Economics 45: 487-500

Verrier E., Colleau J.J., & Foulley J.L.,1993. Long term effects of selection based on the animal model BLUP ina afinite population. Theor. Appl. Genet. 87: 446-454

Wray N.R. & M. E. Goddard, 1994. Increasing long term response to selection. Genet. Sel. Evol. 26: 431-451

Wray N.R. & R. Thompson, 1990. Advances in selection theory. Proceedings of the 4th World Congress on Genetics Applied to Livestock Production, XIII: 167, Edinburgh.

Wright S., 1931. Evolution in Mendelian populations. Genetics 16: 97-159.

Nutrition and feeding of dairy sheep and goats in traditional systems and their effect on typical products' quality

G. Zervas & E. Tsiplakou

Department of Animal Nutrition, Agricultural University of Athens, Iera Odos 75, GR - 118 55 Athens, Greece

Summary

In the Mediterranean region small ruminants farming systems, with dominant extensive grazing situation, produce meat and dairy products with exceptional physicochemical and nutritional properties and original characteristics. These are due to a number of factors like local animal breeds of low productivity, great variability in botanical and chemical composition of grazed herbage, natural environment, very low chemical inputs in animals feed, and very limited animal medication, but also a specific technology for the processed products. These products have beneficial effects in consumers' health, who should always be clearly informed about the origin, traceability and authenticity of the products.

Keywords: sheep, goats, traditional systems, nutrition and feeding, quality products

Introduction

There is an increasing interest for sustainable forms of livestock production systems, which will provide a balanced relationship between environmental, socio-cultural and economic factors. The small ruminant sector is examined with particular focus on the possibilities of improving the sustainability of small ruminant systems and on the possible role of the traditional ones to meet the demand of sustainability. There has also been a trend over the last decade for products associated with lifestyle choices and "process quality", which ultimately justify premium prices (Badertscher-Fawaz *et al.*, 1998). The increase in the importance of products from traditional and /or organic farming in one example of this trend (Nardone *et al.*, 2004).

The aim of this paper is to point out the superiority of milk, meat and dairy products from small ruminant traditional systems of production.

Production systems

The sheep and goat farming systems, which are practiced in the Mediterranean area, can be grouped, according to the level of intensification (breed characteristics, land dependency, herd size, stock feeding origin, performance, volume of production etc.) in four classes: home-fed, intensive, extensive without transhumance and extensive with transhumance (Zervas, 1995).

The most common form of traditional farming today in Mediterranean area is extensive rearing of sheep and goat for milk and meat, with the livestock often herded in mixed flocks, which usually results in a more efficient utilization of pasturelands. The great majority of sheep and goats are kept in less favoured areas (LFA) in relatively small sized flocks /herds.

Dairy sheep and goats are highly variable in their morphology, body size, milking capacity, prolificity, carcass composition and growth rate. However, these animals have a strong constitution and perfect adaptability to the harsh environmental conditions. These sheep and goats of local and regional hardy breeds, are characterized by low productivity, relatively low nutritional requirements and high resistance to diseases. Most of them are based mainly on extensive use of non-fertilized natural pasture resources. Thus, in this respect these extensive sheep and goat production systems are believed to be more close to organic and contribute to the economic and environmental sustainability of LFAs by enhancing their biological diversity (Zervas, 1995).

Nutrition and feeding

Most of sheep and goat nutrition (60-80%) is based on natural grazing, and only during pregnancy and early lactation some supplementary feed (hay, straw and concentrates), is offered which is a higher-cost practice, when grazing cannot meet animals' requirements in energy and protein. No feed additives, antibiotics or other pharmaceuticals for deseases' prevention are usually used in these traditional extensive production systems.

Small ruminants are the most efficient transformers of forage from brush, pasture range, other feeds unsuitable for humans, and renewable plant material produced by solar energy, as low-cost feeds, into high quality animal products with distinguished chemical composition and organoleptic characteristics, and with the least use of fossil fuel (Donald, 1994).

However, grazing dairy sheep and goats, in typical extensive or semi-intensive systems in the Mediterranean area, are often subjected to undernutrition, in relation to seasonal changes in forage or by-products availability (Caballero *et al.*, 1992; Zervas *et al.*, 1996). In general, there is a difficulty in ensuring annual regularity of food supply to small ruminants and this in one of the major constraints for the mainly extensive systems. In this respect, improving pasture and extending the grazing season are important in a forage-based sheep/goat operation. The supplementation of the pasture with legumes (alfalfa, clover), wherever is possible, provides high-quality proteins, reduces the need for nitrogen fertilizers and limits the seasonal nutritional fluctuations.

Grasslands in Mediterranean area are composed mainly of annual species, with great botanical interest, which is sustained by high grazing pressure for a relatively short period. Species counts are high (>6 000), although there is a considerable variation according to space and time, soil fertility and grazing pressure. The swards include mainly species from the plant families *Poaceae* and *Fabiaceae*, while the wooded pastures are dominated by shrubs and trees (e.g. *Medicago, Pistacia, Quercus, Olea* etc.). It should be pointed out here that these grazing areas receive no application of artificial fertilizers, nor agrochemicals and no agricultural management other than grazing. There is, also, no environmental pollution due the long distances of grasslands from main roads and cities.

Quality products from sheep and goats

The grazed forage contains many naturally occurring bioactive molecules with antioxidant and anti-inflammatory properties, like bioflavonoides, phytosterols, oleouropein, hydroxytyrosol, tyrosol etc. due to multispecies composition and climatic xyrothermic conditions.

When sheep and goats are gazing, the conjugated linoleic acid (CLA) content in their milk, meat and dairy products increases, especially when the grass is at an early growth stage (Tsiplakou *et al.*,

2004). CLA may modulate the risk of many diseases, including obesity, atherosclerosis, chronic inflammatory diseases and cancers (Lawson *et al.*, 2001).

Polyphenols, like flavonoids with their powerful antioxidant properties, have anti-carcinogenic and anti-atherogenic implications. Olive tree leaves for instance, which contain a series of phenolic minor compounds, such as hydroxytyrosol and oleuropein, are widely used in sheep and goat feeding and enrich milk in oleuropein (Zervas, unpublished data). Meat quality is also improved by increasing the levels of chemically active natural antioxidants via the feed, because they increase the oxidative stability of meat post mortem.

Thus, the combination of animals' breed, their relatively low milk yield, the wide and exceptional botanical and chemical composition of grazed herbage and the total animals' feeding regime, gives a particularity in sheep and goat milk which is of superior quality, rich in fat, from which niche products, like cheeses, with exceptional physicochemical and nutritional properties and original characteristics can be obtained. This superiority concerns hygienic and satinary, dietetic and nutritional, and gustative and gastronomic quality (Boyazoglu and Morand-Fehr, 2001). About 95% of sheep and goat milk is transformed into typical dairy products that have a regional or local connotation of origin quality (e.g. the cheeses Roquefort, Pecorino romano, Feta, Fiore sardo, Manchego, Serra da Estrela, Indiasabal etc.). The transformation and maturing process of these cheeses and consequently their quality is very much linked to historical and cultural uniqueness right through the production, marketing and consumption chains. These traditional products have always played an important role in the dietary habits of the people where they are produced and cannot be diminished due to globalization. Tradition in the Mediterranean area is part of its cultural heritage.

The nutrition of sheep and goats affects the fatty acid profile of milk fat, which has higher percentage of the short chain fatty acids (Table 1), compared to cow's milk and contributes significantly to final aroma, taste and flavour of milk and milk products. In addition to that it has been shown that the short chain fatty acids and stearic acid of milk do not increase serum cholesterol.

From the consumer point of view, the dairy products from extensively farmed sheep and goat milk have superior nutritive value, due to their high content in fatty acids, aminoacids, peptides, Ca, CLA and other vital compounds, and special characteristics (flavour, taste, aromas) which allow for their designation as products of extra quality and lead, therefore, to high demand (Boyazoglu and Morand-Fehr, 2001).

Sheep and goat meat, particularly for early-weaned lambs and kids, has always been very much appreciated by most Mediterranean populations as meat of high quality.

Conclusions

Typical traditional products of animal origin from small ruminants, such as cheese and meat, in the Mediterranean have a specific originality and authenticity which is due to animal breeds, to the natural feed resources these animals use, to the reduced routine medication, and to specific technology used

Table 1. The fatty acid profile of sheep, goat and cow milk fat.

	C_4	C_6	C_8	C_{10}	C_{12}	C_{14}	C_{16}	C_{18}	$C_{18:1}$
Sheep	4.2	2.0	2.2	6.0	3.1	5.5	16.9	15.8	38.8
Goat	3.1	2.8	3.0	10.1	6.0	12.2	27.2	27.5	25.6
Cow	2.0	2.2	1.1	3.0	2.7	9.0	25.0	13.8	33.0

for the processed products. These niche products, if properly marketed, can get better price, at least from consumers who take a special interest in animals, which have been raised in natural environment and with very low chemical inputs.

References

Badertscher-Fawaz, R., R. Jorin & P. Rieder, 1998. Einstellungen sutierschntzfragen: wirkungen auf den fleischkonsum. Agrarwirtschaft 47: 107-115.

Boyazoglu, J. & P. Morand-Fehr, 2001. Mediterranean dairy sheep and goat products and their quality: a critical review. Small Ruminant Research, 40: 1-11.

Caballero, R., E. Fernandez, J. Rioperez, P.J. Hernaiz & M. Arauzo, 1992. Nutritive status of Manchega ewes grazing cereal stubbles and cultivated pastures. Proceedings of the 43rd Annual Meeting of the EAAP September 1993, Madrid, pp. 356.

Donald, Elg. G., 1994. The role of grazing sheep in sustainable agriculture. Sheep Research Journal, Special Issue 1994, p 37-51.

Lawson, R.E., A.R. Moss & D.I. Givens, 2001. The role of dairy products in supplying conjugated linoleic acid to man's diet: a review. Nutrition Research Reviews, 14: 153-172.

Nardone, A., G. Zervas & B. Ronchi, 2004. Sustainability of small ruminant organic systems of production. Livestock Production Science, 90: 27-39.

Tsiplakou, E., K. Mountzouris & G. Zervas, 2004. Concentration of conjugated linoleic acid in grazing sheep and goat milk. Animal Science Review (In Press).

Zervas, G., 1995. Beef, sheep and goat production systems and extensification in Greece. In: Extensification of Beef and Sheep production on Grassland . M.G. Keane and A. Pflimlin (Eds), Teagasc 1995: pp. 285-304.

Zervas, G., K. Fegeros & G. Papadopoulos, 1996. Feeding system of sheep in a mountainous area of Greece. Small Ruminant Research, 21: 11-17.

Typicity of Mediterranean sheep products: improvement of nutrition and feeding

F. Bocquier[1], C.H. Moulin[2] & P. Hassoun[1]

[1]Agro. M, UMR ERRC INRA PHASE, 2, Place Viala, 34 060 Montpellier Cedex 1, France
[2]Agro. M, UMR ERRC, INRA SAD, 2, Place Viala, 34 060 Montpellier Cedex 1, France

Summary

Sheep production systems of the Mediterranean circle have a large historical background both in their typical use of rangelands and fertile areas and on their lambs and dairy products. These systems are however very diversified and there is no simple and unique way to improve feeding systems and to isolate strong competitive advantages. Their main competitive advantages being firstly to be located close to their traditional markets the second advantage being to have specific products that are well identified and closely linked to their practices on a given territory and in a region. Hence, the main way to give them a stronger advantage is probably to qualify seriously, by geographical specifications, the production system as a whole. With feeding systems that include transhumance, the geographical certification with automated GPS and e-ID (electronic identification), which can be done by the shepherd himself, could be a reliable mean to remunerate the fundamental ecological service they provide. In other systems, such as sedentary ones, this geographical link to the territory can be reinforced by rules that imply a limitation of imported forages and concentrates thus leading to better use of all available local resource. This choice is, however, a strong constraint that can be solved, here again, by means of electronic devices (selective passages) using e-ID, in order to automatically allocate animals for an efficient use of local resources according to their requirements and/or their genetic potential. Hence there is a need for farmers to integrate such tools of "precision husbandry" in order to improve their products quality chain (and traceability) and to efficiently use Mediterranean local resources. On the social point of view, a better access to rangelands, limited conflicts with other rural activities would greatly help these productions to gain competitive advantages. In addition, it must be admitted that farmer incentives have to be a new income to sustain livestock ecological activities (livestock biodiversity interactions).

Keywords: sheep production system, feeding, reproduction, grazing, concentrate.

Introduction

Among livestock production of the Mediterranean areas sheep and goat are well represented (Boyazoglou, 2002). The major fact is that it is in these areas around the Mediterranean basin that is found the large majority of milked sheep, with numerous highly selected dairy breeds. The feeding of these animals is of peculiar importance since, like for all other farming animals, it represents the major cost. In theses areas, the very long term effect of domesticated sheep and goats on landscapes and agricultural systems is very tenuous: these animals, particularly goats, were used for their ability to use shrubs, while sheep were preferably bred in areas were they could find herbages. Even if the sheep and goat were often found together, the tendency is to specialise the production systems according to regional and local feeding resources. However in many cases this small ruminant production is associated with cattle productions. This long history gave a tradition of typical products that were consumed locally either in meat or in dairy products. Hence the typicity of these products comes from local know-how both in the production system and in the processing of animal products. The reputation of some products, especially cheeses (Roquefort, Pecorino Romano, Manchego, Feta, etc.), together with urbanisation, made these products to be easily recognized by a broader number of consumers even out of the Mediterranean circle. Hence it is tempting to make such products out of their traditional areas, in zones where farming conditions are more favourable to forage and/or to grain production.

In fact, on the feeding point of view, many reasons may explain why these breeding systems have been maintained in there traditional areas. But first of all, these ancient systems were capable to evolve and will continue to do so conserving what can be considered as typical and commercially relevant. Furthermore, due to the peculiar role of these animals on the environment and the fact that this may offer a possibility for populations to stay in these places this point has also to be discussed. This paper analyses, with some examples taken in sheep, the importance of feeding systems according to their constraints and their potential improvement on lambs and milk products of the Mediterranean areas.

Mediterranean lambs productions systems

It is not possible to detail the different types of lambs that are found around the Mediterranean basin. Lambs can however be classified according to their physiological age. Some are slaughtered at very light weight, i.e. few weeks after birth (Spain), these lambs are often found in areas were dairy ewes are bred and the competition for milk may explain why such a tradition exist. In others places and close to the systems of production of grass-fed lambs, i.e. temperate areas were grass is continuously found all year round, slaughtered lambs are heavier. According to different places the slaughter weight is around 30 to 35 kg at a mean age of 3 to 4 months. And in the southern coast of the Mediterranean Sea, there is a tradition of slaughtering lambs of more than one year old.

Light lambs of Spain

In the case of light lambs, there is a clear competitive advantage of the Mediterranean area since the cost of producing this lamb is included in the whole lactation revenue of its dam, whose milk is used for cheese making. It is generally interesting to compare the price of lamb with the equivalent milk price: breeder's decision to sell lambs varies according to this ratio, but as consumers want light lambs the margin of carcass weight is weak.

Mid-sized lambs of south of France

In the Merinos d'Arles lamb production system (Molénat *et al.*, 2003), large flocks (mean 700 ewes) are managed in a system were ewes are mated in spring and lambing occurs in autumn. The very peculiar situation is that in the lowland plain, irrigated prairies produce famous hay ("foin de crau") that is sold a good price because it has a PDO since 1997. The sheep production is tightly associated to the hay's production; since ewes are supposed to graze the autumn and winter regrowth of these prairies (until mid-February). This available good pasture is used to feed ewes at the end of pregnancy and during the suckling period. After weaning of lambs, and at the mating period, ewes are driven in the surrounding steppes, and in the rangelands of Alpilles hills. By the mid-June 110 000 ewes of this area are carried for a "Grande transhumance" to the high Alpine mountains where they stay, at an altitude of 1 800 to 2 500 m, until the early autumn. Autumn born lambs, from mid-September to the end of October, are suckling for 9 to 15 weeks. Heavy lambs, i.e. older than one year, are not very common even if they are easily sold for Moslem rites (see example of Tunisian production). Hence most of the lambs are feedlot fattened for 3 to 9 weeks before being sold at an average BW of 33 kg (120 d) for pure breed Merinos and 38 kg (100 d) for crossbred lambs. Solid food is made of hay, straw and concentrates; the mean quantity eaten by a lamb is comprised between 35 to 40 kg, there are practically no lambs that are totally forage-fed. These concentrate are regular ones for pure breed lambs and they are specially made for lambs sold within a label (i.e. crossbred lambs). The crossbreeding with rams of improved breeds (Ile de France, Berrichon, etc.) is often used to obtain heavy carcasses with a better conformation. In fact the main problem is that Merinos d'Arles breed is not well shaped and adult ewes' bodyweight is approximately 50 kg; the lambs sold in 2003-2004 were mostly (72%) classified in O3 and only 26% were in the desired class of R3 (EUROPA scale).

In this typical sheep production system the lambs, fattened with undefined concentrates, are not very different to other lambs produced in France (although local population appreciate this type of lambs). Hence the real typicity of this sheep production is the pastoral system as a whole. For this system based on a complex sequential use of a large variety of local resources is conceived to be in good synchrony with the biomass potential of uncultivated vegetations according to the season. One of the key-factor is that this system use cheap available vegetations, hence his major function is to participate to the control of shrub invasion both in the plains and in the mountains. This system is able to adapt to changing situations and in all occasion shepherds use the available lands to feed their animals. In addition the large body reserves variations observed during the annual cycle (Molénat *et al.*, 1993) is only possible with such a hardy breed that is also capable to transhume in the mountains. The other competitive advantage of such a system is that it sustains many activities both in the lowlands (hay production, preservation of bio-diversity) and in the highlands (maintaining landscapes that was constructed by pastoralism thus favouring rural tourism). If this can be better recognised by specific subsidies that take into account this environmental role; the long-term ecological role of such a system would be ensured. Out of a decision of the farmers that would better define the feeding systems of fattened lambs in order to obtain a PDO, the main improvement that can be proposed is to obtain a GPI certification. With the generalisation of electronic identification (e-ID) in 2008 in Europe, this could be done at a reduced cost for it may be done by the shepherd himself. For this, it can be imagined to build a device that include an e-ID reader coupled with automatic localisation, GPS or Galileo, which certifies the presence of ewes and lambs at a peculiar places of the territory at given dates previously defined in the GPI rules. Treaceability of animals and meat would give to this product its original place on the market that may satisfy the consumer whose feeling is also to participate to sustainable agriculture in difficult areas.

Heavy lambs of north-Africa

In the North-African coast of Mediterranean areas there is an even more difficult environment for the sheep productions. In these regions, aridity and hot temperature have a direct impact on biomasses that are even more seasonal than in the northern coasts. In Tunisia one of the typical sheep breed is the fat-tailed Barbary (Khaldi, 1989). The peculiarities of such a breed has been extensively analysed and the agro-pastoral feeding system well described (Atti *et al.*, 2004). Even if the reproduction of this breed is not very seasonal, the variations of feeding levels and the level of body reserves have a direct effect on the reproductive capacities of ewes. At the spring mating, there is frequently a three months widespread of lambing. This widespread has been directly related to the sate of body fatness at the start of reproduction; the first ewes are those that were in good body condition while lean ewes lasted 3 months before being fertile. This delay is the necessary time for a full replenishment of body reserves (Atti, *et al.*, 2001). The suckling period occurs in autumn and early winter, at a time when there is still some available grass. The weaning of lambs occurs at around six months and they are not fattened with concentrates, for it has been demonstrated that its potential growth is limited while his ability to survive with scarce resources is remarkable.

In such a system, the typicity of lamb is very clear for it's a fat-tailed lamb that cannot be confounded with other types of lambs. This production system is very robust to climatic uncertainly because of the length of the fattening period, thus allowing both for dams and lambs to survive and produce mostly from spontaneous rangelands. On strict feeding point of view the possibilities to improve this breed x system production are limited. The main competitive advantage of this breed is its high ability to survive to effects of severe drought. It has been demonstrated that the fat-tail per se allows surviving to complete starvation of 2 months and that its ability use dietary energy at re-feeding is remarkable (Atti *et al.*, 2004).

Solutions to improve the nutritional status of sheep in such lamb production's systems of the Mediterranean areas are very close to those used all over the world, when feeds and/or nutrients are bought out of the farm and given to animals during the rough periods. In these areas, the real challenge is to find new equilibriums between natural resources availability and the productive cycle of both ewes and lambs (Moulin *et al.*, 2003). In an attempt to offer all year round three month's old lamb, breeders have tempted to use both autumn and spring mating on two halves of their flock (Moulin *et al.*, 2004). The overall mean reproductive rate is increased, because unfertilized ewes can be driven into the second sub-flock but ewes that are suckling during winter are often underfed, due to lack of stored roughages, thus re-enforcing the detrimental effect of lower spring fertility (75%) compared to autumn fertility (90%). And what is observed is that the proportion of ewes from the autumn lambing flock is increasing and breeders have difficulties to maintain such a system which is very time consuming. The final result is that breeders go back to the traditional system of one lambing per year. The lesson of such an evolution is that typical traditional systems are very well adapted to the natural resources availability. Furthermore, the breed uses which are hardy do not respond enough to artificially improved feeding systems to be economically interesting. Hence the alternate solution which consists in changing the breed, in adoption of more intensive feeding systems based on grains and by-products and in the total abandon of the traditional areas of grazing is totally different. In this latter case, the competitive advantage of the Mediterranean regions, compared to other regions of the world with all year round grazing possibilities is weak. But the consumers of lamb's meat are also located around the Mediterranean areas thus maintaining the interest in such a typical production.

Mediterranean dairy sheep production systems

The Mediterranean countries produce almost 50% of the world milk sheep production (FAO, 2003) and are, for sure, the traditional zone for sheep dairy products among which cheese being the main product. Specialized breeds that are found in practically each country and/or regions are often linked with a product, i.e. Lacaune breed and Roquefort cheese, Manchega and Manchego, Sarda and Pecorino Romano etc. (de Rancourt *et al.*, 2005). Dairy sheep production systems are very heterogeneous regarding the land utilization, from zero-grazing to very extensive land use with or without transhumance. Differences also come from milking routines s (manual or machine) and transformation (on farm or in industrial manufactures).

On the feeding point of view, dairy systems have to face two main constraints: seasonally of forage production and collecting period of the dairy industry. Among adaptation to forage availability transhumance is mainly practiced in French and Spanish Pyrenean region ("Pays Basque") and in Greece. This practice is either done to avoid summer drought or to allow the lowland farm to make stocks of preserved forages. Anyway, in most cases there is a necessity to have most of the ewe's lambing synchronized at an adequate period.

Full grazing dairy systems

Full grazing dairy systems are not very common because milk is generally produced during the wintertime when grass is not abundant enough to fulfil animal's requirements. Furthermore, access to pastures is often limited during this period, in Basque Country and in Sardinia, by the climatic conditions that are the main factor conditioning grazing possibilities. Hence, dairy ewes generally full graze only in late lactation stage on either extensive lands (Greece) or, after a transhumance to mountain pastures (Basque Country).

In Sardinia, in the pastime, livestock were fed on natural pastures and winter forage crop (oats, barley and Italian ray-grass). During the last two decades lowlands irrigation has been greatly developed carrying highly-intensified farms (Pirisi *et al.*, 2001). In these systems, lambing occurs in early winter with a winter spring milk production transformed into cheese. However, irrigation allows an important forage production in late spring and summer. Consequently, lambing can be delayed (late winter) and milk production period too. This system allows increasing cheese market opportunities with tourism (Fois *et al.*, 1997). Such a delayed system has been evaluated during 3 years in comparison with indoor feeding and autumn vs winter lambing (Fois *et al.*, 1997). It shows that ewes in grazing system had a significant (P<0.05) higher milk yield, and higher lamb birth weight with higher average daily weight gain. Only milk fat and protein content were lower (67 *vs* 75 g l^{-1} and 55 *vs* 58 g l^{-1} respectively). In a 10 weeks comparison trial between indoor and at pasture (Italian ray grass –IGR) feeding system, Piri *et al.* (2001) also demonstrated that ewes fed on pasture produced more milk and with better microbiological characteristics. Only cheese making parameters were better for indoor system. Logically, grazing system consumed much lesser concentrate and hay. Finally, costs including food, cropping and veterinary expenses are more favourable for the grazing system: the gross margin being twice higher in the full grazing system (Fois *et al.*, 1997).

In such systems, other experiments have also been conducted (Molle *et al.*, 2000, 2003) to improve diet quality at pasture through new forage seeding such as Sulla legume (*Hedysarum coronarum*). Mixed with ray-grass or sown on separate paddocks, results with Sulla were encouraging but experiments need to be carried on. in order to assess the effect of a wider range of using conditions (time allowed for grazing, season impact, milk quality when dedicated to cheese manufacture, economic aspects, etc.)

It is generally expected that systems largely based on grazing (autumn or winter lambing) would perform better results both in term of economics and self sustaining capability.

Grazing and transhumant dairy systems

In the Mediterranean area, the traditional dairy system classified from the land utilization implies three different breeding systems: sedentary, with transhumance and nomadic (Boyazoglu, 1989). The first is increasing in the developing areas whereas the latter is mainly located in the north of Africa and the Near East. The transhumance was used in several countries (France, Greece, Italy Spain) and mainly close to the mountains (Pyrenees, Corsica Island). Nowadays, only the Basque Country on both France and Spain Pyrenean sides is commonly practicing transhumance for dairy sheep (Mandaluniz & Oregui, 2004; Marijuan *et al.*, 2004; Morin *et al.*, 2004). Very few studies were published on the dairy sheep transhumance system. However, out of the economical aspects, it may play an important role from environmental and biodiversity points of view.

In the case of Basque systems, ewes are allowed to restrict grazing during wintertime and are brought to mountainous areas for summertime. As stated before, the synchronization of lambing is of major importance. In figure 1, the consequences of the widespread of lambing among Manech ewes of the same flock (Arranz & Bocquier, 1997) are presented. Late lambing ewes have a shortened lactation (109 *vs* 178 days) when compared to early lambing ewes and because they were dry-off and they did not have enough time to replenish their body reserves at the time of annual mating period (Figure 1: [1]). The level of body fatness at mating has a direct on fertility rate which is lower in late lambing ewes, hence in this type of ewes there is a limited number of ewes that can join the flock (Figure 1; [3], while those in good condition have a high level of fertility (Figure 1: [2] which allows to find the majority of this type of ewes grouped in the next lactation. In the constituted flock of each origin there is neither differences in lactation length (Figure 1 ;[3], 169 vs 163 days) nor in total milk output (149 vs. 139 litres). Hence the ewes that were not fertile at this mating period had to face the same situation as described in the first par of the cycle (Figure 1 ; [1], because they are going to have a late lambing or they left the flock and are to be replaced by ewe-lambs of a lower milk yield. Such a dispersion of lambing has been shown to have negative impact on the fat and protein content of whole milk delivery (Fraysse *et al.*, 1996; Barillet *et al.*, 2002).

The conclusion is that in such a system the feeding use of local resources, i.e. mostly based on grazing (Arranz *et al.*, 1997), either in the lowlands (winter) or in the mountains (summer) is very important. The consequence is, however, that high requirements must coincide with period of high grass availability, if there is some problem of reproduction it has a direct impact on the total milk delivery either because of shortened lactation or by a higher replacement rate of ewes. The competitive advantage of such a system can be found in the extended use of natural resources that prevents from shrub invasion of mountain lands as described for Mérinos d'Arles system and other transhumant system.

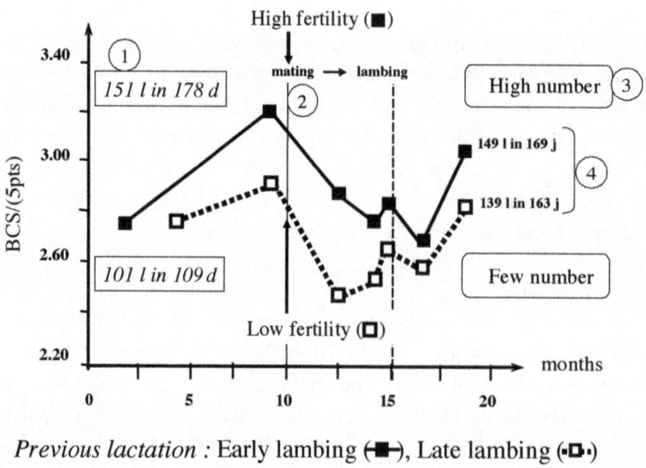

Figure 1. Evolution of Body Condition Score (BCS) of Manech dairy ewes: consequences on reproduction and milk productivity (Arranz & Bocquier, 1997).

Sedentary dairy systems

In the Roquefort Area of southern France, the traditional breeding system is based on the Lacaune breed (Barillet *et al.*, 2000b) and the cheese is protected by a PDO. In such a system, ewes are indoor-fed during wintertime at a period where the major part of milk production occurs. Therefore, the period of maximum requirements has to be fulfilled with conserved feeds. Many years ago the breeders took economical advantage of feeding ewes with a large proportions of commercial feeds of high nutritional density; namely dehydrated Alfalfa, or sugar beet pulps, concentrates, high quality hay (Foin de Crau) and so on… This situation no longer exists since new rules of PDO were changed few years ago, that were aimed to reinforce the relationship between milk and local feed characteristics. One aspect of regulation is to limit the amount of feed that are not directly produced on the farm: the threshold of 200 kg DM / year / ewe has been chosen. Another aspect of the regulation being that the use of some peculiar feeds, or additives, cited in a positive list is forbidden.

There has been a direct effect of these new rules on the evolution of individual milk production which is well illustrated by the data collected in the breeding improvement program of the Lacaune breed (Astruc and Barrilet, 2004). While, on the period from 1981 to 2004, the calculated annual genetic merit of ewes (ΔG) has constantly increased by +6.41 liters/year/ewe, the phenotypic (ΔP) progress was lower: + 5.79 liters/year/ewe and even stabilized on the last year. As a logical consequence (because $P = G + E$) the calculated mean effect of environment (E) is considered to evolve negatively. Although many factors are included in the E component it is tempting to correlate E evolution with the increasing proportion of farmers that passed under the threshold of 200 kg/DM bought out of their farm each year (Figure 2): one can see that these two curves surprisingly well superposed. Although this shortcut interpretation is probably rough, it is obvious that breeders did not change their feeding practices, i.e. ewes of the flock are collectively fed and there is no adjustment of food supply or food density, to the phenotypic level of milk production. As previously reported (Bocquier *et al.*, 1995), the feeding strategy applied to a flock consists of rationing all the ewes

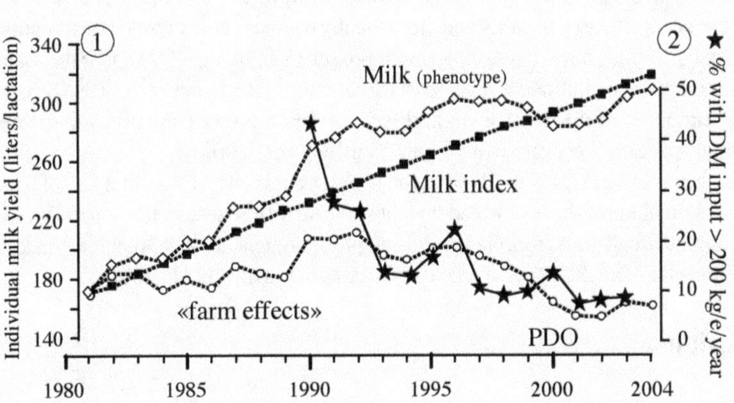

Figure 2. (1) Individual evolution of genotypic and phenotypic progress in primiparous Lacaune ewes (Astruc et al., *2004). (2) Evolution of the proportion of breeder that bought more than 200 kg DM/ewe/year (Morin* et al., *2004).*

above the mean requirements of the group: i.e. above the mean milk production (mMY). This is quite logical, since the majority of ewes that significantly contribute to the filling of bulk tank are producing 10 to 20 % more than the mMY. Inversely, more than 50 % of the ewes of this flock are overfed and receive more than their strict requirements. With the recent PDO constraints to use all the forages produced on the farm, the mean quality of the diet is lowered and the best ewes of the flock are, by now, more penalized. This adverse effect on the milk volume is not *per se* a real problem since the farmers are organized in order to limit their annual milk bulk volume. There is, however, a direct effect on fat and protein contents of milk (Bocquier & Caja, 2001) which are important factors of the cheese process efficiency.

In the Roquefort dairy system, since a long time, there has been a constant effort to reinforce the identity of this product. Out of the cheese processing which is strictly limited to Roquefort caves, the transformed milk is only provided by ewes of the Lacaune breed and, nowadays, the feeding practices are better linked to the territories and some peculiar feeds and additives have been prohibited. Of course, in the field of nutrition, there is still a need to introduce some innovative feeding practices in order to maintain, or even improve, the sustainability of this production. Several propositions can be promoted according to geographical and climatic conditions. When possible, some breeders have already included a larger part of grazed forages in the feeding system. There is however an urgent need to better know the grazing abilities of this ewe in order to optimize grazing systems (Hassoun *et al.*, 2003).

The other main challenge is to make homogenous groups of ewes according to their requirements and intake capacities (Bocquier *et al.*, 1995). Up to now there were numerous technical breakpoints to overcome. Briefly, it seems by now possible to use the electronic identification (e-ID) that is going to be obligatory in Europe to twice a day sort the ewes according to adequate nutritional criterions. This can be done by control access doors that will filter each ewe into authorized zone (sheep barn parks, grazing paddock) where offered feeds are of adequate formulation and prevent other ewes from passing and to be maintained in this group. Inside the sheep barn, one can imagine that all ewes are first introduced in the group with low quality diets, some ewes can move to the next area where the quality of the diet is better and some others ewes can reach the last area were the diet is of the best quality. In such a case low quality forages will be given to low producing ewes without any concentrate, and progressively, the last group will be fed with best forages and adequate doses and composition of concentrates. The main advantage of such a strategy is that the first effect is to optimize the forage use according to its quality and dramatically decrease concentrate consumption. This proposition is realistic since it has been shown, and proven (Vacaresse, 1992), that the major step of this improvement can be obtained with at least three groups (Bocquier *et al.*, 1995). As an example, on a flock composed of 400 adult ewes and 165 primiparous ewes, the division in three groups gives an annual spare of concentrate of 18 tons (Barillet *et al.*, 2002).

These Lacaune systems have largely evolved in the recent years toward a greater use of local resources and high technical know-how of breeders that serve an efficient selection on milk yield (Barillet *et al.*, 2001b), milk quality and morphological and physiological ability to machine milking efficiency (Marie-Etencelin *et al.*, 2002), and mastitis resistance (Barillllet *et al.*, 2001a).

General conclusions

Sheep meat and cheese products of Mediterranean circle are part of the historical and cultural inheritance. These productions systems, that are still well developed, were capable to evolve and adapt to the societal and commercial changes of the recent periods. In term of feeding practices different strategies that were adopted tried to maintain a link with their local or regional origin. In terms of marketing some products are easy to identify and their typicity is quite clear. This is the case of very light or heavy lambs that are not in the common evolution with present conception of breeding objective. The benefit from such a production is only possible because the first one is a co-product of milk production and the second a cheap way of producing meat in an arid zone. Hence these two types are really typical and have a competitive advantage in their region. For mid-sized lamb carcasses, there is a clear problem of typicity for this type of lamb, which is quite easy to produce with improved breeds and low cost diets either grass or cereal based that can be found out of this zone. One can notice, however, that these lambs are issued from a complex breeding system that exerts a specific role on land use of these areas. This peculiar aspect will be discussed hereunder because it is very similar to the role played by transhumant dairy ewes.

Dairy ewes systems also evolved and adapted to the present constraints. Although these conclusions may not apply to all systems, a large movement of certification took place in Europe and many cheeses are by now certified (OPD, PGI) or sold in short circuits to awarded consumers. Some of the cheese products chains were also able to diversify on new products (yogurts, fresh cheeses) in order to maintain one of their most typical products (traditional long maturing cheese). This movement was accompanied by changes in feeding practices including lists of feeds and additives that are no longer accepted and minimum daily time spend grazing (also for animal's welfare aspects). The latter point being in the same line as the whish to reinforce the link with a maximization of on farm produced feeds and/or the direct use of local resources.

In most sheep production systems, either meat or milk, the cost of manpower is the major limiting factor. Considering only this point, there is no competitive advantage of these systems if their ecological role is not measured and economically evaluated. Furthermore, the costs and benefits of various sheep production systems, with their specific practices, must be based on local values and local constraints, causing sustainable practices to be region and culture specific (Tilman *et al.*, 2002). In most of the above described systems there are combinations of intensive and extensive phases that are based on the local calendar of feed resources and animal requirements. As a primary transformer of this biomass into high quality products, sheep, like other ruminants, play an essential role. These systems must in turn be efficient in their ecological service role. Because there is no doubt that appropriately stocked and managed rangelands-sheep ecosystems are efficient methods of producing high-quality protein with minimal environmental negative impacts. This requires however evaluating pastoral practices and proposing improvements that could each contribute to increase nutrient use efficiency and simultaneously acting favourably on the renewal of the resource. The improvement of nutrient use efficiency can be obtained either by supplementation, but for sure, by a good matching of temporal and spatial supply with animals demand. Such precision husbandry can be done in large scale farms and by smallholders: the tools being different. These pastoral systems can also take advantage of services provided by adjacent natural and semi-natural areas and thus be developed to restore some degraded ecosystem. The major problem being a double societal and consumer choice: a good price for the products and a level of subsidies that correspond to the correct payment of ecological service provided by these animals. The political decision to better ensure animal products traceability, by mean of generalized electronic identification, may also be beneficial to the productive processes. New electronic devices may help to built up geographical auto-certification of the whole breeding process by the shepherd himself thus justifying his ecological services. Complementarily, these e-ID developments may reduce on farm time consuming operations (animal sorting) and significantly enhance conserved feed efficiency. Such a demarche is in the direction to use decision tools, and argued advisories, that may greatly help farmers to maintain their economical profitability. At a longer term, an easier access to rangelands and land use adapted rules will favour the fundamental ecological role of these typical Mediterranean productions.

References

Arranz J.M.& Bocquier F., 1997. Relations entre performances et état corporel des brebis laitières en Pyrénées Atlantiques. Renc. Rech. Rumin., 4: 29.

Arranz J.M., Lagriffoul G. & Bocquier F., 1997. L'importance du pâturage dans l'alimentation hivernale des brebis laitières en Pyrénées Atlantiques et son incidence sur la production laitière: approche par enquêtes en fermes. Renc. Rech. Rumin., 4: 156.

Astruc J.M.. & Barillet F.., 2004. Résultats du contrôle laitier officiel. CNBL Edition.

Atti N., Bocquier F. & Khaldi G., 2004. Performance of the fat-tailed Barbarine sheep in its environment: adaptative capacity to alternation of underfeeding and refeeding periods. A review. Anim. Res., 53: 165-176.

Atti N., Thériez M. & Abdennebi L., 2001. Relationship between ewe body condition at mating and reproductive performance in the fat-tailed Barbarine breed. Anim. Res. 50: 135–144.

Barillet F., Marie C., Jacquin M., Lagriffoul G. & Astruc J.M., 2001b. The French Lacaune dairy sheep breed: use in France and abroad in the last 40 years. Livest. Prod. Sci., 77: 17-29.

Barillet F., Rupp R., Mignon-Grasteau S., Astruc J.M. & Jacquin., M 2001a. Genetic analysis for mastitis resistance and somatic cell score in French Lacaune dairy sheep. Gent. Sel. Evol. 33, 397-415.

Barillet F., Such X., Lagriffoul G., Caja G. & Bocquier F. 2002. Evolutions des techniques d'élevage en brebis laitières. In: Nutrition, alimentation et élevage des brebis laitières: maîtrise de facteurs de production pour réduire les coûts et améliorer la qualité des produits. Barillet F. & Bocquier, F. (Eds), Options Méditerranéennes Série B: Etudes et Recherches. 42: 85-115.

Bocquier F. & Caja G., 2001. Production et composition du lait de brebis: effets de l'alimentation. INRA Prod. Anim., 14: 129-140.

Bocquier F., Guillouet P. & Barillet F., 1995. Alimentation hivernale des brebis laitières: intérêt de la mise en lots. INRA Prod. Anim., 8: 19-28.

Boyazoglu J., 1989. La production laitière ovine en systèmes extensifs méditerranéens. Cahiers Options Méditerranéennes, Série A, 6, 141-147.

Boyazoglu J., 2002. Livestock research and environmental sustainability with special reference to the Mediterranean basin. Small Ruminant Research 45: 193-200.

FAO, 2003. Faostat, FAO, Rome, Italy.

Fois, N., Ligios, S., Molle, G., Sitzia, M., Decandia, M., Sanna, L., 1997. Dairy sheep farming in Sardinian irrigated lowlands. Cahiers Options Méditerranéennes 38, 219-222.

Fraysse J., Lagriffoul G., Bocquier F. & Barillet F., 1996. Brebis laitières: impact de la structure du troupeau et autres facteurs d'élevage sur la composition chimique du lait livré. INRA Productions Animales, 9: 201-210.

Hassoun P., Autran P., Marie-Etancelin C., Barillet F. & Bocquier F., 2003. Effect of three levels of herbage allowance on voluntary intake and milk production of dairy ewes at pasture: preliminary results. In Proceedings of the 54th Annual Meeting of EAAP, Rome, Italy, 30 Augusy-3 September 2003.

Khaldi, G., 1989. The Barbary sheep. In Small Ruminants in the Near East. Volume III: North Africa. FAO Animal Production and Health 74, 96-135.

Mandaluniz N. & Oregui L.M., 2004. Diet quality of sheep and goats grazing in the Gorbiea Natural Park (Basque Country). Cahiers options méditerranéennes Série A 61, 179-184.

Marie-Etancelin C., Casu S., Aurel M. R., Barillet F., Carta A., Deiana S., Jacquin M., Pailler F., Porte D. & Tolus S. 2002, New tools to appraise udder morphology and milkability in dairy sheep. Cahiers Options Mediterranéennes, CIHEAM, 55: 71-80.

Marijuan S., Mandaluniz N., Ruiz R. & Oregui L.M., 2004. Utilisation of mountain pastures by dairy ewes: Eastern Basque Country situation. Cahiers options méditerranéennes Série A 61, 185-188.

Molénat G., Dureau R., Fabre P. & Lambertin M., 2003. Les « herbes » des troupeaux ovins transhumants de Crau. Multiples dimensions d'une gestion pastorale et fourragère. Fourrages, 176, 43-461.

Molénat G., Lapeyronie P., Vincent M. & Gouy J., 1993. Variations de l'état corporel en système méditerranéen transhumant. In Pratiques d'Elevage Extensif. Identifier, modéliser, évaluer. INRA Etudes et recherches sur les Systèmes Agraires et le Développement, 27: 123-136.

Molle G., Decandia M., Fois N., Ligios S., Cabiddu A., Sitzia M., 2003. The performance of Mediterranean dairy sheep given access to sulla (*Hedysarum coronarium* L.) and annual ryegrass (*Lolium rigidum* Gaudin) pastures in different time proportions. Small Ruminant Research 49, 319-328.

Molle G., Sitzia M., Decandia M., Fois N., Scanu G. & Ligios S., 2000. Intake and performance of dairy ewes grazing Mediterranean forages either as pure or mixed swards. Cahiers Options Méditerranéennes 52, 187-192.

Morin E., 2004. Base de données d'appui technique OPTISUD, Institut de l'Elevage (antenne de Toulouse, France).

Moulin C.H., Blanc F., Ezanno P. & Bocquier F., 2003. Modelling as a tool for the teaching of livestock dynamics. Anim. Res. 53, 439-450

Moulin C.H., Pluvinage J. & Bocquier F., 2004. Les relations entre agrandissement des troupeaux et changements de conduite: exemples d'élevages d'ovins allaitants en Crau. Renc. Rech. Rumin. 11: 145-148.

Pirisi A., Piredda G., Scintu M.F., Fois N., 2001. Effect of feeding diets on quality characteristics of milk and cheese produced from Sarda dairy ewes. Options Méditerranéennes: Série A. Séminaires Méditerranéens 46, 115-119.

Rancourt (de) M., Fois N., Lavin M.P., Tchakerian E. & Vallerand F., 2005. Mediterranean sheep and goats production: An uncertain future. Small Ruminant Research (in press)

Tilman D., Cassmann K.G., Matson P.M., Naylor R. & Polasky S., 2002. Agricultural sustainability and intensive production practices. Nature 418: 671-677.

Vacaresse C., 1995. Essai de conduite en lots virtuels pour l'alimentation des brebis laitières. Pâtre. 424: 18-20.

Traditional cheese production systems in Mediterranean areas

G. Licitra[1], I. Piccitto[2] & P. Campo[2]

[1]*Università degli Studi di Catania, Dipartimento di Scienze Agrochimiche e delle Produzioni Animali della Facoltà di Agraria,Catania, Italy*
[2]*Consorzio Ricerca Filiera Lattiero-Casearia, Regione Siciliana, S.P. 25 km 5 97100, Ragusa Mare, Ragusa, Italy*

Summary

Cheese making through the centuries gradually evolved in different streams. The first was the liquid fermented milks such as yogurt. The second through allowing the milk to acidify to form curds and whey. The nomad founded the whey drinkable and the curds edible. Further, the whey could then be drained either through perforated earthenware bowls or through woven reed baskets or similar material. Later on, to extend shelf life of the curd, the addition of salt and the dehydration process was introduced. Since ancient times, salted cheeses are wide spread around the world for the important role not just of protein, fat, but also for salt content in the hot regions.

Of the countries studied, 47% of fifty-one dairy products evaluated were made through acidification process using the ancient techniques. This simple technique applicable at the household level, was diffuse because in the arid region the need of healthy drinkable liquid was crucial to survive. The use of sun drying the curd was another guarantee of food supply in the time of famine.

Traditional dairy products are the result of many biodiversity attributed to the territory, the man, the culture, the nature and are definitely testimonial of heritage and history of each ethnic human resource. Unfortunately, most of the developing countries try to plan for the future by looking to the developed countries and eliminate their own rural culture.

Keywords: traditional dairy products, Mediterranean area, biodiversity.

Introduction

The first historical references to cheese production using milk in the cheese making process from priests, is a Sumerian Frieze from Mesopotamia, today's Iraq, from about 3000 B.C. which is preserved at the British Museum of London. Remnants of cheese were found in the tomb of a king of the Egyptian First Dynasty (c. 3000-2800 B.C.). Other Egyptian tomb murals of 2000 B.C. show butter and cheese being made, and other murals show milk being stored in skin bags suspended from poles demonstrating knowledge of dairy husbandry at that time (Fox & McSweeney, 2004).

A consequence of "The Agricultural Revolution" which occurred in central Asia in the steppe, in Mesopotamia, in Anatolia, in the Middle East, was the domestication of animals and plants. Man changed from hunter to farmer. Archaeological surveys have established that in the Neolithic age, the ancient Sumerian and Mesopotamian cultures of Tigris-Euphrates basin raised animals and engaged in dairy production as early as 7000-8000 B.C.

A legendary story of "how the cheese making process was discovered" is the one that describes a fortuitous combination of events by an unknown Arab nomad. He filled a saddlebag with milk to sustain him on a journey across the desert by horse. After several hours riding, he stopped to quench his thirst, only to find that the milk had separated into pale watery liquid and solid white lumps. Because the saddlebag, which was made from the stomach of a young animal, the milk had been effectively separated into curds and whey by the combination of something unknown in the stomach (the rennin enzyme), the hot sun and the galloping motions of the horse.

It is important to consider that during the Greek Age the technology to transform milk became common and the production of different types of cheeses came about, not only fresh and soft, but also hard cheese to be grated, which was made by pressing the curd with a heavy weight. The art of manufacturing and cheese making, especially with the use of vegetable rennet, was improved by the Etruscans, shepherd populations, which handed the art of cheese making down to Romans. The Romans improved the cheese manufacturing of the Greeks and produced dairy products from sheep, goats and cows which they consumed fresh or aged. After centuries the monk's culture in the Middle Ages was the most important factor contributing to the development of cheese technology and to the evolution of cheese varieties. Monasteries and feudal estates became the sole oasis of religious devotion, culture and enlightenment. In addition to their role in the spread of Christianity and in the preservation and expansion of knowledge during the Dark Ages, monks from all over Europe became cheese supporters (Montanari, 2000).

The Age of Enlightenment, arising out of the Renaissance, was a time of tremendous inventiveness and entrepreneurial spirit and cheese makers created new and exciting varieties.

Dairy products in the Mediterranean area and Europe

The evolution of cheese making through the centuries was very different between Europe and the Afro-Mediterranean countries. In Europe, the industry of dairy products is very well developed and produces many specialties to meet the consumer need of fresh dairy products and its economics importance is more than 50% of the overall dairy market. We must also underline the extraordinary importance of more than 2000 traditional cheeses in Europe and 140 of these have the Protected Denomination of Origin. In the developing countries, including the Mediterranean region, almost all the dairy products are household operations using traditional preparation systems. Livestock farming and milk and milk products play an important socio-economic role. The extent of dependence of the farmer on traditional milk products varies from country to country and within countries. Climate, development of roads and transport within a country, and the level of industrial milk processing are some of the factors in determining the importance of these products.

FAO (1990) reports that traditional dairy products are associated mostly with the native cattle keepers and the amount of milk involved in traditional processing may be taken to be at least 80% of the total milk produced, i.e. total milk less 10% that is produced in the commercial sectors and another 10% which is marketed from the traditional sector via milk processing plants. The fact that only a small fraction of the milk produced is marketed by commercial enterprises implies that nearly all traditionally produced dairy products are marketed through traditional, informal marketing channels. Most of them are marketed through inter-household sales and exchange, rural trading centers and the common weekly or bi-weekly rural market days. In the majority of cases, the women who process the milk are also the ones involved in marketing the dairy produce with occasional assistance from other female members of the family or young boys.

The nutritional importance of traditional milk products in the diet is related to the overall availability of milk to the population of each country, and within a country to the consumption of milk and milk products in rural and urban areas. Since much of the preparation of traditional milk products takes place in rural areas - many of them isolated by lack of transport - their importance from a nutritional point of view is greater for the local population than for that of the urban areas.

Technologies of the main traditional milk production

In the developing countries studied, there encompasses a wide range of climatic conditions ranging from the hot humid coastal areas, dry and semi-arid grass hinterlands to high altitude highlands with subtropical to temperate type climates. Over the centuries technologies have evolved within different communities resulting in products. Table 1 shows the dairy products of the Mediterranean basin.

Of these countries the mean diurnal temperatures may range from as low as 15-17°C in the highland areas to as high as 35°C in semi-arid and arid areas. These high ambient temperatures coupled with the general lack of refrigeration facilities imply that the milk, often containing high initial numbers of bacteria, becomes sour in 12 to 24 hours. The standard of hygiene applied to milk production in developing countries is poor and as a result the quality of milk is poor. Fermentation of milk to control the growth of spoilage bacteria and some pathogens is the most common aspect of technology in the preparation of traditional milk products. Further prevention of the spoilage of milk is based on boiling, or on souring, simple condensing with sugar addition or immediate processing, used in traditional milk products. The use of natural fermentation as being the most important means of achieving the necessary souring either in the formation of a fermented milk product ready for consumption or in an acidification or souring intermediate stage in a product's preparation. The value of lactic fermentation not only affects the shelf-life of the product but affects the quality and characteristics of the product. Natural acidifications end up in the whey and in the curd. Acid whey is a pleasant, refreshing drink for immediate consumption while the curd is consumed fresh or stored for future use. Table 2 shows that of the countries studied 47% of dairy products were made through

Table 1. Dairy products (n° = 51) evaluated per countries of the Mediterranean basin.

Country	Incidence %	Total no.	Most diffuses name of the dairy products
Egypt	14	7	Daani, Domiati, Karish, Mish, Rahssr, Keshda Mosakhana, Zabady
Cyprus	10	5	Trachanas, Graviera, Anari, Kefalotyri, Paphitico
Turkey	10	5	Ayran, Beyaz Peyniri, Kasar Peyniri, Tulum Peyniri, Mihalic Peyniri
Lebanon	8	4	Umbris, Fresh cheese, Karichee, Kanafeh Bil Jbin
Jordan	8	4	Naboulsi, Jemid, Djamid, Rayeb
Syria	6	3	Shenglish, Mesanarah, Medaffarah
Mauritania	6	3	Caravane, Beurre traditiuonnel, Lait caille
Morocco	4	2	Jben, Leben
Algeria	4	2	Aoules, Takammart
Tunisia	2	1	Raibb
In more than one of the above countries	29	15	Wagashi, Shankalish, Laban, Labneh, Kishk, Akawieh, Kashkaval, Halloumi, Zebdah/Zibd, Feta Country, Samneh, Shmena-semma, Smen, Yogurt, Ayib

Table 2. The coagulation systems used in the processing of dairy products (n° =51) in the countries of Mediterranean basin.

Coagulation systems	Incidence %	Total no.	Most diffuses name of the dairy products
Rennet	45	23	Caravane, Shankalish, Mesanarah, Medaffarah, Beyaz Peyniri, Kasar Peyniri, Tulum Peyniri, Akawieh, Jben, Naboulsi, Kashkaval, Fresh cheese, Halloumi, Graviera, Kefalotyri, Paphitico, Daani, Domiati, Rahssr, Takammart, Feta Country, Kanafeh Bil Jbin, Mihalic Peyniri
Acidification	47	24	Wagashi, Shenglish, Laban, Labneh, Ayran, Kisk, Leben, Jemid, Umbris, Trachanas, Karish, Mish, Zebdah/Zibd, Raibb, Djamid, Samneh, Beurre traditiuonnel, Keshda Mosakhana, Shmena-semma, Smen, Lait caille, Zabady, Yogurt, Rayeb
Heat/Acid	8	4	Anari, Aoules, Ayib, Karichee

Table 3. The typology of dairy products (n° = 51) in the countries of the Mediterranean basin.

Typology	Incidence %	Total no.	Most diffuses name of the dairy products
Fermented milk	19	10	Leben, Trachanas, Laban, Zabady, Yogurt, Rayeb, Ayran, Samneh, Beurre traditiuonnel, Smen
Soft cheese	35	18	Wagashi, Caravane, Labneh, Akawieh, Jben, Umbris, Fresh cheese, Anari, Daani, Domiati, Karish, Mish, Keshda Mosakhana, Zebdah/Zibd, Karichee, Shmena-semma, Lait caille, Ayib
Semi hard cheese	16	8	Shenglish, Beyaz Peyniri, Naboulsi, Kashkaval, Halloumi, Feta Country, Raibb, Kanafeh Bil Jbin
Hard cheese	18	9	Shankalish, Kasar Peyniri, Tulum Peyniri, Graviera, Kefalotyri, Paphitico, Rahssr, Djamid, Mihalic Peyniri
Sun dried cheese	12	6	Mesanarah, Kishk, Jemid, Aoules, Takammart, Medaffarah

acidification process using the ancient techniques. This technique applicable at the household level, was diffuse because in the arid region the need of healthy drinkable liquid is crucial to survive. The use of sun drying the curd was another guarantee of food supply in the time of famine. These soured milks, as well as being liquid milk products in their own right, are the basis for the production of unsalted butter, ghee and curd in the household, the local village processing unit or the industrial scale dairy factory.

Most of the cheeses studied in these countries are fresh and soft (Tables 3 and 4). Considering the hot climate and low technology, we were expecting to find much more aged cheeses, but after further evaluation it appears reasonable that the main use of the milk is the day by day self consumption or through fermented products with short shelf life.

It is difficult to consider production for future need if first they are not able of satisfy the daily need. In the period of abundance they age the cheese by salting, sun drying or under oil or using spices and other ingredients.

The use of spices and other ingredients is testimony to the unique ethnic culture that makes the relationship with the territorial resources very strong. The spices used were in 18% of the products: Shenglish (thyme, cumin seed, chilly powder, pepper and salt), Shankalish (thyme, aniseed, paprika, nigella and cumin), Mesanarah (nigella grains), Laban (garlic), Samneh (turmeric), Shmena-semma (roasted piece of ram horn or leaves of plants or seeds), Smen (rosemary and thyme), Medaffarah (black cumin, nigella sativa is added to the hot paste), Mish (red and green pepper), Shenglish (thyme, cumin seed, chilly powder, pepper and salt). Ingredients used were in 10% of the products: Kishk (ground wheat), Naboulsi (mahlab, prunus mahaleb and mastic, pistacia lentiscus), Raibb (yeast flavoured), Kanafeh Bil Jbin (sugar, flower water, ground biscuits cooked in ghee), Beurre traditiuonnel (honey and juice).

Mediterranean traditional dairy products vs. European PDO cheeses

It is necessary to underline that in these figures we are not taking into consideration the European industrial dairy products but only their PDO cheeses. In figure 1, it is evident that in the Mediterranean countries 47% of the dairy products are consumed within two weeks; in Europe only 14%. If we

Table 4. The aging time of dairy products (n° = 51) in the Mediterranean basin.

Aging time	Incidence %	Total no.	Most diffuses name of the dairy products
None ripening	16	8	Laban, Ayran, Leben, Trachanas, Samneh Beurre traditiuonnel, Smen, Ayib
7-14 days	31	16	Keshda Mosakhana, Anari, Fresh cheese, Akawieh, Jben, Karichee, Kanafeh Bil Jbin, Shmena-semma, Lait caille, Karish, Wagashi, Zabady, Yogurt, Rayeb, Raibb, Umbris
< 60 days	14	7	Caravane, Shankalish, Feta Country, Kashkaval, Shenglish, Daani, Zebdah/Zibd
60-180 days	16	8	Kasar Peyniri, Paphitico, Rahssr, Mihalic Peyniri, Graviera, Kefalotyri, Tulum Peyniri, Beyaz Peyniri
181-360 days	5	3	Naboulsi, Halloumi, Djamid
Variable ripening sun dried or pickled	18	9	Mesanarah, Kishk, Jemid, Aoules, Takammart, Medaffarah, Labneh, Domiati, Mish

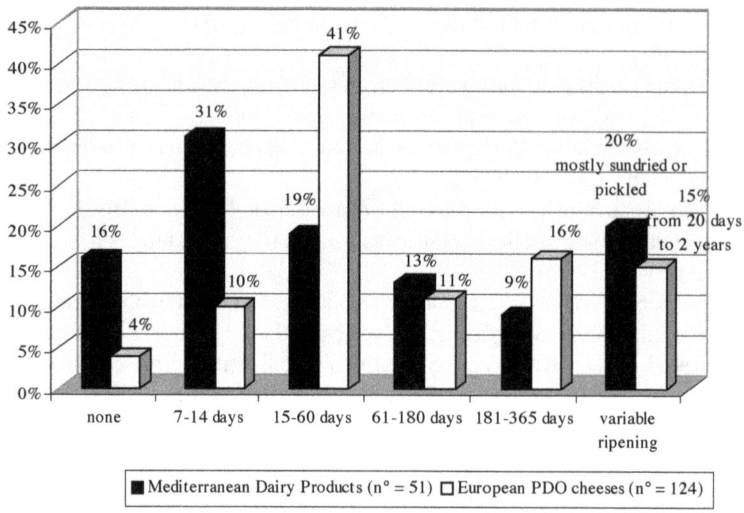

Figure 1. Ripening period of traditional Mediterranean dairy products and European PDO cheeses.

consider the products aged less than 60 days, the European show the consumption of 55% of the total PDO produced versus the 66% of the Med countries. In Europe the tradition of aging the cheeses is much more diffuse than in the Mediterranean, where the cheese is not aged for long periods (only 9% over 6 months) but are saved by sun drying or under oil, 20% (Figure 2), or adding spice, sugar and other ingredients, 28%.

Figure 2 shows that in the Mediterranean countries 54% of the dairy products are fermented milk (acidified and sour milk, clarified butter, butter milk, yogurt), and soft cheeses, only 16% are semi hard cheeses and 18% hard cheeses.

If we consider the differences between Mediterranean traditional dairy products, reported in table 1, and Italian and French PDO cheeses, Italy produces much less soft cheese, only 20% compared with 73% of France. In alternative Italy more semi hard cheeses are produced, 60% vs. 27% in France. Further, in Italy 20% of the cheeses have a hard texture while no cheeses have a hard texture in France. Similar values are reported for the hard cheeses in the Mediterranean countries (18%), but as we mentioned before they appear to be related to dehydration or pickled process.

Biodiversity factors of traditional cheeses: the case of Ragusano cheese

The major components of biodiversity expressions that give a specific identity to the products are: the production area; the geographic environment with the natural resources; the production locations and also the structural and architectural aspects; the human factors with emphasis on the role of women; the farming systems and breeds; the quality of raw milk and the traditional technologies of cheese making and aging; the preparation of the finished product responsible for the relative shelf life; the uses and customs tied to the relationship with the market and consumer.

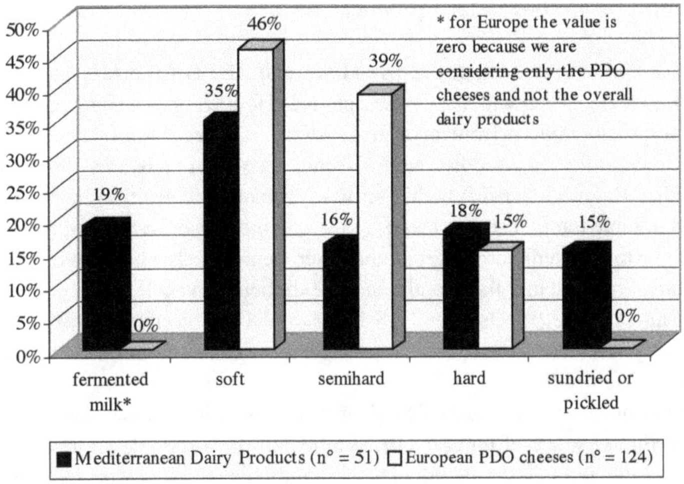

Figure 2. Typology of traditional Mediterranean dairy products and European PDO cheeses.

Production area

Traditional cheese is deeply tied to the territory of origin from which it gets its most important peculiar characteristics tied to the environment, the weather, the field arrangement, the farms ambient for cheese making and aging the cheeses, the landscape, the variety of native pasture and breed and all of the cultural and archaic aspects that are done for the historic cheeses.

High quality raw milk

A microbiologically healthy milk is rich in pro-dairy bacteria, due also to the elevated number of species present, which allows for in improvement in the processing, being indispensable in the principle enzymatic processes that contribute to acidification of the milk and/or of the curd and define the aromatic, lipolytic and proteolytic profiles of dairy products.

Man, their values and cultural journey on traditional cheese making

Every productive system is characterized by a sequence of biologic events, each being marked by its own natural rhythms. Man, who must manage this series of biologic processes, support them, but at the same time, control them to give shape to the milk and to obtain a high quality product, an expression of nature. Technology is tied to the processing location, the native micro flora, and the tools, still wooden or of copper, where man's imprint, his culture and his traditions still have an inestimable value (Licitra *et al.*, 1998). The wooden tools historically have been considered essential elements in making good cheese, bringers of the native pro-dairy micro flora characterized by the local environment and therefore, as an expression of itself (Corallo *et al.*, 2002). The porous structure of wood, particularly in the coagulation phase of the vat, improve the settlement of niche bacteria communities, biofilm, that are left in less than 10 minutes, from the wood to the milk, allowing acidification of the milk and then maturation of the curd in the making of Ragusano cheese.

Animal nutrition and healthy properties of dairy products

The goal is to show that with traditional production systems and with animals free to graze on pasture, the level of the main health components of traditional cheeses, produced following the natural process is significantly higher with respect to products from innovative systems. We use the Ragusano P.D.O. cheese as an example. The availability of countless natural forage essences in pastures of Iblean territory are crucial for the final Ragusano quality, both in terms of aromatic and health components (Carpino *et al.*, 2004b). Of notable interest are the results relative to the properties of conjugated linoleic acid, which was shown to have anti-carcinogenic, anti-atherogenic effects, alterations on the distribution of nutrients and lipid metabolism, and can cure diabetes (type II) by reducing hyperglycemia. In Ragusano cheese, the isomer C18:1, t11 and CLA content, 29.95 and 9.38 mg FFA/g fat respectively, has resulted by twice comparing it to a significant sample of Italian commercial cheeses (Banni & Martin, 1998).

A preliminary screening, done on several historical Sicilian cheeses at different ages, made with animal on pasture showed the presence of natural antioxidants, all trans retinal (vitamin A), α-tocopherole (vitamin E), squalene and β-carotene, at levels slightly higher with respect to other Italian and foreign cheeses. β-carotene is present in Ragusano cheese with values of about 450 μg/100g (Rosato *et al.*, 2003). With regard to the aromatic markers, it has been shown (Carpino *et al.*, 2004a) that Ragusano presents eight unique markers that are linked to the pasture: 2 aldehyde (E,E-2,4-octadienal, dodecanal), 1 alkane (hexadecane), 2 esters [(geranyl acetate, (E)-methyl jasmonate)] and 3 terpens [1-carvone, L(-)carvone, citronellol]. The complexity of the odors of these cheeses is always higher with respect to c.d. total mixed ration cheeses and, therefore, should become an economic opportunity for these products.

Coagulation system

The system of milk coagulation represents one of the main factors of biodiversity. Table 2 and its relative discussion demonstrate the biodiversity of dairy products in the Mediterranean countries. In Europe due to the much lower yield of natural acidification, it is not part anymore of the cheese making process system where, instead, rennet from young animals and vegetable rennet are widespread.

Aging system

Aging takes place in fresh and ventilated areas with geologically natural walls (i.e. where the presence of natural micro flora contributes in a significant way to the best aging of the product). The ancient "caves" have made the history of world wide recognized European cheeses as Roquefort, Cabrales, Castelmagno, Formaggio di Fossa, Gorgonzola, and Ragusano. Just as it is for the cheese making, it is historically "the man" who directs the aging of the cheese, with his attention, examining the state of aging and the individualization of defects.

Conclusion

As long as traditional stock of low milk production potential continues to be the major source of milk for the rural peoples of African-Mediterranean regions, traditional milk processing will continue to be a household activity revolving around natural fermentation of milk, traditional butter churning methods and ghee preparation. To improve the bargaining power of the traditional milk producer processors, products with a longer shelf life such as ghee and aged cheeses should be encouraged.

While improvements to existing household level technologies may be made, the introduction of more efficient milk processing techniques that respect the traditional processes but improve the efficiency and the quality of their production system, will require the setting up of milk processing units at the village (community) level.

The consumption of traditional milk products is increasing due to growth of the population, and to peoples demand and their awareness of the nutritional importance of these products in their diet. There must be a positive increase on production quantity and quality to satisfy the self consumption but also to open commercial opportunities in the urban area and in the international market.

"Economic growth – including that caused by globalization – brings urbanization that endangers rural traditional societies". "Globalization seems to jeopardize traditional values, in short, a threat to cultural values and identity" (Stiglitz, 2001).

The identity of the developing countries is their strength in survival or they will lose their roots and remain economically poor. The development of less favored countries should go through the transformation of the advanced societies, educating them to have greater tolerance and cultural integration, to understand that diversity and biodiversity are valuable and not something to defend oneself from.

Every project should starting from the demands of the base productivity and not bringing down from high pre-packaged project that resolve more problems in developed countries than in those developing.

To reach this goal it is necessary to encourage the cooperation within these countries, working directly in their systems, laying down rules, responsibilities and motivations for the local institutions.

The west can contribute to improving the international context if it chooses to develop projects that give value to the local communities, and their unique institutions.

Our idea is that economic development should occur in these societies without denying their diversities, traditions and roots.

Science and technology within the farming communities much demonstrate that production within natural systems, according to local traditions, is not only a memory, but a journey of natural biologic processes that bring out aromatic, organoleptic and health properties that only these systems are able to guarantee the consumer.

Consumers in the developed countries must be educated in the cultural value of the foods from the developing countries which includes production methods and sensory characteristics of the foods. The consumer of the developed countries, must know how to distinguish, by informing them on the cultural value of food and educated to develop his own critical and sensorial capacity. Biodiversity is a point of excellence of Mediterranean populations and traditional cheese products represent an excellent social-economic resource beyond just an opportunity to reinforce the details of different ethnicities.

Globalization impose a concentration and standardization of production, a lowering cost of production, which looks to produce enormous quantities and standardized qualities to allow an easy positioning in the market at a low price. In theory it was to feed the world, the same motivation for producing GMOs, but the fact is, it is only speculation for large holding companies who have patented various native produce, creating financial dependence on the native populations. Monopoly of the market based on competition, prices and service, and not on quality, even less on respecting the environment and the farm communities.

Development asks for a social-cultural transformation of the society. The needs must start from the bottom. Shared prosperity will set out across a social, cultural and human partnership, in order to develop human resources, to favour the understanding among cultures and the exchanges among civil societies.

One will understand how difficult it is, but if real development is desired in the less fortunate countries, the objective must be changed from economic to social.

It is only right to end with a voice of hope, because after Seattle, the W.T.O. was forced, by the less developed nations that finally came together, to put in their agenda, not only commercial aspects, but respect for the environment, eco-sustainability, and the ban on the GMO. Edgar Morin in "Mother Country" (Morin & Kern, 1994) and "The Human Identity" (Morin, 2002), argues that we live in a world in which "the destiny community" is evident and presents his idea of "Mother Country". It deals with going towards a universal society based on the genius of diversity and not on the lack of genius in homogeneity. A world that cannot be governed by the interests of a few and Morin even suggests the necessity of a "planetary government", if we want to truly govern human destiny in an equal and joint way.

References

Banni, S. & J.C. Martin, 1998. Conjugated linoleic acid and metabolites. In: Trans fatty acids in human nutrition, J. J. Sebedio and W.W. Christie (Editors), Oily Press, Dundee, Scotland, pp. 261-302.

Carpino, S., S. Mallia, S. La Terra, C. Melilli, G. Licitra, T.E. Acree, D.M. Barbano & P.J. van Soest, 2004a. Composition and aroma compounds of Ragusano cheese: Native pasture and total mixed rations. J. Dairy Sci. 87: 816-830.

Carpino, S., S. Mallia, G. Licitra, P.J. Van Soest & T.E. Acree, 2004b. Aroma compounds of some Hyblean pasture species. Flavour Fragr. J. 19: 293-297.

Corallo, L., R. Gelsomino, P.S. Cocconcelli, P. Campo, S. Carpino & G. Licitra, 2002. Molecular characterisation of lactic acid bacteria of Ragusano cheese. Abstracts. J. Dairy Sci. Vol. 85, Suppl.1. J.W. Fuquay (Editor), American Dairy Science Association, Savoy, Illinois, pp. 330.

FAO, 1990. The technology of traditional milk products in developing countries. Part A. FAO animal production and health paper 85. Rome, Italy.

Fox, P.F. & P.L.H. McSweeney, 2004. Cheese an overview. In: Cheese: Chemistry, physics and microbiology, P.F. Fox, P.L.H. McSweeney, T.M. Cogan & T.P. Guinee (Editors), 3rd Ed., Vol. I, Elsevier Academic Press, Amsterdam, pp. 1-18.

Licitra, G., G. Portelli, P. Campo, G. Longombardo, G. Farina, S. Carpino & D.M. Barbano, 1998. Technology to produce Ragusano cheese: a survey. J. Dairy Sci. 81: 3343-3349.

Montanari, M., 2002. Latte e Formaggi nel Medioevo. Parma economica 4: 9-20.

Morin, E, 2002. Il metodo 5. L'identità umana. Trans. Susanna Lazzari. Raffaello Cortina Editore, Milan, pp. 291.

Morin, E. & A.B. Kern, 1994. Terra-Patria. Trans. Susanna Lazzari. Raffaello Cortina Editore, Milan, pp. 194.

Rosato, M., S. La Terra, G. Di Rosa, J. Horne, S. Carpino & G. Licitra, 2003. CLA and natural antioxidants in historical Sicilian cheeses. Proceedings Cheese Art 2002. CoRFiLaC Editor.

Stiglitz, J. E., 2002. The globalization and its discontents. W.W. Norton & Co., New York, pp. 282.

Evolution of production system of Iberian pigs in Spain

A. Daza[1], J. Ruiz[2], A.I. Rey[3], A. Olivares[3] & C. López Bote[3]

[1]*Departamento de Producción Animal, Universidad Politécnica de Madrid*
[2]*Departamento de Zootecnia, Universidad de Extremadura. 10071 Cáceres*
[3]*Departamento de Producción Animal, Universidad Complutense de Madrid*

Summary

Total number of Iberian swine sows in Spain is close to 230 000, giving rise to around 2 millions finished Iberian pigs in 2003. The Iberian swine constitutes a singular breed, strongly adapted to the ecological conditions of the "Dehesa" ecosystem. This breed includes a number of strains, all of them showing low prolific and growth rates, very early maturing fat deposition, and an excellent meat quality, which is valuated in the market according to the fatty acid profile (palmitic, stearic, oleic and linoleic acids) of subcutaneous fat. This paper mainly deals with available feeding sources of the "Dehesa" for Iberian pigs, the productive features of this breed and the evolution that feeding and production systems have experienced last years.

Keywords: "Dehesa" ecosystem, feeding systems, production systems, quality, performances.

Introduction

Around 2 million Iberian pigs were slaughtered in Spain in 2003 (MAPA, 2004), from which about 15% were fattened under free-range conditions, in which is called "montanera" (fed on acorns and grass in the "Dehesa"), and the other 85% were reared in confinement and fed on mixed diets. Based on these figures, the number of Iberian sows can be estimated in nearly 230 000, which means an increase in the number of reproductive animals of more than 100% compared to the year in which Spain joined the EU (1986). The Iberian swine sector includes a national association of farmers (AECERIBER) and three Protected Designations of Origin (Dehesa de Extremadura, Guijuelo and Jamón de Huelva) which control the trade of high quality meat products (dry-cured hams and forelegs) elaborated under strict norms of animal production and industrial manufacture. Traditionally, carcasses of Iberian pigs were paid to farmers as a function of the genetic features of the animal (pure Iberian, crossbreeding of ¾ Iberian with ¼ Duroc, crossbreeding of ½ Iberian with ½ Duroc) and feeding system during the fattening, from 100 to 160kg of live weight, including three different systems:

1. "montanera", in which animals are fed outdoors and consume exclusively acorns and grass;
2. "recebo", (half the time in "montanera" and half time fed on mixed diets); and
3. fed on mixed diets in confinement.

There is nowadays a Quality Norm for the production of Iberian hams, Iberian forelegs and Iberian loins (BOE, 2004) establishing that it is compulsory the use of pure Iberian sows if the meat products are commercialized as ibérico. In addition, such Quality Norm establishes the existence of products of different qualities depending upon the feeding background of the animals ("montanera", "recebo" and on mixed diets in confinement). Moreover, since a few years ago, the payment in the trade of slaughtered Iberian pigs considers the fatty acid profile of subcutaneous fat, the range for each fatty acid within each commercial class is officially published each year.

The "Dehesa" ecosystem and the Iberian pig

The "Dehesa" located an agroforestry system covering more than 9 millions ha located the Southwest and the Centre of the Iberian Peninsula. The available feeding sources for Iberian pigs in this ecosystem are the grass (from October to June), by-products of cereal crops and horticulture during summertime, non harvested figs and olives and mainly the acorns from trees of the genus *Quercus*, which are available from November to February. It is estimated that Spain has 2.9 millions ha of evergreen oaks and 0.4 million ha of cork trees, of which 2.1 and 0.3 millions ha respectively are forming part of the "Dehesa" ecosystem (MAPA, 2001). Of all this area, the surface used for Iberian pig production only represents around 450 000 ha, the rest being used for bovine and ovine production, for hunting, etc.

Acorn production is highly variable, depending upon the year, the healthiness of the tree, its age, the tree management, the variety of tree and the tree itself (Vázquez, 1998). The "Dehesas" containing trees show an average density between 20 and 40 trees per ha, and the average annual production per tree is close to 15kg, although in some cases the density is much higher (80-100 trees per ha) (Esparrágo *et al.*, 1994). The whole acorn (shell, skin and endosperm) is not totally used by Iberian pigs. The shell and the skin, which constitute around 21-23% of the total weight of the acorn, are removed by the animal while ingesting. The endosperm of the acorn from evergreen oaks shows a 67% of dry matter (81.5% of nitrogen free extractives, 4.7% of crude protein, 6.3% of fat, 5.7% of crude fibre and 1.75% of ash on a dry matter basis) and is very rich in oleic acid (C18:1 n-9) (66% of total fatty acids), linoleic acid (C18:2 n-6) (14.7%), palmitic acid (C16:0) (12.6%) and stearic acid (C18:0) (3.2%) (Rey *et al.*, 1997), although its chemical composition is variable depending on the year and season. According to a recent experiment in which data from three consecutive years were recorded, protein, fibre, fat and oleic acid content of acorns increase during the months the "montanera" takes place (from October to February), while the stearic acid content decreases and that of the linoleic acid kept constant during such period of time (López Carrasco *et al.*, 2005). The grass in autumn and winter, together with the acorns, are the main natural feeding sources on which meat quality of Iberian pig is mostly sustained. Grass consumption is a very important source of protein (137g/kg dry matter), linolenic acid (C18:3 n-3) (450 g/kg of total fatty acids) and α-tocopherol (171 mg/kg dry matter) (Rey *et al.*,1997).

It is accepted that the Iberian pig is a single breed, but it includes a number of lines or strains with either red or black skin ("Negra Entrepelada", "Negra Lampiña", "Valdesequera", "Retintas", "Torbiscal", "Alentejano", etc.), all of them being perfectly adapted to the ecological environment of the "Dehesa". The average adult live weight of boars and sows has considerably increased in recent years, being nowadays around 140-170 kg for females and 180-190 kg for males. The first mating for sows usually takes place when they are 9-10 months old with 80-90 kg in weight, while males reach puberty at 10-11 months of age. The productive life for sows in the traditional system lasts around 6 to 8 parities, while that of males is around 2-3 years. Iberian breed is not highly prolific (7-9 piglets per farrowing), the litter size increasing with the farrowing number, reaching the maximum between the 5th and the 8th one (Vázquez *et al.*, 1994). There are several factors that may influence this variable, such as season (i.e., temperature) or feeding availability. Benito *et al.* (1992) found the highest prolificity in summer births, while Dobao *et al.* (1983) in winter. Suárez *et al.* (2002) observed that the strains "Manchado de Jabugo" (whose origin is the crossbreeding of Iberian breed with British breeds) and "Torbiscal" showed the highest prolificity, followed by the "Entrepelado" and

"Retinto" ones, the "Lampiño" showing the poorest reproductive features. Crossbreeding between different Iberian breed strains improves the prolificity in around 0.5 piglets per litter (Silió *et al.*, 2001) and crossbreeding with the Duroc breed in 2-3 piglets (López-Bote, 1998).

Feeding systems: quality effects

Growth and development of Iberian pig includes different periods, named "cria", "recria", "precebo" and fattening or finishing.

During the "cria" phase (lactation and weaning) the piglet is nursed by the sow from birth (1.2-1.5 kg) to weaning, which is carried out between 28 days (6-7 kg) and 56 days (13-15 kg), depending upon the production system. From weaning to 90 days old (23-25 kg), piglets are usually fed in confinement, with mixed diets containing approximately 3 100-3 200 kcal ME/kg, 18-19% crude protein and 1-1.2% lysine. During "recria" (23 to 60 kg) and "precebo" (60 to 100 kg) previously castrated pigs are reared and fed in different ways depending on the system in which they are going to be fattened (outdoors in "montanera" or fed on mixed diets in confinement). Pigs intended for "montanera" are allowed to carry out physical exercise in farmyards or even using the available feeding sources of the "dehesa", and are daily supplemented with 1.5-2.0 kg of mixed diets with 2 800-2 900 kcal ME/kg and 2.0-2.2% lysine per Mcal of ME (López Bote, 2001), while those animals intended for being finished indoors on mixed diets, are usually fed in confinement and on a higher daily nutritive level, even *ad libitum*. The feeding level during the "recria" and "precebo" phases affects carcass quality and fatty acid profile of subcutaneous fat, and also to the content of tocopherols in muscle. Thus, Daza *et al.* (2005a) observed a compensatory growth during the finishing outdoors in "montanera", which mainly comprised fat, and higher concentrations of oleic and linoleic acids and of α- and γ-tocopherol in those pigs subjected to lower levels of feeding during "precebo". The recommended feeding strategy for Iberian pigs during the growing phases, despite the animals are aimed for being finished outdoors in "montanera" or in confinement, is a low feeding level, with mixed diets containing low fat, low linoleic and saturated fatty acid concentrations and a moderate high crude fibre levels, contributing to the development of the digestive system (López Bote, 2001).

The finishing or fattening phase (100-160 kg) is carried out either outdoors in "montanera" or in confinement with mixed diets. Genetic factors, and other aspects such as sex, age at the beginning of the fattening, age and weight at slaughter, environmental conditions, and so on, have marked effects on productive results of this phase and on carcass quality. According to results shown in Table 1, the "Torbiscal" strain shows the best carcass quality characteristics, while the "Negra Lampiña" one shows the worst, although the meat from this latter strain possesses an extraordinary quality, mainly as a consequence of its high intramuscular fat content and distribution (Benito *et al.*, 1998; Muriel *et al.*, 2004).

In production systems for growing-fattening in confinement, the crossbreeding of Iberian pigs with the Duroc breed improves average daily gain, feed conversion and carcass quality (Serrano *et al.*, 2005a b), although meat quality may be adversely affected. Crossbreeding of Duroc with the "Torbiscal" strain improve carcass leaneness, hams, forelegs and loins yield. The average daily gain during the finishing phase is not affected by Duroc crossbreeding (Dobao *et al.*, 1987). The acceptable productive features (average daily gain and feed conversion) and good carcass and meat quality characteristics of the "Torbiscal" strain makes this type of animals one of the most valorated genetic line for boars in Iberian productive systems, as an alternative of the Duroc breed.

The feeding of Iberian pigs during the fattening phase is the most important factor affecting carcass, meat and fat quality. Pigs finished outdoors in "montanera" and exclusively fed on acorns and grass, show significantly poorer carcass, hams, forelegs and loins yields than animals fed on mixed diets,

Table 1. Carcass characteristics in Iberian pig varieties.

Variable	Varieties					
	LAMP	RE-M	RE-S	RET	ENTRE	TORBIS
Carcass weight (kg)	159.4	134.1	136.4	139.4	143.2	157.5
Carcass (%)	86.9	83.8	81.5	83.6	83.8	82.7
Hams (kg)	22.4	20.5	21.3	21.1	21.3	25.0
Hams (%) (1)	14.0	15.0	16.0	15.0	15.0	16.0
Forelegs (kg)	13.6	13.9	14.2	14.1	14.2	16.3
Forelegs (%) (1)	9.0	10.0	10.0	10.0	10.0	10.0

LAMP = Lampiño; RE-M = Retinto mamellado; RE-S = Retinto Silvela; RET = Retinto; ENTRE = Entrepelado; TORBIS = Torbiscal.
Source: Barba *et al.* (2000).

Table 2. Carcass characteristics and major fatty acids percentage of subcutaneous backfat outer layer according feeding system.

Variable	Feeding system	
	Free- range	Formulated diet[1]
Carcass weight (kg)	118.9^a	130.0^b
Carcass (%)	78.4^a	82.0^b
Left longissimus dorsi muscle (kg)	1.7^a	2.1^b
Psoas major muscle (kg)	0.5	0.5
Ham 1 (kg)	10.4^a	11.8^b
Ham 2 (kg)	10.4^a	11.9^b
Foreleg 1 (kg)	7.0^a	7.6^b
Foreleg 2 (kg)	7.0^a	7.6^b
Longissimus dorsi area (mm^2) (2)	1976.6	2056.5
Backfat area (mm^2) (2)	5210.5^a	4207.6^b
Backfat thickness (mm) (2)	59.3^a	46.3^b
C16: 0 (palmitic acid) (%)	18.9^a	20.2^b
C18: 0 (stearic acid) (%)	8.7^a	9.8^b
C18: 1 (n-9) (oleic acid) (%)	52.6^a	46.9^b
C18: 2 (n-6) (linoleic acid) (%)	10.3^a	12.5^b
C18: 3 (n-3) (linolenic acid) (%)	0.6^a	0.4^b

[1]In confinement (2) at site of the last rib. Different superscripts in columns means they differed $P<0.05$.
Source: Lopez - Carrasco *et al* (2003).

but the fatty acid profile is most appropriate from a technological and commercial point of view (higher concentration of oleic acid and lower of palmitic, stearic and linoleic acids; Table 2) and the muscle tocopherol content is higher (Daza *et al.*, 2005). The weight gained as a consequence of acorn and grass feeding, that is, the length of the outdoor rearing in "montanera", also influences the fatty acid profile of subcutaneous fat, and particularly that of oleic acid (Montero de Espinosa *et al.*, 1992), although more scientific knowledge is needed on this topic in order to increase the number of animals yearly fed outdoors in "montanera", allowing optimization of natural resources.

As previously explained, around 85% of Iberian pigs slaughtered in Spain are finished in confinement and fed mixed diets. Three general orientations of feeding practices in the last 50 years can be differentiated: traditional feeding, fake feeding and promising feeding.

Traditional feeding was based in the use of mixed diets rich in carbohydrates (barley and wheat) and poor in fat, which leaded to a high consistency in pig fats, which as a consequence of feeding composition were rich in saturated fatty acids, showing melting points above 30ºC. This latter criterion was used as a method for identifying the feeding background (either "montanera" or in confinement) after slaughter. Subsequently, since last decade of the 20's, mixed diets were enriched in unsaturated fats in order to avoid endogenous synthesis and achieve lower melting points, similar (and even lower) to those of fat from pigs fed outdoors in "montanera" on acorns and grass (26-27ºC). Feeding sources rich in linoleic acid were frequently used in feed formulation, such as corn, barley, wheat and soya bean. This feeding strategy, which is fake in its purpose, led to an increase in the susceptibility of the meat to lipid oxidation, to an enlargement in the ripening process of hams and forelegs and to a marked decrease in the sensory quality of meat products (López Bote, 2001).

In recent years there is a trend to include ingredients rich in oleic acid in mixed diets for Iberian pig fattening (lard, oleins, high oleic sunflower oil, peanuts, rapeseed oil, etc.), with the purpose of achieving an adequate fatty acid profile and a meat sensory quality similar to that of meat from pigs fed in "montanera". Furthermore, in order to substitute the antioxidant effect of tocopherols form acorns (γ-tocopherol) and grass (α-tocopherol) found in the meat of pigs fed outdoors in "montanera", mixed diets used for feeding Iberian pigs in confinement are supplemented with 100-200 mg/kg of vitamin E (α- tocopheryl acetate). In different experiments in which linoleic acid was replaced by oleic acid in the mixed diets without changing the saturated fatty and the linolenic acid contents, and feeding was supplemented with supranutritional levels of vitamin E, an increase in oleic acid and a decrease in linoleic acid in subcutaneous and intramuscular fat, together with a decrease in the susceptibility of meat to lipid oxidation (due to an increase in muscle and microsome α-tocopherol), a lower Σ(n-6)/Σ(n-3) fatty acids ratio, and an improve in some rheological properties of subcutaneous fat, such as consistency, elasticity, cohesiveness and adhesiveness (López Bote *et al.*, 2002; López Bote *et al.*, 2003; Daza *et al.*, 2005b) was achieved. As expected, the influence of vitamin E supplementation in these experiments led to a reduction in the susceptibility of pigs tissues to lipid oxidation. It has been also observed an interaction between the level of vitamin E in the feeding and the ratio oleic/linoleic acid, so that the lower the vitamin E inclusion in the feeding and the lower the MUFA/PUFA ratio, the lower the muscle microsomes α-tocopherol concentration, leading to an increase in the prone of the meat to get oxidized. (López Bote *et al.*, 2003)

Supplementation of feeding with copper may decrease the proportion of saturated fatty acids and increase that of MUFA. In fact, supplementation with only 35mg of copper per kg of feed was enough to modify the fatty acid profile of hepatic tissue from Iberian pigs fed in confinement with mixed diets. The concentration of lauric (C14:0), palmitic (C16:0) and margaric (C17:0) acids in neutral lipids decreased and that of oleic acid and unsaturation index increased, although the latter two effects were not significant (Lopez Bote and Rey, 2001).

These feeding strategies are promising for the Iberian pig sector as a whole, since are aimed not only to improve the profitability of farmers, but also the technological quality (industrials) and the sensory, nutritional and health features of the product (consumers)

Production systems

The use of biological (both animal and vegetal), labor, technical and financial resources have experienced notable changes in the last 50 years. The traditional free range system for rearing Iberian pigs, which has nowadays almost disappeared, was characterized by:

- The use of pure Iberian breed.

- Reproductive plans involve farrowing during May-June and November-December for adult sows, and during February-March for gilts born in November-December. Non-pregnant sows mated in summer, and that consequently will not deliver in November-December, will mate again during autumn, together with the gilts. Those sows delivering in February-March will mate again in May-June.
- Weaning of piglets at 8-9 weeks of age.
- Feeding of reproductive animals based in the available sources of the "Dehesa" and supplemented with mixed diets during the last month of pregnancy and lactation. According to Benito *et al.*, (1986), the free range feeding of Iberian sows in the "Dehesa" allows in saving close to 30% of energetic needs for gestation and 20% for lactation.
- Birth in open pens, which led to a high rate of mortality in piglets due to crushing.
- Free rearing of pigs during the "recria" and "recebo" phases, in farmyards with feeding supplementation (1.5-2.5 kg/day) and finishing in "montanera", with an animal density of 0.6-0.7 pigs/ha. Also "recria" and "recebo" phases in montanera and finishing with mixed diets.

Such a traditional productive model gives rise to meat and meat products with the highest sensory quality.

The free range system was subsequently changed into a confinement one in different extent, through the design of pens and buildings similar to those for selected pig genotypes, thus increasing of the productive rates, the reduction of the weaning age, the use of crossbreed Duroc x Iberian or pure Duroc sows (with better prolific and productive features) (Table 3).

At the moment, the Official Quality Norm establishes (despite the reluctance of some important members of the Iberian swine industry, such as dry cured ham processors) that the sows should be pure Iberian breed, allowing the boar to be either pure Duroc or crossbreed Duroc x Iberian. Nevertheless, some farmers claim that prohibiting the use of pure Duroc or Duroc x Iberian sows

Table 3. Performances of the production systems of Iberian pig.

Genetic type used	Extensive Iberian (1) x Iberian (2)	Semi extensive Iberian (1) x (Duroc x Iberian) (2)	Intensive Iberian (1) x Duroc (2)
Farrowings/ sow/year	1.7-1.9	2.0-2.2	2.37
Sow effective life (3)	7	6	5
Born piglets	7-8	9-10	11-12
Liveborn piglets	6-7	8-9	10-11
Piglets mortality birth-weaning (%)	10	9	10
Weaning age (days)	50-60	40-50	28
Weaned piglets/sow/year	5.9	7.7	9.4
Postweaning mortality (%)	2	3	3
Fattening mortality (%)	1	2	2
Produced pigs/sow/year	10.3	15.5	21.2
Feed conversion kg/kg (postweaning period)	2.7 (14-23 kg) (55-90 days)	2.4 (11-23 kg) (42-90 days)	2.0 (7-23 kg) (28-90 days)
Feed conversion kg/kg (fattening period) (23-150 kg)	5.0	4.4	4.0
Slaughter weight (kg)	150-170	150-170	150-170

(1) male (2) female (3) number of farrowings per sow.
Source: Daza, 2001.

could be an economical drawback for Iberian pig farms, due to a significant decrease in productive traits (Table 3).

The future of Iberian swine sector is nevertheless promising. The extraordinarily social prestige of Iberian meat products, the progressive increase of consumers income (which stimulates consume of highly rated products), the notable effort carried out in the organisation of the productive sector and in the homogenization, typifying, traceability and hygienic trustfulness of the derived meat products, and the exportation to countries of the European Union, and to America and Japan, could be exponents of such future.

References

BOE, 2004. Real Decreto 1781/2004 de 30 de julio.

Barba, C., J.V. Delgado, E. Dieguez, & P. Cañuelo, 2000. Caracterización productiva de las variedades del cerdo ibérico basada en el comportamiento en matadero. Archivos de zootecnia. 50: 125-132.

Benito, J., A. Fallola, G. López de Torre & J.L. Ferrera, 1986. La línea Valdesequera de cerdo ibérico. In: Conservación y desarrollo de las dehesas portuguesa y española, P. Campos and M. Martín (Editors), MAPA, Spain, 189-218.

Benito, J., C. Vázquez, C. Menaya, A. Fallola & J.L. Ferrera, 1992. Estudio sobre la estacionalidad de las producciones del cerdo ibérico. Jornadas Técnicas sobre Obtención de Productos Ganaderos Naturales en el Ecosistema de la Dehesa. Feria Internacional Ganadera del Quinto Centenario, Zafra, Badajoz, Spain. Tomo II: 70-77.

Benito, J., C. Vázquez, C. Menaya, J.L. Ferrera, J.M. García-Casco, L. Silió, J. Rodrigañez & M.C Rodríguez, 1998. Evolución de los parámetros productivos en distintas líneas de cerdo ibérico. IV Sympósio Internacional do Porco Mediterránico. Évora. Portugal. 26-28 November.

Daza, A., 2001. Sistemas de explotación. In: Porcino ibérico: aspectos claves, C. Buxadé & A. Daza (Editors), Ed Mundi Prensa Libros SA Madrid, Spain, 151-175.

Daza, A., A. Mateos, A.I. Rey & C.J. López Bote, 2005a. Feeding level in the period previous to the late fattening phase influences fat composition at slaughter in free- ranged Iberian pigs. Arch. Anim. Nutr. (in press).

Daza, A., A.I. Rey., B. Isabel & C.J. López Bote, 2005b. Effect of dietary vitamin E and partial replacement of poly- with monounsaturated fat on fatty acid patterns of backfat and intramuscular fat in heavy pigs (Iberian x Duroc). Journal of Animal Physiology and Animal Nutrition. 89: 20-28.

Dobao, M.T., J. Rodrigáñez & L. Silió, 1983. Seasonal influence on fecundity and litter performance characteristics in Iberian pigs. Livest. Prod. Sci. 10: 601-610.

Dobao, M.T., J. Rodrigáñez., L. Silió., M.A. Toro & E. De Pedro, 1987. Crecimiento y características de la canal de cerdos ibéricos, duroc x ibérico y jiaxing x ibérico. Investigación Agraria. Producción y Sanidad Animales. 2 (1): 9-24.

Espárrago, F., F. Cabeza de Vaca., F. Brito & M. Burzaco, 1994. La producción de porcino ibérico. In: La agricultura y ganadería extremeñas en 1993. Ed Caja de Badajoz, 229-240.

López Bote, C.J., 1998. Sustained utilization of the Iberian pig breed, Meat. Sci. 49: S17-S27.

López Bote, C.J., 2001. Alimentación del cerdo ibérico con piensos compuestos. In: Porcino ibérico: aspectos claves, C. Buxadé & A. Daza (Editors). Ed Mundi Prensa Libros SA, Madrid, Spain, 247- 272.

López Bote, C.J. & Rey, A.I, 2001. Susceptibility of hepatic tissue of Iberian pigs is enhanced by free-range feeding and reduced by vitamin E supplementation. Nutr. Res. 21, 541-549

López Bote, C.J., B. Isabel & A. Daza, 2002. Partial replacement of poly- with monounsaturated fatty acids and vitamin E supplementation in pigs diets: effects on fatty acid composition of subcutaneous and intramuscular fat and on fat and lean firmness. Anim . Sci, 75: 348-358.

López Bote, C.J., B. Isabel., J. Ruiz & A. Daza, 2003. Effect of vitamin E supplementation and partial substitution of poly- with monounsaturated fatty acids in pigs diets on muscle, and microsome extract α- tocopherol concentration and lipid oxidation. Arch. Anim. Nutr. 57: 11-25.

López Carrasco, C., J.M. Contreras., A. Daza., A.I. Rey & C.J. López Bote, 2003. Fatty acid composition and carcass characteristics of Iberian pigs fed in either free-range or in confinement with acorns and grass or a formulated diet. EAAP Congress, Rome, September, 2003

López Carrasco, C., T. Muñoz de Luna., A. Daza., A. Rey & C.J. López Bote, 2005. Variaciones inter e intraanuales de la calidad de bellotas de encina en una dehesa de Castilla-La Mancha. XLV Reunión Científica de la SEEP, Villaviciosa (Asturias) Mayo de 2005 (In press).

MAPA, 2001, Anuario de Estadística Agroalimentaria.

MAPA, 2004. Boletín mensual de Estadística Agraria, 6.

Montero de Espinosa, V., E. Osorio., E. Sabio & J. Benito, 1992. Efecto de diferentes reposiciones en montanera en la composición en ácidos grasos del tejido adiposo subcutáneo de jamones frescos de cerdos ibéricos. Jornadas Técnicas sobre Obtención de Productos Ganaderos Naturales en el Ecosistema de la Dehesa. Feria Internacional del Quinto Centenario. Zafra, Badajoz, Spain, Tomo II: 100-105.

Muriel, E., J. Ruiz, J. Ventanas, M.J. Petron & T. Antequera (2004) Meat quality characteristics in different lines of Iberian pigs. Meta Sci., 67: 299-307.

Rey, A.I., C.J. López Bote & Sanz Arias, R., 1997. Effect of extensive feeding on α- tocopherol concentration and oxidative stability of muscle microsomes from Iberian pigs. Anim. Sci. 65: 515-520.

Serrano, M.P., D.G. Valencia., A. Fuentetaja., R. Lázaro & G.G. Mateos, 2005a. Efecto del sexo, el peso al sacrificio y la línea paterna sobre la productividad de cruces retinto ibérico x duroc en sistemas intensivos. ITEA, Vol Extra, 26, Tomo I: 264-266.

Serrano, M.P., D.G Valencia., D. del Pico., R. Lázaro & G.G. Mateos, 2005b. Efecto del sexo, el peso al sacrificio y la línea paterna sobre la calidad de la canal de cerdos ibéricos x duroc destinados a la industria de productos curados, ITEA, Vol Extra, 26 Tomo II: 828-830.

Silió, L., C. Rodríguez., J. Rodrigáñez & M.A Toro, 2001. La selección de cerdos ibéricos. In: Porcino ibérico: Aspectos claves, C Buxadé & A. Daza (Editors). Ed Mundi Prensa Libros SA, Madrid, Spain, 124-149.

Suárez, M.V., C. Barba., J. Forero., J.R.B. Sereno., E. Diéguez & J.V. Delgado, 2002. Caracterización reproductiva de varias razas porcinas de origen ibérico. I Análisis descriptivo. Archivos de Zootecnia. 51:245-248.

Vázquez, C., C. Menaya., J. Benito., J.L. Ferrera & J.M. García Casco, 1994. The influence of parity and season on the prolificacy and maternal ability of Iberian sows. Pig News and Information. 15 (4): 121- 124.

Vázquez, F. M., 1998. Semillas de Quercus: biología, ecología y manejo. Ed Secretaría General Técnica. Consejería de Agricultura y Comercio. Junta de Extremadura, pp. 234.

The role of pasture improvement in the rehabilitation of the "montado/dehesa" system and in developing its traditional products

D.G. Crespo

Fertiprado, 7450-250, Vaiamonte, Portugal

Summary

The "montado/dehesa", an old agro-sylvo-pastoral system, which predominates in the south-western semi-arid zone of the Iberian Peninsula, is formed by a beautifull open forest of cork and/or holm oaks, the soil being used by a long rotation of cereals and fallow natural pastures, which tend to be invaded by shrubs. Although the system is not highly productive on a per hectare basis, it produces a diversity of valuable traditional products, and provides feed and shelter for such a great number of domestic and wild animals, that it is considered one of the most interesting ecosystems in the world. However, the system is endangered by recent technological, economic and social changes which affected the rural society, particularly during the last 50 years. As a result large areas are being degraded, some have already disappeared, being urgent to find a sustainable solution to preserve the system. As most of degradation is associated with the abandon of cereal cultivation, accompanied by a shrub invasion of the natural pastures, and the reduction of feed resources for animal production, a sustainable solution to rehabilitate the system could rely on pasture improvement. A model of low cost permanent pasture improvement has been developed and widely demonstrated in Portugal and consists of establishing biodiverse legume-grass mixtures, the seeds of legumes being previously inoculated with specific *Rhizobia*, and the soil regularly fertilized with the missing nutrients, particularly phosphorus. These permanent legume-rich pastures have a great effect on soil fertility and on the carrying capacity of the land, which results in a considerable increase of the production of traditional products. At the same time, they contribute to arrest shrub invasion and facilitate the management of the system, thus reducing the costs of maintenance and increasing the income of the system.

Keywords: "montado/dehesa", traditional products, pastures, legumes, biodiversity.

Introduction: definition, origin and characteristics of the "montado/dehesa"

The "montado/dehesa" is a Mediterranean savana-like agro-sylvo-pastoral system which predominates in the semi-arid areas of the Southwestern part of the Iberian Peninsula, although with some similitudes in other areas of the Mediterranean basin, particularly in North Africa (Northern parts of Morocco, Algeria and Tunisia), Southern Italy (specially the "pascolo arbolato" in Sardinia), and in some niches of French and Spanish Catalonia. The system consists of an open forest of cork oak (*Quercus suber*) and/or holm oak (*Q. rotundifolia*) with a number of adult trees per hectare varying from 20 (or less, in degraded areas) to 60 (or more in young tree populations), the key number for a good "montado" being circa 40 adult trees/ha. The trees, which provide shelter, shade and feed, are pruned periodically at intervals of 6 to 12 years, to increase acorns and cork production.

The soil, in general acid and shalow, tends to be covered by a shrubby and herbaceous vegetation but traditionally it is also cropped with cereals (mainly wheat and oats) in a long rotation with natural pastures (Peco, 2002).

The historic origin and evolution of the "montado/dehesa" seems to go back to the times of the Roman Empire and its domination in Iberia. Indeed, the Romans brought to this territory three main systems of land use: *ager*, cultivated land with annual crops (cereals and pulses); *silva*, forest land to supply wood for building, firewood for heating and charcoal for cooking, as well as game for hunting; *saltus*, open spaces in the forest to grow pastures and other feed resources for raising grazing animals (pigs, sheep, goats, cattle, horses).

The civilization which developed after the Romans has converted oak forests in "montado/dehesa", which includes cropping land, thus integrating in the same space the three former systems of land use. This was achieved through the action of man, who selected and pruned the best oak trees, particularly cork and holm oaks, eliminating the less desirable (*Quercus coccifera, Q. faginea, Q. pyrenaica*, etc.) and, at certain intervals, ploughed the land to remove undesirable shrubs (mainly *Cistus* spp., but also, depending on soil/climate conditions, *Cytisus* spp., *Retama* spp., *Lavandula* spp., *Erica* spp., *Myrtus* spp., *Arbutus* spp., *Viburnum* spp., *Ulex* spp., etc.), in order to cultivate cereals and restore the herbaceous natural pasture cover, formed by a great diversity of species of the *Poaeae, Asteraceae, Fabaceae, Geraneaceae, Cruciferae, Plantaginaceae, Umbeliferae, Liliaceae, Ranunculaceae, Borraginaceae,* etc. families.

The system is therefore mainly used for cropping and animal production, as well as to produce cork and firewood. In the past, cork was used only to construct beehives and handicraft goods for various uses, but during the past two centuries it has been progressively valued to produce stoppers for wine botling, and insulating pannels. The production of cork, which is stripped every 9-11 years, is very variable, as it depends on the presence, density, size and state of health of the cork oak trees; some "montado/dehesa" are formed only by holm oaks, and therefore do not produce any cork, but others dominated by adult cork oaks at circa 40 trees/ha, may attain or exceed 3 000 kg of cork/ha/every period of 9 years, which in this case becomes the main economic resource of the system.

Cereals never provided great economic returns, since the respective yields are low (400-1 200 kg/ha) due to the poor conditions of the soil, which is often deficient in organic matter and in some nutrients, particularly phosphorus. Indeed, cereal cropping in the "montado" has been rather regarded as a source of subsistence food (wheat for the local populations) and feed (oat grain, stubbles and straw for livestock) than an important source of income. The main argument in favour of the cultivation of cereals has allways been the role of the ploughed bare fallow in controlling shrubs and in regenerating the herbaceous pastures at regular intervals, which may vary from 6 to 12 years, acording to soil type, tree density, degree of shrub invasion, etc.. As a rule, the year of ploughing to remove the shrubs coincides with that of pruning the trees, so that the cereal crops which follow in the rotation may benefit from more favourable conditions of light and moisture.

Natural pastures and other feed resources (acorns, leaves of trees and shrubs, cereal stubbles) and the associated animal production, have always been a major source of economic return, as it may be deduced from the meaning of the Portuguese name "montado" and the Spanish "dehesa". Indeed, these terms seem to have a convergent meaning, both related with the use of pastures by grazing animals: a space which is defended ("dehesa") from public grazing, those wishing to use it with their animals must pay a tax/head ("montado"). Since long and until 3 or 4 decades ago, it was common to associate the grazing of the "montado/dehesa" with the transhumance of the animals (particularly sheep), which normally were moved to mountain pastures from early Summer to middle Autumn, but today this practice is almost abandoned.

One of the important economic returns of the system is the Iberian grazing pig, which, like the closely related wild boar (both part of the same species, *Sus scrofa*), is able to live all the year from the grazing resources, finding in the autumn/winter period an opportunity to be fattened on acorns produced by holm and cork oaks. The yields of acorns are very variable with plant size and density, and also with climatic conditions of the year, incidence of diseases, particularly insect attacks, mainly those of the genera *Tortrix*, *Lymantria* and *Balaninus* (Montoya Olivier, 1993), etc.. A study carried out in Spanish Extremadura on areas dominated by holm oaks (Garcia *et al.*, 2005) indicates a 3 years average yield variation of the acorns, going from 280 to 540 kg/ha, able to fatten 0.48 to 0.95 pigs/ha from October to January/February. The meat of this pig is highly valued for its quality, particularly in the form of traditional cured meat (smoked hams and various types of sausages).

Other important source of income are sheep and goats, which produce meat (lamb and kid), fiber (wool and skins), and milk, this one entirely transformed in cheese and other traditional co-products. Cattle, generally of local races, is also present in some "montado/dehesa" systems, but traditionally was raised, together with horses, donkeys and mules, mainly as a source of draft power and means of transportation, although today beef cattle (pure and cross-bred) has replaced most of them. Poultry (hens, turkeys, geese and ducks) and honey bees are other traditional sources of income produced on the feed resources of the "montado/dehesa".

Many wild animals are also living in the "montado", and they may constitute another source of income, mainly those valued by hunting activities, such as deer, wild boar, hare, rabbit, partrige, duck, pigeon, dove, thrush, etc.. Even other existing wild animals, such as the mammals lynx (rare), genet, wild cat, badger and fox, and the birds black and white storks, cranes, voltures, eagles, etc., which today call the attention of wild life conservationists and tourists, may contribute to increase the income of the system.

Various edible wild mushrooms (*Agaricus campestre*, *Lepiota procera*, *Boletus* spp., *Marasmius oreades*, *Pleurotus eringii*, etc.), may develop well in the "montado/dehesa" (Montoya Olivier, 1994), particularly in areas of higher soil organic matter, and if properly exploited can provide some significant additional income to the system.

In spite of the great diversity of products, the total carrying capacity of the grazing resources available in the traditional "montado" is relatively low, between 0.1 and 0.5 equivalent cattle unit (ECU)/ha/year (1 ECU=1 head of cattle with 400 kg live weight). Indeed, most of the natural pastures yield between 250 and 1 250 feed units (FU- feed energy contained in 1 kg of barley grain) of a very variable quality, often defficient in protein and rich in fiber, particularly during the dry season.

The reason for the low productivity of the natural pasture resource is multiple, but it is mainly associated with the relatively low contribution of herbaceous legumes to their composition, and the poor performance of legumes in the traditional system. Such an argument seems difficult to be understood, if we consider that the "montado/dehesa" is part of the Region which is the "center of diversity" of Mediterranean pasture and forage legumes (Bennett & Cocks, 1999). However, the situation becomes clear, if we consider the following four main reasons justifying the poor contribution of legumes:

The first reason is related with the generalized low phosphorus (P) content of the soils and the relatively high requirements of legumes for this nutrient. The adaptation to low P content of the existing legume species and ecotypes may also explain their relatively poor response to generous applications of fertilizer P (Crespo *et al.*, 2004).

The second is related with the ploughing phase of the "montado", which is practiced periodically to eliminate shrubs in one year, and to sow cereals (wheat and oats) in the next two. These repeated ploughing operations (traditionally with a moulboard plough) bury the seeds of the annual herbage legumes too deep, thus reducing drastically their chance to survive and dominate.

The third reason relates with traditional grazing management practices. Indeed, the pastures of the "montado/dehesa" are grazed mostly during late autumn, winter and spring, as in summer and early autumn, the animals, if not trashuming to the mountains, preferentially graze on the cereal stubbles. This practice prevents the removal of the dry pasture (standing hay) resulting from the pick of spring growth, which accumulates on the soil, affecting the future germination of the legume seeds and the regular development of the new seedlings.

A final reason justifying the difficults faced by legumes in the natural pastures of the "montado/dehesa" is the tendency of shrubs to become dominant over the herbaceous component, particularly over legumes, since these are particularly sensitive to shade and moisture stresses induced by the shrubs.

In conclusion, inspite of the relatively low productivity of the traditional "montado" on a per hectare basis, it produces a considerable number of high quality traditional products, and holds such a great diversity of plants and animals, that it is considered one of the most attractive systems in the world, and therefore it must be preserved to continue providing a beautiful landscape, able to produce quality rather than quantity.

The degradation of the "montado/dehesa" and its causes

The "montado/dehesa" and their traditional products, have been suffering some serious degradation derived from deep changes in the technological, economic and social parameters of the rural society which for centuries has been taking care of, and living from it.

First of all, the farm mechanization which emerged and developped in the 20th century has weakened the oak tree component, through the use of deeper ploughs which affected the root system of the trees. Many trees have died of this effect and today we find large areas of "montado" which, having been dense in the past, hold today only a sparce tree population.

The natural regeneration of the "montado" through seedlings and young trees, basis of the long term persistence of the system, has also been severely affected by the use of wide ploughs, disc harrows and scarifiers, which, contrary to the traditional man-driven narrow and shallow plough pulled by oxen or mules, blindly destroy all the young plants on their pass, thus putting in risk the future of the system.

The generalised use of the mechanic chain-saw, instead of the traditional ax and hand saw in the pruning operations, has been responsible for the modification of the shape of the trees, with some negative impact on the yields of acorns and eventually also on the cork.

An excessive or even total removal of trees, has been verified during periods of scarce fuel oil for heating and cooking, which caused a supervalorization of the fire wood and charcoal, thus contributing to the elimination of many trees, then converted into a valuable but unrepeatable cash crop. During war periods or other critical times, areas around densely populated towns were the main victims of this process of desertification.

Another severe attack to the system was induced, in the second and third quarters of the 20th century, by economic policies favouring the cultivation of cereals combined with heavy mechanization. The trees of large areas of "montado", particularly those in best soils, were uprooted to facilitate cropping operations, including the use of combine harvesters, and simultaneously, deep ploughing was erroneously advised with the aim of increasing wheat yields. Unfortunatelly, the results were much different, as soil erosion and the exhaustion of the accumulated soil fertility rapidly led to the ruin of the system. Today, the European Union is providing considerable subsidies to those who decide to reforest these lands and recreate the "montado/dehesa" sytem, an operation that, apart from being costly and long time taking, is not allways well succeeded.

Other problems arrived with the decline of wool prices, and the propagation of African swine fever in the sixties, which being accompanied by the prohibition of grazing pigs, have stimulated the destruction of significant areas of holm oak "montado", promptly converted into *Eucalyptus* plantations, at the time providing better economic returns.

Another serious menace lies on insect attacks, particularly xilophagous (e.g. *Platypus cylindrus*) as well as on fungal diseases, which have caused severe damages in the last twenty years. Among the problems caused by fungus, particularly severe is the difusion of *Phytophtora* spp., which apparently is causing the rapid decline and death of many cork and holm oaks. The real causes of these problems are far from being entirely understood, but it is thought that the excess of cultivation in the soil, particularly with disc harrows, and the consequent damage in the root tissue, together with certain changes in the management practices (abandon of good practices of pruning, lack of removal and burning of the dead parts of the trees and other wooden residues lying on the soil, etc.), may be responsible for most of these problems.

A last (but not the least) cause of degradation is the profound tranformation of the rural society in the last 5 decades, particularly derived from the migratory movement of the populations to the cities, initiated after the 2nd world war. This has induced a severe scarcity of labourers to look after flocks and herds, milk the sheep and goats, shear the sheep, manage the grazing pigs, prune the trees, strip the cork, keep the beehives, etc.. This new situation led to the abandonment of some areas where, in the absence of grazing animals, shrubs grow freely. The construction of fences and water points have helped to reduce the amount of labour required by the traditional systems of animal management, but the investments required to build and keep them in function are often not compatible with the poor income derived from the low stocking rates which are mantained by the natural pasture resources.

Another matter of concern is the recent devaluation of cork due to the inovation and production of cork substitutes for wine bottling (plastic, aluminium, even when in combination with cork discs and aglomerates, etc.) and insulation pannels. This is creating some concern among people depending economically from cork oak, although it is believed that cork production will overcome the crisis, particularly due to its unreplaceable quality.

More recently, some changes in the Common Agricultural Policy converting the direct aids to cereal areas and grazing animals into "historic income" to farmers, together with the degradation of cereal and meat prices, due to policies of market globalization, have also contributed to increase the difficulties for keeping the traditional system. The tendency has been to eliminate the cropping phase and to reduce or abandon grazing, but if no alternative is found, not only the feed resources of the "montado" will degrade due to the desappearence of stubbles and straw and the degradation of the pastures, but the entire system itself will run into a big risk, caused by the invasion of shrubs, thus converting the "montado" into an easy prey of ravaging fires.

Pasture improvement- the key for rehabilitating the "montado/dehesa"

With cereals out of the "montado/dehesa", an alternative must be found to control the shrub invasion and keep the system productive. Some farmers who have already abandoned cereals and grazing animals on their cork oak areas, are controlling shrubs through disc harrowing at 4-6 years intervals, combined or not with operations of shrub cutting. Such a method is not sustainable since, apart the significant costs involved, it will continue affecting the root system of the trees, and particularly will enhance soil degradation. Indeed, disc harrowing will increase soil erosion and simultaneously will reduce soil fertility by facilitating the oxidation of organic matter and the leaching of nutrients. The method is also not compatible with keeping an economic animal production, as feed resources tend to become shorter.

The only sustainable alternative consists in improving the pasture yield and quality, in order to keep a much higher stocking rate, able to control the shrubs, improve soil fertility, and facilitate the entire management of the "montado". In fact, among the causes of invasion of shrubs in the traditional "montado", are the low grazing pressure supported by the natural pastures conjugated with the poor levels of soil fertility. Indeed, in various "montado" areas, it has been observed that the invasion of *Cistus* spp. and other shrub species, increases as soil fertility degrades, and that a good way of controlling the situation is to enrich the soil in nutrients and organic matter, thus stimulating the development of the herbaceous component, which, if properly grazed, will stop shrub invasion.

However, to be efficient, any model of pasture improvement for the "montado" must involve low costs, which is only possible if the pasture is long lasting (at least 6 years, preferably 10 years or more), and if the nitrogen required to increase yields and quality (protein) originates from biological fixation.

Inspired on the Australian use of Mediterranean legumes for pasture improvement and ley farming rotations, a model for improving the pastures of the "montado/dehesa" has been developed and largely diffused in Portugal, from where it passed to Spain and more recently to Sardinia (Italy). This model has a few fundamental elements which will be briefly presented and discussed below.

Elements for efficient pasture improvement

If compared with a motor car, the proposed pasture improvement model must consider the following essential elements:

1. Legumes, the motor.

   In any low cost pasture, productivity shall depend entirely from an abundant presence of legumes, which shall never fall below 35% of the total dry matter (DM) yield, the ideal being to keep a balance of 50-65% legumes to 50-35% other species. New established pastures, normally are much richer in legumes (up to 90%), but as soil fertility raises with time, the presence of nitrophilous plants (grasses and some other non N fixing plants) will increase.

   Apart their capacity to fix nitrogen by using sun-light energy, and therefore to build soil fertility, legumes are also responsible for improving the quality of the pasture, since protein, a limiting nutritional factor in the natural pastures, is quite abundant in legume-rich pastures. The animal intake is also much higher in legumes than in other plants with identical level of DM digestibility. Therefore, legume-rich pastures assure higher levels of ruminant animal production. Even monogastrics, such as pigs and poultry can use legumes with great efficiency, and this enhances the value of legumes in the "montado", as they may contribute to develop some of their most traditional products, such as grazing black pig and free range poultry. During their long period of flowering, most legumes provide abundant "grazing" for the honey bees, the honey produced being of excellent quality.

   The legume species which so far have been used in the improvement of pastures of the "montado/dehesa" are annuals with moderate to high levels of hard seeds and among these we count as most important Subterranean clovers (*Trifolium subterraneum* spp. *subterraneum*, spp. *brachycalycinum* and spp. *yanninicum*), Arrowleaf clover (*T. vesiculosum*), Persian clover (*T. resupinatum spp. resupinatum*), Balansa clover (*T. michelianum*), Cupped clover (*T. cherleri*), Rose clover (*T. hirtum*), Gland clover (*T. glanduliferum*), Medics (*Medicago polymorpha, M. sphaerocarpus, M. murex, M. truncatula,*) Biserrula (*Biserrula pelecinus*), Serradellas (*Ornithopus compressus, O. sativus*) and Lotus (*Lotus subbiflorus*). Of all these species there are more than 60 cultivars available, which differ in the content of hard seeds, pest and disease resistence, length of cycle, adaptation to soil and climate conditions, habit of growth, etc..

2. Rhizhobium, the starter.
   As the motor of a car needs a starter device to start running, so the legumes need specific bacteria of *Rhizobium* to nodulate well and fix nitrogen.
   As mentioned before, the natural legume population of the "montado" is often relatively poor and so is the natural *Rhizobia* occurring in the soil. To solve this problem, the model provides an efficient method of seed inoculation, using specific effective strains of *Rhizobium* for each legume species, protected by a firm seed coating through a pelleting operation.
   When well nodulated, legume-rich pastures can fix high quantities of N, which may reach or exceed, in favourable plant/rhizhobium/soil/climate conditions, 200 kg N/ha/year. However, the most common figures vary from 30 to 150 kg N/ha/year, the higher rates being attained when appropriate seed inoculation with selected specific and effective *Rhizobium* strains is practiced, and the soil pH and plant nutrients are adequate. (Materon, 1986).
3. Grasses, the catalytic agent.
   Legumes, through their ability to fix atmospheric nitrogen, can create excessive quantities of N in the soil, which, if not timely removed, will not only induce soil acidification but also will increase the level of nitrates in underground water. This is today an important environmental problem, which requires a prompt, safe and economic solution. In this context, grasses are the ideal component to clean the system, as they can uptake great amounts of N, contributing to increase DM yields, and to balance the quality of the pasture in terms of energy and protein. Indeed, grasses are, in general, deficient in protein for the amount of energy they can supply to animals; on the contrary, legumes contain an excess of protein in relation to the energy they have. Therefore adding grasses to legumes provides a well balanced feed regime for animals and at the same time the environment is kept in perfect conditions. If grasses are absent from legume-rich pastures, there is also a risk of invasion by nitrophilous weeds, such as *Urtica,* spp., *Carduus* spp., *Malva* spp., *Erodium* spp., etc., which may affect the yield, quality and persistence of the pastures.
   Among the grasses to be associated with legumes, the annuals *Lolium rigidum,* and *L. multiflorum*, and a few summer semi-dormant cultivars of the perennials *Dactylis glomerata, Lolium perenne* and *Phalaris aquatica* are the most important. However, grasses, particularly the perennial species, require high soil fertility to grow and persist well, condition often only achieved after 3-4 years of legume dominant pastures.
4. Plant nutrients, the fuel.
   As mentioned before the majority of the soils of the "montado" are acid and poor in phosphorus, but a few may also be deficient in some other macro (potassium, calcium, magnesium, sulphur) or micro-nutrients (molibdenium, boron, zinc, copper, cobalt, manganese, etc.). Legume-rich pastures are sensitive to missing nutrients, which may affect both production and persistence. Therefore it is fundamnetal to pay attention to the soil content of the various nutrients, through regular soil analysis, correcting all the deficiencies eventually found.. Phosphorus is a key element which deserves special attention being applied at sowing time and top-dressed in each subsequent year. Once an appropriate level is applied at sowing, maintenance will require a low input in the order of 5 to 10 kg of P/ha/year. The same principle applies to potassium and magnesium if needed. Micro-nutrients, when required, are applied at longer intervals (3 to 6 years). When pH is below 5.3 (in H_2O), liming must be applied, but this operation shall be in moderate amounts (1 to 3 t/ha), being preferable, if required, to repeat it after 3 or 4 years. With reference to this subject, it should

be kept in mind that it is more important to choose plants adapted to soil pH than to adapt soil pH to pre-chosen plants.

5. Plant biodiversity, the suspension.
   The use of complex seed mixtures formulated with various species and cultivars of legumes and grasses, well adapted to local soil and climate conditions, offers various advantages over simple mixture of one legume with one grass. Indeed, an adequate and well chosen biodiverse pasture will produce more than one based on a single or double species or cultivars. Also, it will withstand better eventual rainfall variation from year to year (leading to shorter or longer growing seasons), and will assure more regular pasture cover in divergent soil conditions (niches) of the same paddock, such as differences in pH, drainage, depth, texture, etc.. Indeed, biodiversity will act like a suspension, introducing more flexibility in pasture management, and increasing stability and persistence of a pasture exposed to variable conditions of climate and soil.
6. Animal, the user.
   Provided grazing management is adequate, animals are not only the users of the pastures, they are also the maintainers of their productivity. In fact, there are no good pastures without grazing animals, as they play an important role in the maintenance of yields and quality, and also on the control of herbaceous and shrubby weeds, and this may be of particular interest in the "montado", where there is a tendency of shrub invasion after long periods of absence of grazing animals.
   Animals should not overgraze the pastures, as this will affect production and persistence, but undergrazing should also be avoided, as the pastures may deteriorate and fail to persist. Pasture economy is also affected by both over and undergrazing. Particular attention should be given to grazing management during the dry season, when a considerable quantity of seeds may be ingested by the animals. The removal of the dry pasture through grazing is essential, as the ingestion of seeds increases the nutritive value of the pasture and improves the germination patterns of hard seeds lying on the ground. Indeed, aproppriate dry season grazing is essential to position the seeds in the soil and simultaneously remove the dry standing pasture in order to eliminate its insulating effect, thus increasing the thermic amplitudes on the ground, which facilitate hardseededness breakdown. Summer grazing will also eliminate the shading effect, therefore facilitating the exposure of the new seedlings to sunlight after their emergence with the autumn rains. However, too hard grazing can lead to the exaustion of the seed bank and this must be avoided. The most important for the correct use of a pasture is to determine which is the adequate stocking rate that can be maintained in a pasture along the year, keeping it as long as possible, until the productivity of the pasture changes, then requiring new adjustments.
7. Conserved forage, the insurance.
   Pasture growth is not regular along the year; it changes continuously, according to plant composition, temperature, soil moisture, day length, light intensity, etc. Pasture quality also varies with time as plants develop and mature. However, grazing animals have to be fed regularly every day, and although they can withstand considerable changes in the nutritive value of the herbage for certain periods of time, they require a minimum quantity and quality of feed to be kept in acceptable conditions of production.
   Roughly, a Mediterranean legume-rich pasture distributes its annual yield and quality pattern by producing circa 1/4 in the autumn with very high quality, 1/4 in the winter with high quality, and 1/2 in the spring with high to medium quality, which degrades after complete maturation of the seeds and with the progress of grazing in Summer. The double production in spring will allow two important facts: the production of seeds to regenerate the seed bank, and the feed availability during the summer when pasture growth is nil.

A well managed pasture shall provide feed to cover the great majority of the animal requirements, and this implies that the stocking rate shall be adjusted to average pasture production. However, there are some periods in the year, particularly in the transition from summer to autumn, and sometimes also during winter, where suplemental feed may be required. The irregularity of rain distribution in Mediterranean climate also introduces frequent inter-annual variations in the pasture growth, which may create some abnormal short pasture supply problems. To solve these situations, feed reserves of conserved forage (hay, silage, haylage) should be produced on the farm, otherwise they have to be bought from the market at a higher cost (straw, industrial by-products, concentrates, etc.). Taking in consideration that the cost of one FU in conserved forage is normally twice or thrice of that in the grazed pasture, and 4 to 6 times more expensive in straw and concentrates, attention should be paid to the quantity and quality of forage to be conserved on the farm. Indeed, conserved forage should be considered as an insurance which must cover not only normal periods of feed scarcicity, but preferably also some more irregular climatic accidents affecting pasture production. As the "montado" can assist in covering some more difficult or irregular periods of pasture availability, by providing twigs and leaves of the oak branches, grazed *in situ* after being cut down, annual needs for forage conservation may be limited to 1 FU per kilogram of animal live weight grazing in the "montado", that is circa 1/6 of the total annual feed requirements. Also the yield of acorns, if not used by pigs, can assist ruminant nutrition in eventual prolonged autumn droughts but this shall be taken with care as in dry autumns the acorns may delay their maturing period, falling down green and with a high tanin content, able to block protein in the rumen, thus affecting the N nutrition of the grazing animals. Pigs are not affected by this problem.

8. Manager, the driver.
   Man is the most important element. Indeed he must take every decision concerning the "montado", pastures, animals, trees, fodder crops, bees, etc. Taking in consideration the complexity of the system he may need a holistic approach for his decisions. Particularly, he has to pay attention to pasture establishment and management, deciding on soil preparation, sowing time, seed mixture and sowing rate, sowing method, fertilizers and soil ammendments, weed control, type and productive function of the grazing animals, stocking rates and grazing management, conservation of fences and water points, fertilizers topdressing, areas for forage conservation, method of forage conservation, animal management, animal health, accounting and sales of products, etc.. He has also to decide on the complex management of the trees such as prunning, cork stripping, using the acorns, etc..

Results of pasture improvement in "montado/dehesa" farms

Many examples of results obtained through this model of pasture improvement in "montado/dehesa" farms could be presented. However, there is one, whose evolution has been oriented and accompanied by the author for more than 30 years, that deserves particular attention. This is the farm "Herdade dos Esquerdos" (in Vaiamonte, Monforte, Alentejo, Portugal), a 285 ha cork/holm oak "montado" at 30-40 trees/ha, in a region of 550 mm annual rainfall, soils acid (pH 5.3- 6.2) and shallow derived from granite and gneiss, originally very poor in P (less than 2 ppm), and soil fertility (organic matter 0.7-1.2%).

Until 1973 the farm was exploited through a traditional 9 years rotation, including a ploughed (with mules) bare fallow-wheat-oats-6 years natural pasture, which was progressively invaded by *Cistus salvifolius* and *C. crispus*. The yields of cereals were very low (400-800 kg/ha) and were only justified by the need to control the invasive shrubs.

The acorns were grazed every year by an average 120 fattening pigs, from October to January/February, and the natural pastures by 350 merino sheep, arriving in the farm by late February and remaining on pastures until late June, when they were moved to the 60 ha of wheat and oats stubbles, being fed there until early August. By this time they have finished the stubbles and were moved out to another farm of the same proprietor. In total (including the grazing pigs fed on acorns) the stocking rate of the farm was estimated in 0.87 sheep equivalent/ha/year.

In 1973, following the crisis of the African Swine fever, and the fall in wheat prices, aggravated by difficulties in finding labourers to continue with the traditional system, it was decided to initiate a programme of pasture improvement and intensification of sheep production, which included the sowing of biodiverse legume-rich pastures, on one well fertilized (superphosphate 18% P_2O_5 at 500 kg/ha) paddock per year, accompanied by fencing and provision of water points. At the same time the rotation with cereals was progressively discontinued and from 1987 onwards 88% of the agricultural area was occupied with permanent improved pastures, the remaining 12% (the best soils) being continuously cropped with annual legume-rich forages for hay.

Today the farm is an organic farm involved in sheep milk production for organic cheese, keeps a stocking rate of 8 sheep equivalent/ha/year, including 40 Iberian sows and 4 males to produce pigs for fattening on the acorns (50-80 organic fat pigs /year) and young pigs for the market. Apart from that, some recent demonstration trials carried out on the farm has shown a good potential for developing the production of high quality poultry meat on pastures.

On the other hand, the effects on the soil were remarkable. Soil organic matter raised considerably to 1.5-4.4%, and a similar situation occurred with the levels of P, which raised to 12-40 ppm, the lowest values being only found on the continuously cropped land for hay. The elevation of the soil organic matter, apart representing a considerable sequestration of carbon from the atmospheric CO_2, has also improved the water holding capacity of the soils, which reflects in a further increase of pasture productivity. Soil erosion, in the past a matter of great concern, has been totally arrested. The difficult and costly control of shrubs in the past, is now completely and naturally resolved, as the improvement in pasture productivity and the correspondent high grazing pressure on the land, have contributed to the total eradication of the shrubs, today only found outside the fences, along the road sides. The trees have an excelent vegetative aspect, the yields of acorns are more regular, and the cork thickness seem also to have improved, although some more time is required to conclude on this matter. As a final result, landscape has improved and fire risks are much smaller and easier to control.

On the other hand the farm management is much facilitated and the income has increased considerably. Apart from the annual forage hay crop which requires sowing, fertilizing, cutting and haymaking operations every year, the remaining area is now just topdressed whith rock phosphate 26,5% P_2O_5, at the rate of 250 kg/ha every second year. The sheep and pigs are kept on the paddocks during the entire year, although receiving some concentrate feed during lactating and late pregnancy periods. The management of the sheep flock is facilitated through individual electronic chip identification, associated with an adequate soft-ware programme (Ovigest) specially developed in cooperation with an electronic-informatic engineer.

Today, the farm receives a great number of interested people, including farmers, researchers, students, congress participants, etc., and the author and his sons, as well as a dedicated team of collaborators are all proud of the results achieved.

References

Bennett, S.J. & P.S. Cocks (Editors), 1999. Genetic Resources of Mediterranean Pasture and Forage Legumes. Kluwer Academic Publishers, Dordrecht/Boston/London, pp. 241.

Crespo, D., A.M.C. Barradas, P.V. Santos & J.P. G. Carneiro, 2004. Sustainable improvement of Mediterranean pastures, In: Grassland Science in Europe, 9: 840-842.

Garcia, D., S. Ramos, F.M. Vázquez, J. Blanco, A.B. Lucas, J.J. Barrantes & M. Martinez, 2005. Estimación de la producción de bellotas en los encinares de la Comunidad Extremeña en la Campaña de 2004-2005. Consejeria Agricultura y Medio Ambiente, Merida, Badajoz.

Materon, L., 1986. Maximizing biological nitrogen fixation by forage and pasture legumes in semi-arid areas. In: Nitrogen Fixation by Legumes in Mediterranean Agriculture, Martinus Nijhoff Publishers, Nederlands, pp. 33-40.

Montoya Oliver, J.M., 1993. Encinas y Encinares, Ediciones Mundi-Prensa, Madrid.

Peco, B., 2002. Grassland diversity under dry conditions: the role of management in nature conservation. In: Grassland Science in Europe, 7: 675-881.

Animal production and farm animal genetic sources utilization in Turkey

M.I. Soysal

Trakya University Agricultural Faculty of Tekirdag, Department of Animal Science, Tekirdag, Turkey

Summary

Turkey is situated in the Northern hemisphere and is a gateway between Asia, Europe and Africa. The area of the country is 780 576 km² and is surrounded by Black Sea in the north, Aegean Sea in the west and Mediterranean in the South. The European part of Turkey is referred to as Trakya. The Asian part of Turkey is called Anatolia peninsula and is crossed by two mountain chains from west towards the east. The chains are causing three different climate zones, namely Mediterranean, the Interior Steppe and Eastern Anatolian climates.

Even though Turkey is among the leading countries in the number of livestock, the animal productivity sometimes can be quite low due to high number of low producing unimproved native animals. Nevertheless, the adaptation capacity of these indigenous animals to the local conditions is higher than improved western breeds. Differences in climate, topography and soil lead to great diversity in the utilization of natural resources in Turkey in general and in the Mediterranean mountain part of Turkey in particular.

In general, as a consequence of development in animal husbandry on the better land and climate areas, indigenous breeds have either been completely replaced by high yielding animals or replaced by massive upgraded western breeds that respond better to improved feeding and management and intensive production conditions. Sheep production is heavily dependent on pasture, late spring, summer and autumn. In the winter and early spring, flocks are kept in barns with some supplementary concentrates and without moving to the pasture due to snow and rainy conditions. This is called traditional winter system feeding which is mostly not sufficient for the nutritional requirements of the animals. Generally, lambing and consequently milking occur in late winter or mostly at the beginning of the spring.

Keywords: Turkey, nomads, nomadic animal husbandry.

Introduction

Turkey has approximately eleven million heads of cattle, roughly 27 million heads of sheep and 7 million heads of goats. Native cattle and sheep breeds represent 40% and 95% of total cattle and sheep breeds respectively (Table 1).

Native cattle breeds consist mainly of Native Black (Yerli Kara), Eastern Anatolian Red (Dogu Anadolu Kirmizisi), Grey Steppe (Boz step) and Southern Red and Yellow (Güney Sari Kirmizisi). Native sheep breeds consist primarily of Red Karaman (Morkaraman), White Karaman (Akkaraman), Daglic, Awassi (Ivesi), Kivircik, Karayaka and Tuj. The most peculiar indigenous goat breeds of Turkey are basically Angora goat (Tiftik Kecisi).

Table 1. Livestock numbers in Turkey by year (heads)[1].

Years	Total sheep	Hair goat	Angora goat	Cattle		
				Total	Pure+cross	Native
1971	37 008 000	15 042 000	4 127 000	12 939 000	-	-
1975	41 366 000	15 216 000	3 547 000	13 751 000	-	-
1980	48 630 000	15 385 000	3 658 000	15 894 000	-	-
1985	42 500 000	11 233 000	2 103 000	12 466 000	-	-
1990	40 553 000	9 698 000	1 279 000	11 377 000	4 683 000	6 694 000
1995	33 791 000	8 397 000	714 000	11 789 000	6 478 000	5 311 000
1996	33 072 000	8 242 000	709 000	11 886 000	6 704 000	5 182 000
1997	30 238 000	7 761 000	615 000	11 185 000	6 405 000	4 780 000
1998	29 435 000	7 523 000	534 000	11 031 000	6 428 000	4 603 000
1999	30 256 000	7 284 000	490 000	11 054 000	6 608 000	4 446 000
2000	28 492 000	6 828 000	373 000	10 761 000	6 544 000	4 217 000
2001[2]	26 972 000	6 676 000	346 000	10 548 000	6 474 000	4 074 000

[1]State Statistic Institute, 2001, Turkey.
[2]Permanent.

Animal husbandry based on the usage of pasture in our country is very common. Two kinds of traditions exist in this tendency. The first one is practiced by nomadic people, especially in the eastern and south-eastern part of Turkey.

The nomadic animal husbandry

The term nomadic stands for moving together with flocks according to the level of vegetations in different cities and pastures around the country. The distance can be in some instances about 300-400 kilometres. Although numbers of such farmers are decreasing drastically, there are still some tribes famous for their high skilled level of sheep husbandry.

There are social support programs in order to encourage the nomadic people to settle in some villages and to make them sedentary by means of giving them land and buildings. However, they insist on practicing the traditional nomadic way of sheep husbandry. Most of them voluntarily abandoned traditional life by being sedentary in urban areas.

The decrease of pastureland in the area has decreased the number of farmers practicing the nomadic culture. The so called "Göcebe" or "Göcer" in Turkish Nomadic stands for the farmers who earn their life from very old times as a tradition as to move among the pasture in high plateau and residence places for winters according to the conditions of nature, seasons and sort for their herds. This is in fact even now a very common practice in old Turkish tribes in central Asia.

The people that earn their life by this way of sheep husbandry are also called "Yörük". This Turkish term can be defined as a person that was a fast walker and lived in woven tents from black hair of goats in old times.

They were also called Turcoman Nomads which survived mainly in some Mediterranean mountain area and in some Southeast and East Anatolia and have nearly completely vanished. In South-eastern and Eastern Anatolia the Göcers still practice the nomadic herding, winter their flocks in the valley and drive them to the high mountain pastures in the summer. The flocks are invariably mixed some sheep and some goats. Most of the milk is produced while stocks are on summer pasture in high mountains. The milk is processed into cheese, butter and yogurt for the long haul to market.

In south-eastern Anatolia, farmers known as "Beritans" practicing nomadic herding, winter in Urfa Province. They start to move in April to the south and pass through Maden, Palu district and Karliova (Bingöl). Other farmers called "Savuk" winter in Tunceli (Cemisgezek) and Elazig and they move in April to Bingöl, Erzincan and even to Erzurum province. They start to come back in autumn (September) for wintering in lowlands and valleys. It is reported by State Statistic Institute that by May 1986 almost 1 million head of cattle 3 million head of sheep and 1.5 million head of goats lived in upper Euphrates region. The herds arrive in December to the place where they winter (Bayder, 1986).

The nomadic herding farmers so called Yörüks have been named and identified according to their way of movement and colour of tents. There are famous nomadic tribes practicing nomadic herding such as Kara Kecili (black goat) and Sari Kecili (yellow goat). The Sari Kecilis are travel by camels. But the tribe of Bahsis travels by horses. It is necessary that every family of tribes keep distance in their placement in high plateau pastures called "yayla" as 500 meters in order not to mix the herds.

There are still Sari Kecili tribes consisting of 200 families practicing the nomadic herding in Mersin province of the Mediterranean cost. They winter in Silifke, Gülnur and Anamur coast. They move in summer months to Beysehir and Seydisehir districts, which are the pastures of Konya province. Nomadic herding farmers always lend the pasture from the sedentary farmers of the places where they move. Yörüks called Sari Kecili, Kara Tekeli, Karakoyunlu, Gebizli still keep the way of nomadic herding within the days. They settle their tents in the low land area but herdsman move the herd for grazing to upland pasture (yayla) and come back in the night. It is practiced every year but they do not winter in the same place due to control of ticks. Nomad tents in Turkey can be seen throughout the coastal region of Turkey.

Nomadic herders, the Yörüks who come down to the pastures walking in winter and in the summer when the sun shrivels the vegetation on the coast travels up into the mountains to the so called "yayla", the high mountain plateaus and valleys where there is sufficient grass and foods for the animals until autumn rains again regenerate the pastures on the coast. In other parts of Turkey, there are larger number of these nomads who carry everything with them on donkeys and camels along with their flock of sheep and goats. To control the large flocks of sheep and protect herds against the wolves and guard their property, they also have a large breed of dog. However, the life of the Yörüks is becoming less nomadic in Modern Turkey and they are losing many of their tradition.

They still make superb rugs (kilim) with patterns and colors particular to the clan and region. The problems of scarcity of pasture area let them to diminish their numbers. According to the law numbered 2510 that passed from the parliament in 1950, the government encouraged them to become sedentary. In 1970 in Konya, Samsun, Eregli respecively 139, 168, 100 families are settled to the villages as sedentary by giving them land without any loan. After 1970, other laws have been passed and they were given some subsidiary to settle in the village of Hatay, Burdur, Diyarbakir, Elazig, Mardin provinces, comprising of 754 families (Kücükcakar, 1989). Another practice in Erzurum is the so-called fattening in yayla (upper pasture) or fattening in Mountain area of Erzurum. About 90% of our total livestock population is being fattened only at pasture and meadow areas at a period from early spring to winter. Turkey has 21 million hectares of pasture and meadowland in which 15-18 million tons of hay was produced annually up to 1987 (Baydar, 1987).

In Black Sea Region, a second type of nomadic herding is practiced identical to the Mediterranean region but not so common as in the South. In this type of nomadic herding, transitionally only in some areas, sheep herds with herdsman or shepherds called "Coban" are moved from sedentary places of village of the farmers by walking to the high plated pasture called yayla where there is sufficient amount of grass and favorable climate for sheep herds.

They usually start to move to the yayla region in middle of May and then come back in mid-August. Generally they winter in villages located lower areas than yaylas (Atasever, S., 1994). Some breeds don't practice the usage of pasture in upper Mountain pastures (yayla). The sheep are milked twice a day. Rams are allowed into the herd once a year depending on the area in early September or late August. The free ram joining has being practiced. Flushing is practiced before the joining of rams. The lambs are not given supplementary foods. In pastures sheering has been done via scissors or clippers. The nails, tails and horns of the breeds are not cut in the herds.

Today the Yörük nomads are confined mainly to the Toros Mountains with their high pastures. Until 19th centuries thousands of Yörük nomad lived in Ottoman territories of the Balkan and throughout the western southern and southern eastern Anatolia. They live summer on the high pastures, autumn at lower altitudes and winter months in the sheltered encampment with their flocks. These pastoral communities or tribes or nomads began to settle on the land. Actually, they have been forced to settle the land in Ottoman times and this process gained momentum under Turkish republican period. In the beginning of such settlement years Yörüks resisted permanent settlement

Today most Yörük communities live permanently in their winter villages, so the sight of Yörük tents on the high pastures in summer is becoming increasingly rare. Yet there will always be a few Yörüks who continue to make annual migration in to Toros mountain drawn by the lush grazing, fresh mountain air, ice cold spring water and above all by the freedom of mountain life. They start to go to the yayla in May where weather and grazing conditions are favourable for Yörük and their sheep populations.

During the summer months in the mountains milk from sheep and goats is used to make butter and cheese which are taken to the market along with wool and surplus livestock. After weather gets cold, Yörüks move dawn the mountains to the autumn camps where they make butter and cheese for their own use during the winter. The winter villages are usually situated on the coastal plains where the weather is mild and there is plenty of rich winter grazing and firewood. Their felt tents known as Kara cadir (Black tent) made of goat hair were their only homes. The shepherd called also from Adriatic Sea to the chine wall "Coban" are protected from the cool of night by felt cloaks (kepenek) was the typical wearing on. The migration of yörük nomads is an ancient part of their culture with its own unique tradition. Nowadays tribes or communities called "honamli", "Karakoyunlu", "Gebizli", "Hayta", "Yeni Osmanli" yörük still spend their summer on the yayla between the provinces of Isparta, Konya,Antalya,Burdur, of interior part of Mediterranean region. Apart from small number of yörük belonging "Sarikeceli" and "Bahsis" tribes most of Yörüks now live in settled life. But these who persist in their nomadic tradition do not show any inclination to give it up.

Sedentary animal breeding

Sedentary farming people that live in the villages own most of the remaining native livestock. These herds are predominately of cattle and water buffalo. They are triple purpose beasts used for work, meat and milk.

The remaining and most part of the country where nomadic herding is not applied use exotic breeds or so-called western breeds especially in cattle husbandry. However, most of and principal sheep breeds are native (indigenous) breeds. There are still native or so-called auto chanteuse cattle breeds. Indigenous cattle breeds are mostly named according to the coat colour and geographical location. Grey Steppe cattle are grown primarily in Thrace Region near Greek and Bulgarian borders. They are fine large beasts somewhat resembling Brown Swiss. The Native Black breed is raised in central Anatolia. They are for the most part very small rather refined cattle somewhat like the size of Jersey but even smaller. The Eastern Anatolian reds live mostly in Eastern Anatolia. The reds and yellows of south and south-eastern Anatolian are raised in southern Turkey. Both are tall slim and long legged resembling the very thin Guernsey. Rather good milk producing specimen have been selected out of all four breeds but little selection is practiced in most village herds.

Native cattle breeds are kept mainly for meat production. They are distributed in mostly semi-mountain and mountainous areas under poor feeding and management conditions. Improved cattle breeds or so-called modern breeds are used mostly for milk production. Majority of them are kept by small-scale family enterprises. In fact 90% of all herds contain 0-5 cows, which shows that they are managed as traditional extensive cattle production system and also primary income of farmer is not obtained only from dairy cattle husbandry.

Every day early in the morning, all flocks of the farmers in the same village are unified and moved by the management of one herdsman in order to utilize the common pasture of their village. In the evening the total herds return back from the common pastures of the village and cattle are distributed to the owner farmer. In general, traditional extensive sheep production systems are applied for centuries. Some intensification measures such as improved management of natural pastures, use of roughage, use of silage and concentrates in winter had been developed during the general improvement course of animal husbandry in Turkey.

Sheep production is heavily dependent on pasture, during late spring, summer and autumn. In the winter and early spring, flocks are kept in barns with some supplementary concentrates and without moving to the pasture due to snow and rainy conditions. This so called traditional winter system feeding which is mostly not sufficient for the nutritional requirements of the animals. Generally lambing and consequently milking occur in late winter or mostly in the beginning of spring.

Another tradition in Turkey was a kind of modified form of Nomadic way of life called temporarily move from permanently settled village or cities in winter to the more cooler yayla for more comfort and healthy life or more available grasses for the animals in summer. This tradition is now become fashion for the people lives in industrialized cities as a summer houses. This was mainly and originally practised for the animal breeders bringing the sheep population more productive pasture and environmental area.

With the first warm days of spring in the villages all around the Turkey families begin to think about moving their summer residence on cool summer pasture called "yayla".

The migration from winter village to the summer housing is legacy from the days when Anatolia was mostly populated by nomadic and semi-nomadic peoples. Generally in the southern part of Anatolia they use their own felt tents made by goat hair. In Black sea region summer settlement are made of permanent structures with each family moving into their own wooden chalet each year.

Today this is become more popular even for urban people as agro-tourist point of view. Today efforts has directed to avoid the encroachment of modern life on the "yayla's" so that essential characters of this way of life is preserved. Up to now general characteristic of animal husbandry from the native animal genetic resources" points of view were given. As it can be easily postulated from the possibility of improving traditional system some alternatively source of income is needed. At this point some case studies from Turkey will be given. The examples are coming from felt making and cream from water buffaloes milk for a famous Turkish dessert. Felt making is nowadays practiced in very few areas of Turkey. In spite of the unique, healthy characteristic of fibre from wool or goat hair, the wool and hair doesn't make money in the market. This is mainly due to being not as competitive as synthetic fibre from the price point of view. Now government is encouraging the felt makers to create new area of usage as decorative material in order to make the product of wool and hair itself voluble. It seems there is a great variety of potential area mainly concentrated in the field of handicrafts. Government now organizing open air exhibitions, workshop, TV programs etc., opportunity to express them are also one way of alternatives opening their way. Felt making nowadays remained very few areas of Turkey such as Urfa, Balikesir, Selcuk, Tire, Afyon, Ödemis, Selcuk, Sandikli. Felt making, one of the traditional handicrafts is trying to survive and challenging with synthetic fibre. Felt is a tissue obtained by joining animal fibbers especially wool under heat, dampness and pressure with help of soap, oils, and acids. Kicked felt making which is one of the oldest of Turkish handicrafts was brought to Anatolia in the 11th century by Turks migrant from central Asia and still survives today. Felt is produced kicking method in family scaled small workshop rather then factory making. Besides lamb wool, camel, rabbits, mohair and goat bristles are all used. Natural colours (black and white, browns) are used for the surface and design are applied with felts coloured with natural colours and dyes. Most of designs has geometric shapes but figures from natures can also been seen. Socks, pillows, carpets, coverings, shepherd felt cloaks known as "kepenek" are made in various family sized small workshop with regional design and colour.

Today, very few families produce felt in the mentioned areas. Governments are trying to make their usage more popular as decorative objects, big pieces of felt decorated in colourful styles are used to adorn horses, covering them from head to tail.

A second example can be given for water buffaloes, used as a chance for improving traditional systems. As the level of agricultural mechanization increased, number of water buffaloes decreases drastically. On the contrary, a lack of information and of precise information on their products made water buffaloes and their product less valuable. It is estimated that 57 000 water buffaloes are bred. Nowadays it is argued that due to their unique nutritive value water buffaloes are getting more and more popular. Especially its non substitutable in traditional products, i.e. the famous Turkish dessert produced with the cream called "kaymak", making the cake popular again.

According to the traditional summer production system of feeding, the sheep herds are moved temporarily from the villages where the owners live (which are located mostly in lower altitudes and have hotter climatic conditions) to the common pastures (which are located mostly in higher altitudes and have cooler climatic conditions and more green fresh roughage sources of pasture). This is called a kind of nomadic sheep grazing system.

In order to increase the productivity of the period of 1980-1990, several projects were initiated mainly in the western part of the country and a total 280 000 heads of pure breed exotic pregnant heifer were imported, mostly from European Countries and from the USA.

During the last few decades modern breeding techniques have greatly improved the production potential of most farm animal species. Thus trends have led to use intensive crossbreeding of native breed by exotic imported western breeds.

In conclusion while the percentages of modern breeds were increased the percentages of native farm animals breeds came to the danger zone of extinction. This trend of loss of variation narrows the genetic base of species. The genetic basis of any specific animal genotypes has consisted of several genetic components each responsible for different traits, which leads to high variable genetic content. Holstein-Friesian breeding stockbreeders union was established in 1995. The organization is under way and seems to contribute the systemic improvement of dairy cattle breeders in the country. In general, as a consequence of development in animal husbandry on better land and climate areas, indigenous breeds have either been completely replaced by high yielding animals or replaced by massive upgraded western breeds that respond better to improved feeding and management and intensive production conditions. The genetic farm animal sources of Turkey and the main purpose of breeding numbers and features are explained in this study.

One of the still remained traditional handicrafts practised in Turkey (e.g Urfa, Balikesir, Afyon, Ödemis and Tire, Selcuk, districts) is felt making. Felt is a tissue obtained by joining animal fibbers, especially wool, under heat, dampness and pressure with the help of soap, oils and acid.

Kicked felt making, which is one of the oldest of Turkish handicrafts, was brought to Anatolia in the 11th century by Turks migrating from Central Asia, and still survives today. Felt is produced by the kicking method or in factories.

Besides lambs wool, rabbit, camel, mohair and goat bristles are all used. Natural colours (black, white, brown) are usually used for the surface, and designs are applied with felts collared with natural synthetic dyes. Most of the designs have geometric shapes.

Felt is also decorative. Big pieces of felt, decorated in colourful styles, are used to adorn horses, covering the animal from head to tail. The horse carrying the bride after the wedding was traditionally decorated with a harness of felt.

Shepherds (called choban in Turkish) still use the felt cloak known as "kepenek" however. Today, very few families produce felt and those who do make only one or two pieces.

The process of felt making can be summarized as using the material wool that has been combed so that all the fibres lie in the same direction for felting. Biting the wool by thick stick cause swell then the wool spreads on cotton layer on the dried reed mat on the workbench or ground according to the pattern, size, and shape of desired end product.

Until obtaining a square of wool with all the fibres going in the same direction on top of the wetted surface then gone through the same process again on a second layer if (top and bottom coverage layer) required. Rule is placing the wool so the fibres are at right-angles to the previous layer. The more layers of wool you put down like this, the thicker, firmer and stronger the finished felt will be.

The layer of combed wetted wool can felted by hard kicking,(biting the wool layer rolled onto wooden pole by legs), rolling, folding, refolding, wetting with water with melted soap continuously approximately half hour. After that, second time same operation is operated approximately one hour. This can be traditionally done by rolling the layer on the surface of wooden pole. Felting happens because every strand of wool is covered by tiny scales. The primary mechanism of felting is abrasion. As the individual hairs get rubbed together the scales catch on each other, and the effect is that the whole thing shrinks in an irreversible way to make a mat. For some decoration it has been placed some different colours of wool on top of obtained layers. After obtaining the desired design according to end product the material are ready to start felting. With -hot water liberally sprinkle the wool so that it is wet all over. Spinks soapy water into the wool to thoroughly saturate it. Add more water if you need it and if it doesn't appear to be very soapy then sprinkle a few drops of washing up liquid directly. When it is fully rolled you are ready for the physically tough bit. Pushing down hard, roll onto the wool. The wool should now miraculously be a piece of felt. More precisely it is called pre-felt at this stage.

The second stage traditionally called cooking or ripening the pre-felt manufactured in sauna like Turkish bath (called keche hamami) designed for especially for felt making.. This stage now is practised in workshops of the manufacturers. Discard the layer cotton made layer wrap and place the felt directly onto the surface. Again use with very hot water and sprinkle the felt liberally. Add more washing up liquid if it is not looking soapy enough. The last stage is practised 15 minutes in more dry conditions. The last stage is called last kicking (biting by legs) the felt rolled on to pole as 15 minutes approximately.

References

Anonymus, 2001. State Statistic Institute. Türkiye.

Atasever, S., 1994. Samsun'un Bazi Ilcelerinde Yayla Koyunculugunun Genel Yapi Analizi. Ondokuz Mayis Üniversitesi. F.B.E., (Master Thesis).

Baydar, M.C., 1986. Daglar ve Yaylalarda Servet Meralari. T.O.K.B. Dergisi.

Baydar, M.C., 1987. Dag Besisi veya Yayla Besisi. Hasad Dergisi.
Dulkadir, H., 1993. Yörükler. Icel Kültür Dergisi Sayi 29–30 Kasim
Kücükcakar, H., 1989. Sari kecililer artik göcmeyecek. T.O.K.B. Dergisi. (38), 50–51.
Sayici M., A.Topbas, 2003. Anadoluda Kececilik; http://www.akmb.gov.tr.

Improvement in general traditional livestock management practices in South Lebanon through different interventions

M.T. Farran[1], C. Kayouli[2], A. Soukehal[2], R. Metzger[2], M.L. Hayek[2], I. Nahhal[2], Z. Jaber[2], N. Fahd[2] & V.M. Ashkarian[1]

[1]American University of Beirut, P.O. Box 11-0236, Riad El-Solh, Beirut 1107-2020, Lebanon
[2]TCP/LEB/2905, FAO Representation, Lebanon

Summary

The total number of dairy cows and goats in Lebanon is 86,200 and 428,200 heads, respectively. Nabatieh region alone, accounts for 7 300 cows and 66 700 goats. Only 40% of the dairy herd is of pure imported breed and the goats mostly belong to the local "Baladi" breed. Dairy animals and goats under traditional feeding practices and poor hygiene management conditions result in low milk production, low fertility rate with several metabolic diseases. Most farmers are ignorant in proper roughage feeding and concentrate feeding is arbitrarily practiced. The presence of only one government veterinarian limits the services to few farmers.

The TCP/LEB/2905 was formulated to provide the animal small holders with technical assistance in the fields of milk hygiene and milk processing, general farm management, artificial insemination (AI), forage production, and animal nutrition in South Lebanon. Proper milking, filtering, agitating, and quick cooling of fresh milk to preserve quality were emphasized. Use of strip and teat cups for milk testing and iodine teat dipping were introduced. Proper milk pasteurization procedures were practiced and proper techniques of cheese and labneh manufacturing were introduced. Special emphasis was put on the practice of record keeping, and on balanced concentrate-roughage feeding programs. Twelve young farmers were awarded AI training certificates. Alfalfa plot resulted in 5 annual cuts of 5 tons hay/cut per hectare. Rain fed hay crops of oats or barley in association with vetch was preferred over field corn due to scarcity of irrigating water. The yield obtained from this intervention varied between 8 – 10 tons per hectare. A mixture of carob-olive pulp (1:1 wt/wt) top dressed with 2 and 100 kg/ton of salt and wheat bran, respectively, was ensiled and fed at a rate of 4 kg/cow per day to substitute 1 kg of straw and 2 kg of concentrate. The silage had no obvious palatability problem according to the farmers concerned. This observation was confirmed by the trial data that showed no significant change in milk production over a 6-week trial period. Feeding carob-olive silage could result in a saving of about US$ 120/cow per lactation. A 117-day field trial on goats fed concentrate diet 2 weeks prior to parturition at a rate of 320 g/day, resulted in kid birth weight averaged across both genders higher ($P<0.05$) than that of control. Also, kids born to does in the concentrate group were fed at the age of 60 days, a daily concentrate-barley supplement (180 g) after grazing for 37 days. During this period their corresponding mothers were milked twice a day as opposed to once for the control does. Kids in both groups were allowed to nurse and remain with their mothers for one hour after milking. Both genders in the kid-concentrate group were weaned 11 days earlier and had a corrected weight of 1.8 kg at 75 days heavier than the control ($P<0.05$). Based on the difference in daily doe milk production (780g concentrate vs. 350g control) and body weight gain of kids, feeding concentrate to both does and kids showed a net profit of US$ 4.50 per doe during the trial period.

Keywords: dairy cattle, forages, carob-olive pulp, goats, milk processing.

Introduction

The animal production sector in South Lebanon has seriously suffered during the Lebanese civil war of 1975-1991 and more specifically because of the Libano-Israeli conflict through the years of 1982-2000. Nabatieh, East Saida, and Marjayoun-Hasbaya were the most critically affected areas. The main occupation in these regions is mainly olive production and when the displaced rural populations began to return progressively back to their original home villages, they found their orchards unattended with aged, non productive trees. According to the Ministry of Agriculture (MOA, 2004), 23% of the total area of 57 600 hectare olive groves in Lebanon are in Nabatieh region mostly concentrated in the Hasbaya-Marjayoun area and have never been satisfactory to sustain living. Farmers, who continued living under occupation so they could tend the few animals they possessed, turned to dairy production with few facilities to survive. They have never received any national or international development aid to improve their existing primitive farming conditions. With coming home of the displaced, they took the opportunity to increase the number of their herd and process popular milk products like laban (yoghurt) and labaneh, to cope with the present situation of population boost in the region. According to the latest statistics (MOA, 2004) the total number of dairy cows and goats in Lebanon is 86 200 and 428 200 heads, respectively. Nabatieh alone accounts for 7 300 cows and 66 700 goats. Only 40% of the dairy herd is of pure imported breed and the goats mostly belong to the local "Baladi" breed. Fifty six percent of the goat herd is milking with 12% annual mortality at less than one year of age among calves and kids.

Dairy animals under traditional feeding practices and poor hygiene management conditions result in low milk production, low fertility rate with several metabolic diseases. Dairy diets consist mainly of straw, concentrate, and left over vegetables when available. Grazing is minimal and limited to around two months in the Spring season. Similar feeding regimens are practiced during the cold season (November-February) for goats raised at an altitude up to 800m. Grazing takes place for the rest of the year. On the other hand, goat flocks at higher altitude undergo transhumance to coastal areas during cold and snowy season that extends from November to March. Unfortunately, most farmers are ignorant in roughage feeding like silage and hay and their impact on the general performance of animals. Moreover, concentrate feeding is arbitrarily practiced and frequently results in metabolic diseases in addition to increased production cost.

The presence of only one veterinarian along with one artificial insemination (AI) technician in the Regional Service Office of MOA for the whole South Lebanon area limits the services to few farmers in their immediate vicinity. Farmers in remote areas are frequently not aware of animal vaccination programs and do not benefit from such governmental services that puts them at the mercy of private veterinarians, who often charge them with outrages fees.

In addition to the low milk production which often does not reflect the true genetic potentiality of the imported animals, the farmers are facing milk hygiene problems related to the mode of milking, cooling, transportation, and processing. Cases of Malta fever in humans have been reported in the media not only in the South but also in other areas of Lebanon.

The TCP/LEB/2905 was formulated in order to address the above mentioned issues and provide the small holders with technical assistance in the fields of milk hygiene and milk processing, general farm management, AI, forage production, and animal nutrition, in the regions of Nabatieh, East Saida, and Marjayoun-Hasbaya in South Lebanon.

Material and methods

Several meetings were held with officials from the Regional Service Offices of the MOA, FAO, municipalities, and cooperatives in the designated areas of Nabatieh, East Saida, and Marjayoun-Hasbaya in South Lebanon. The objectives were to obtain information pertaining to the animal sector and public services rendered to farmers in each region. Consequently, three competent agricultural engineers from the area were appointed as national consultants in the fields of animal nutrition, milk production and processing, and extension services. Based on the field visits of these engineers and their individual meetings with animal producers and milk processors, existing impediments were identified. In the presence of Minister of Agriculture, FAO representative, public officials, and local authorities, two launching workshops were held in Nabatieh and Marjayoun for 125 and 300 attendants, respectively to explain the objectives of the project. Afterwards, pilot unit groups among beneficiaries who share the same problems in each sector of each region were formed. The three groups consist each of 20 representative pilot members closely supervised by the three engineers in collaboration with three FAO consultants in the fields of nutrition and animal husbandry, milk technology, and agricultural economics throughout the project. Ad-hoc consultants were hired to assist in solving unforeseen problems.

Milk hygiene and milk processing, and general farm management

Extensive visits to the representative dairy and goat farms were made and drawbacks related to milk production and general farm management practices were recorded. These include among others different stages of milk production, handling, and processing, feeding system, reproductive efficiency, calf rearing, hoof trimming, sanitation and vaccination programs etc.

Artificial insemination

Fifteen young, active farmers from each pilot group were selected to undergo a three day intensive AI training session. A competent, well trained veterinarian from the general directorate of animal resources of the MOA was entrusted to train the young farmers. Three culled cows were used for demonstration in each session. The best of the trainees from the three groups were selected to further undergo an advanced training session. Nine complete AI sets were procured to be placed in cooperative centers.

Forage production

The importance of silage and hay production and their use in animal feeding as substitutes to straw, was emphasized in series of group workshops and individual meetings with farmers.

Consequently, two demonstration plots of 0.5 hectare each were selected in Marjayoun valley and planted with field corn for silage and alfalfa for hay production in Spring 2004. In addition, grains of barley, oats, vetch and ray grass (2, 2, 2, and 0.5 tons, respectively) were planted by the pilot beneficiaries in a total area of 50 hectares in the Fall 2004 and in different locations. The project also procured equipment and implements to manage the new proposed hay crops such as self-propelled and hand-driven mowers, rotary rake, and round baler.

Silage of crop residues and dairy feeding trial

Olive and carob pulp refuse, byproducts of olive oil and carob syrup industries, are being disposed off arbitrarily creating an environmental problem in the region. Banana stems and leaves are abundant in the coastal area. The project introduced a new silage making technique in the region, to provide added values to these residues. A mixture of freshly produced olive and carob pulps (1:1 wt/wt) supplemented with feed grade salt and wheat bran at a rate of 2 and 100 kg per ton of olive-carob pulp, respectively, was prepared. All ingredients were thoroughly mixed and compacted to release air trapped in the mixture and hermetically covered with large plastic sheets for a period of 6 weeks for proper anaerobic fermentation process. The same procedure was applied using either olive or carob pulp separately. Another type of silage was prepared containing 50% chopped banana stem and leaves, 50% carob or olive pulp top dressed with 2 and 100 kg of salt and wheat bran, respectively. A stationary fodder chopper was procured to cut banana stems and leaves.

The silage mixtures were prepared in different locations at the farmers' sites and feeding trials using the mixture olive-carob pulps were conducted on dairy cows in 3 separate farms. The initial individual milk production was recorded daily for two weeks. Afterwards, the cows in each farm were divided into two groups of 3. One group was fed the regular ration (control) and the second received 4 kg of silage per day as a substitute of 2 kg of concentrate and 1kg of straw. Daily milk production was recorded for a period of 6 weeks. The available data were collected and statistically analyzed using a two way ANOVA with interaction and means were separated by the Duncan's method (SAS, 1992).

Goat feeding trial

A pioneer goat farm consisting of 230 milking does was selected in Nabatieh region at an altitude of about 650 m. In addition to their regular ration, a corn-barley-soybean diet, formulated to meet the nutrient requirements of goats (NRC, 1981), was fed to 50 ear-tagged does, at a rate of 320 g/head per day, two weeks prior to parturition. The parturition season of experimental goats extended for a period of 22 days and no cases of twin birth were recorded. Also, 50 control goats were selected from single birth animals during the same 22-day kidding period. Birth weight and gender were recorded and kids were ear tagged. Mortality was recorded as it occurred. Concentrate feeding for the experimental goats continued for a period of 117 days, whereas that of their respective progeny started at an average age of 60 days and lasted for 37 days. The total amount of concentrate for kids and does was measured. Milk production of control and experimental goats was recorded collectively during the last 34 days of the trial. During the first 4 days milking was performed once per day (PM) and afterwards only the experimental goats were milked twice; the control continued to be milked in the evening according to the regular practice of the owner. Experimental kids were allowed to nurse only for one hour after AM and PM milking, and all kids to graze separately from their mothers. Individual body weight for all kids at weaning was recorded and the corrected body weight at 75 days was computed according to the following formulas:

$$\text{Daily Growth Rate}\,(\text{g/day}) = \frac{\text{Weaning Weight}\,(\text{kg}) - \text{Birth Weight}\,(\text{kg})}{\text{Weaning Age}\,(\text{days})} \times 1000 \qquad (1)$$

$$\text{Corrected Weight at 75 days}\,(\text{kg}) = \frac{\text{Daily Growth Rate}\,(\text{g}) \times 75\,\text{days}}{1000} + \text{Birth Weight}\,(\text{kg}) \qquad (2)$$

Data were analyzed using the T-Test of SAS (1992) and a preliminary feasibility study for the whole intervention was made based on an average annual price of concentrate (US$ 200/ton), goat milk (US$ 0.66/kg) and live weight of young kids (US$ 2.00/kg).

Results and discussion

The outcome of meetings with public officials and local authorities along with field visits to farmers and producers revealed that the dairy production in the South could be qualified as an intensive system in which practically all feed ingredients including the straw are purchased. The average annual price of concentrate and straw is US$ 200 and 133 per ton, respectively. The average market price of cow and goat milk is US$ 0.33 and 0.67 per kg, respectively. Almost all farmers are ignorant about the importance of silage and hay feeding and their dairy herd suffers from metabolic diseases such as acidosis, indigestion with displaced abomasums, hoof problems with lameness, low fertility rate, retained placenta, and mastitis etc.

Milk production, hygiene and processing

Several workshops and individual field visits demonstrations were made to emphasize the importance of proper milking per se, filtering, agitating, and quick cooling of fresh milk in order to preserve its good quality. Strip and teat cups were also used in these demonstrations for milk testing and iodine teat dipping after milking, respectively. The use of food grade almasilium jars for milk cooling and transportation was introduced as a substitute of plastic non hygienic jars. Ten electrical chilled water tanks for quick cooling of milk were procured and placed in strategic stations of the project areas to be used by neighboring farmers.

Proper milk pasteurization procedures were demonstrated through the use of locally made gas-heated pasteurizers at home processing operations, and in different locations. Furthermore, processors acquired the proper techniques of cheese and labneh manufacturing through the use of cheese presses and blenders, respectively. The importance of refrigerating the end products was emphasized. Moreover, four milk processing pioneers participated in a one-week intensive training session at "Centre Fromager de Carmajane", France. Ten other young milk processors followed similar training sessions at the creamery of the Agricultural Research and Education Center of the American University of Beirut.

General farm management

Instructions were given to pilot farmers either collectively in workshops or through individual field visits on the importance of record keeping. Special forms for daily milk record and feeding schedule, along with charts pertaining to heat detection, insemination, and calving were handed to the farmers. To cut down on production cost and to minimize cases of metabolic disorders, special emphasis was put on feeding programs based on balanced concentrate-roughage rations, taking into consideration the requirements of the animals during the different stages of production. In order to avoid cases of diarrhea, frequent among young calves, special attention was given to proper colostrum and milk feeding with emphasis made on appropriate sanitary and clean utensils. Farmers acquired hoof trimming techniques and use of appropriate tools through demonstration sessions given at different farms. The dairy animals belonging to pilot farmers were vaccinated twice a year against foot and mouth disease in the three regions. Vaccine against enterotoxemia was also administered to dairy animals, goats and their progenies. Farmers were encouraged to vaccinate herds collectively to minimize on the medication cost.

Artificial insemination

The three AI training sessions held in the different regions were successful and resulted in the selection of 12 young promising farmers capable to acquire the AI technique. Those farmers underwent an advanced intensive AI training session and were awarded AI training certificates at a special ceremony held in Nabatieh and in the presence of Minister of Agriculture, FAO representative, and concerned agricultural cooperative representatives. The fresh graduates were entrusted with complete AI sets placed in cooperative centers to be used in collaboration with a veterinarian designated by the regional service office of the MOA.

Forage production

Farmers attending the workshops that emphasized the importance of silage and hay feeding were eager to embark in the production of corn silage and alfalfa hay. The 0.5 hectare demonstration plot planted with corn for silage in Marjayoun valley in Spring 2004 resulted only in 20 tons/hectare due to unavailability of irrigation water towards the end of the trial in spite of planting a short season variety. It was decided to chop the corn at the early milk stage and feed it as green chopped forage to dairy animals. This action saved some irrigation water that was subsequently diverted to the alfalfa plot which resulted in 5 cuts averaging 5 tons of hay/cut per hectare. Based on these facts, it was decided to concentrate on the production of rain fed hay crops like barley and oat vetch association. This new intervention was successful (8–10 tons/hectare) in stone-free land where equipment could operate. The inaccessible plots were left for grazing. The implements designed to operate in small, hilly, unfavorable terrain were tested in the presence of several farmers and technicians from the MOA emphasizing the importance of proper seed bed preparation. The equipment was handed over to MOA for future operations in collaboration with cooperative centers in the region.

Silage of crop residues and dairy feeding trial

The change in milk production of cows fed the control ration versus the carob-olive silage over a period of six weeks is presented in Table 1. The slight drop in milk production with time was expected and comparable between the control and the silage groups (P>0.05) at the end of the trial. The new silage had no obvious palatability problem according to the farmers concerned. This observation was confirmed by the trial data that showed no significant drawback on milk production. A simple calculation based on the average price per ton of concentrate (US$ 200), straw (US$ 133), and the new silage (US$ 33), shows that a saving of about US$ 120/cow per lactation can be made.

Goat feeding trial

One kid from the concentrate diet group and 3 kids from the control died at a very early stage of the trial. Therefore, the data presented in Table 2 are for 47 kids that were raised according to the traditional practice of the owner and 49 kids that were fed the concentrate diet for 37 days. The male to female ratio was around 1 to 1. Average birth weight of male kids born to does consuming the

Table 1. Effect of olive-carob pulp silage on daily milk production change of dairy cows over a period of six weeks.

	Initial milk production (kg/day)	Milk production change[1] (kg/day)			
		First 2 weeks	Second 2 week	Third 2 week	Cumulative
Feed					
Control	20.1	-0.59	-0.63	-0.91	-0.72
Silage	16.8	-0.24	-0.41	-0.42	-0.37
SEM[2]	1.12	0.781	0.795	0.986	0.818
Farm					
A	18.0	0.98[a]	0.65	-0.05	0.52
B	18.0	0.52[a]	-0.87	0.13	-0.08
C	19.3	-2.75[b]	-1.35	-2.08	-2.07
SEM[2]	1.37	0.956	0.974	1.208	1.001
		(ANOVA *P*-values)			
Variable					
Feed	0.0570	0.7605	0.8466	0.7320	0.7637
Farm	0.7355	0.0343	0.3495	0.3839	0.2035
Feed×Farm	0.5115	0.8092	0.1485	0.4994	0.4115

[1]Initial milk production was taken as a starting point.
[2]Pooled standard error of mean.
[a,b]Means within a column in each comparison group with no common superscripts differ significantly ($P<0.05$).

Table 2. Effect of concentrate supplement on performance of male and female goat kids for 37 days prior to weaning.

		Treatment		
		Control	Concentrate	SEM[1]
Males	Birth weight (kg)	3.50[a]	3.65[a]	0.081
	Weaning age (day)	108[a]	100[b]	2.1
	Weaning weight (kg)	21.4[a]	21.9[a]	0.58
	Corrected weight at 75 d (kg)	15.8[b]	17.3[a]	0.37
Females	Birth weight (kg)	2.65[b]	2.95[a]	0.074
	Weaning age (day)	109[a]	94[b]	1.4
	Weaning weight (kg)	17.5[a]	16.1[b]	0.31
	Corrected weight at 75 d (kg)	12.9[a]	13.5[a]	0.20
Both gender	Birth weight (kg)	2.85[b]	3.26[a]	0.070
	Weaning age (day)	108[a]	97[b]	1.3
	Weaning weight (kg)	18.4[a]	19.0[a]	0.51
	Corrected weight at 75 d (kg)	13.6[b]	15.4[a]	0.33

[1]Pooled standard error of means.
[a,b]Means within a row with no common superscript differ significantly ($P< 0.05$).

concentrate diet 2 weeks prior to parturition was numerically higher but not significantly different than that of control; however those of females and the two genders combined were significantly heavier. Both genders in the concentrate group were weaned 11 days earlier and had a corrected weight at 75 days 1.8 kg higher than those in the control group ($P<0.05$).

The total amount of concentrate consumed by the 50 does was 1 872 kg throughout the trial. Their kids consumed 225 kg of concentrate and 100 kg of barley in 37 days. The total milk production for the control and the experimental goats during the 34 days milking period was 592 and 1 326 kg, respectively. Taking into consideration the average annual price of concentrate, barley, milk, and live weight of kids, a simple calculation based on the difference in milk production and the gain in body weight shows a net profit of US$ 4.50 per doe.

Conclusion

1. Proper milking and milk handling procedures were demonstrated through workshops and individual field visits. Food grade almasilium jars for milk cooling and transportation were introduced as substitutes to plastic non hygienic jars. Electrical chilled water tanks for quick cooling of milk were placed in strategic stations of the project areas. Processors acquired the proper techniques of cheese and labneh manufacturing. Milk processing pioneers participated in intensive training session at "Centre Fromager de Carmajane" in France and at the Agricultural Research and Education Center of the American University of Beirut.
2. Instructions were given to pilot farmers on the importance of record keeping, heat detection, insemination, vaccination, hoof trimming, calving, and colostrum and balanced concentrate-roughage rations feeding.
3. Twelve young farmers were awarded AI training certificates. The fresh graduates were entrusted with complete AI sets placed in cooperative centers to be used in collaboration with a MOA designated veterinarian.
4. Alfalfa plot resulted in five 500 kg alfalfa hay/cut. Due to scarcity of irrigation water, emphasis was made on the production of rain fed hay crops like barley and oat vetch association.
5. The carob-olive silage trial showed no significant decrease in milk production and resulted in a saving of about US$ 120/cow per lactation.
6. Average birth weight of kids (two genders combined) born to does consuming the concentrate diet 2 weeks prior to parturition was significantly higher than the control. Both genders in the concentrate group were weaned 11 days earlier and had a corrected weight at 75 days 1.8 kg higher than those in the control group ($P<0.05$). Based on milk production of does and body weight gain of kids, feeding concentrate to both does and kids resulted in a net profit of US$ 4.50 per doe during the trial period.

Acknowledgement

This work was supported by the Animal Production and Health Division of FAO. The authors thank AGAP Personnel/FAO Rome, Mr. Abdessalam Ould Ahmed, FAO Representative in Lebanon, the staff of FAO Representation in Beirut, Ad-hoc consultants, and employees of MOA/Lebanon who actively participated in this project.

References

Ministry of Agriculture, Republic of Lebanon/FAO, 2004. Agriculture in Lebanon 2003. Directorate of Studies and Coordination, Beirut, Lebanon.

National Research Council, 1981. Nutrient Requirements of Goats: Angora, Dairy, and Meat Goats in Temperate and Tropical Countries. National Academy Press, Washington, DC.

SAS Institute, 1992. SAS User's Guide: Statistics. Version 5 Edition. SAS Institute Inc., Cary, NC.

Session 4. Quality and traceability of typical products

Main papers

Pork meat and processed products deriving from Mediterranean pig production: Quality and origin as sustainability conditions

F. Casabianca

INRA SAD LRDE, Quartier Grossetti, 20250 Corte, France

Summary

Typicality is not always the same than quality: we take the example of salty taste of Corsican ham: when reducing the salt content, local people don't yet appreciate as typical. A 6 hams comparison from 3 Mediterranean Europe countries show big chemical and sensorial differences but also several clusters of consumers rank their preferences on contrasted ways. New notions are introduced to explain these results: archetypicallity and stereotypically.

Keywords: dry cured hams, quality, typical products, consumer preferences, typicality.

Introduction

In several points of view expressed on typical animal products deriving from Mediterranean area, quality is considered as one of the major comparative advantages. But what is quality of a typical product? And who is defining quality? The producer with his conception of the local product or the consumer with his hedonic perception? As far as scientists are concerned by these questions, we must take into account that there is a debate on such notions.

We assume that «typical» is not exactly the same meaning that "quality" and design technical, sensorial and cultural identity of the product. The fact to be typical is referred to the knowledge of the local populations and quality has nothing to do with this identity. In fact, some thoughts must be conducted on defining what is considered as "typical".

We propose to mobilise some typical animal products to improve our definition, with the example of a highly symbolic product: the dry cured ham from Mediterranean Europe. First we present some investigations done on Corsican ham and how to improve his quality without decrease his typicity? Then, we compare 6 hams from several Mediterranean regions on a chemical point of view, a sensorial assessment by experts and a test on consumers' preferences. Finally, we propose some new definitions of "typical" according to the diversity of consumers' preferences.

Identity of Corsican ham

Corsican ham is deriving from a very special extensive production system. Pigs are conducted in an outdoor rearing system using mainly the natural resources that the animals are able to pasture directly on the territory.

Breeders are also procesors of the meats they obtain, in a very simple way: long salting period, long aging process in order to reach the genuine taste of the Corsican ham. In a frequent way, these hams are really salty with chloride contents till more than 12% in the fresh matter after curing. We supposed these high contents may affect the commercial interest of these products and we decided to study the possibility to reduce salt content of Corsican hams (Casabianca & Coutron, 1998).

How to reduce the content of salt?

We manage this reduction using several technologies as short salting period, then a resting period (supposed to distribute the salt within the joint) and respecting long ageing for reaching the same local characteristics of hams. We thought that, reducing this chloride content, more people are able to appreciate Corsican hams and to pay a good price for having some. In table 1, we report the technical results of an experiment done with traditional joints and in very similar conditions than local producers have.

But it changes a lot the sensory traits of what is considered as a Corsican Ham

We made a sensory assessment with an expert panel comparing the two types of salting. Results presented in table 2 shows a reduction of salty taste as expected, but also big changes in several other sensory traits. In particular, rancid taste, fatty taste are increased by the short salting.

Is the less salted still «typical»? When these two types of hams where assessed by local Corsican connoisseurs, they considered that some of the less salty hams are quite strange and they rejected them as non typical. And when preparing their file for recognize Corsican ham as PDO, the applicants deeply reflect on values of chloride content to be expected: not only a maximum (such as for example 10% in raw matter) but also a minimum (6%) that is properly unique in such demands. We concluded that reduce the salt content of the local hams above this minimum is a way to improve its quality but, at the same time, affects its typicity.

Table 1. Effect of type of salting on some traits of Corsican dry cured ham.

Type of salting	Long-term salting	Short-term salting	Significance * = 0.05
Dry matter	25.2	23.8	*
Total chlorides	7.3	4.7	**
Free chlorides	2.0	1.5	**

Table 2. Changes on sensory traits according to the type of salting of Corsican ham.

Type of salting	Long-term salting	Short-term salting	Significance* = 0.05
Brightness	9.3	10.8	*
Salty	12.6	10.8	*
Rancid	8.6	9.2	*
Fatty	7.8	8.6	*
Buttery	6.2	6.7	*

A comparison between six types of Mediterranean hams

During an AIR project on the scientific basis of quality of dry cured hams in Mediterranean Europe, we compared 6 types of hams: From Italy, Parma ham (PA) and a Country style ham (IT); from Spain, Iberian ham (IB) and Serrano (SE); from France, Corsican ham (CO) and Bayonne ham (BA).

Several chemical identities

As evidence (Monin *et al.*, 1996), the characteristics of each type was quite specific (see Table 3) and several differences appears as obvious: IB and CO were really fattier, with a high amount of intramuscular fat and a higher proportion of monounsaturated fatty acids. The salt content shows a higher value of Corsican ham, related to a high content in dry matter (CO was issued from a long ageing just like IB). This is a confirmation of big differences in technological objectives of each country.

Several sensory identities detected by the panel test

When these 6 types of hams were assessed by an expert panel, several sensorial identities were clearly identified (Rousset-Akrim *et al.*, 1996). We can see in table 4 that the chemical composition is giving quite a good anticipation of these properties: IB and CO are more marbled, red and rancid. They show fewer odours of fresh meat and with SE more concretions of tyrosine (long ageing marker). But CO is still the less soft of all types.

But the consumer test shows contrasted clusters

We compared the answers of 3 consumer panels in each country. These panels are composed by local people used to consume hams, knowing well their local hams. And we ask them to rank the

Table 3. Chemical composition of the 6 Mediterranean hams: some values in Biceps femoris.

Hams	IB	SE	PA	IT	BA	CO
Lipids % DM	23	12	10	8	7	16
Mono FA%	57	52	52	50	48	55
NaCl % DM	12	18	18	20	17	18

Table 4. Sensory traits of the 6 Mediterranean hams.

Hams/criteria	IB	SE	PA	IT	BA	CO
Marbled	4.5	2.9	3.7	2.5	3.3	4.1
Red	6.9	4.5	4.5	3.8	4.2	6.5
Tyrosine	3.8	3.0	0.7	0.3	0.6	2.7
Fresh meat	1.7	2.3	2.9	2.9	2.9	1.7
Rancid	2.5	1.3	1.1	1.1	1.3	2.5
Soft	4.8	4.2	5.3	4.5	5.0	3.6

types according to their preferences, without any consideration of price and any information on the origin of the hams.

We can identify some clusters of consumers preferring each type of ham in each country. In table 5, we report the results of ranking of 4 clusters named A to D, and, excepted for the cluster B in Spain, and for the cluster A in Italy, all clusters are represented in the 3 countries.

If we try to understand the different logics of ranking, we can see that CO is recognised as preferred by cluster A, well ranked by cluster D, and rejected by the two other clusters. That is to say that a strong identity as CO's cannot be appreciated by all people, but meet some expectations even far from its original area (the Spanish panel), if local people is able to give value to its unique properties. Making a cartography of consumers preferences, PA, BA and IT were very closed, IB and CO were far from this focal point, SE being intermediate.

What is typicality?

Several definitions for «typicality»

In a project called «TYPIC», G. Giraud (2005) distinguishes two ways to be typical:

- "First, a typical product is an archetype, produced in a traditional manner with a high level of authenticity. It is an on-farm or craft processed product with a clear and identified traceability, sometimes certified. An archetype is very different from its category. It has singular attributes, it is atypical and its market positioning is based on difference face to other competing products. Among consumers, those who appreciate archetypal product are often connoisseur, sometimes they can be expert, with a deep knowledge of the given product. This first understanding of typical product as an archetype is farmers' and small producers' approach. In this way, a typical product can reach niche and regional-oriented markets.
- The second understanding of typicality is based on resemblance. In this way, a typical product is a stereotype. It is sometimes an industrial and often a branded product. It is representative of its category. In this way, a stereotype is designed or displayed in order to be salient. To be representative of its category, such a typical product must be perceived in a central position by means of attributes of similarity. Among consumers, those who appreciate stereotypical products are often brand prone and have a broad or fuzzy knowledge of the given product. Such consumers can be found mainly in mass marketing distribution and also within tourists who are curious but not necessarily connoisseur. This second understanding of typicality is very different from the previous one. But the two are not opposite, they are complementary."

  We can summarise this double-face identity: typical product can be
- an archetype, a unique product which needs a deep learning process to have a good perception of archetypicallity;
- a stereotype, which doesn't need cognitive efforts to be identified as stereotypicallity.

Table 5. Clustering the 3 consumer panels according their ranking.

Cl/Hams	N	It	Sp	Fr	IB	SE	PA	IT	BA	CO
A	89	1	84	4	7.8	3.7	4.1	3.0	3.0	6.1
B	109	66	0	43	2.1	5.2	6.5	5.8	6.8	2.4
C	43	31	9	3	6.3	5.3	7.1	4.0	4.8	3.3
D	24	8	7	9	3.3	2.6	5.1	4.5	5.9	5.0
Tl	265	106	100	59	4.8	4.5	5.6	4.5	5.0	4.0

Consumer decision-making process

We can read a lot of affirmations on the preference of consumer, considered as a homogeneous figure. But what are consumers' expectations? Cluster analysis shows a large variety of behaviours, and in the reality, each of us can be conscious of that, for his own decision making. When we are looking for resemblance, stereotype is searched: it needs less efforts and it ensure risk minimisation. But when we are looking for difference, archetype is searched: it suppose that we show an expert ability and that is not always possible, only when we have cultural proximity with the product we are looking for.

Is PDO label useful?

Additional information such a PDO label can help a lot when product is less known, and when we are in search of a stereotype. Halo effect is still possible and we need information to make us sure in our choice. And PDO label becomes a passport to travel easier out of original area. But in case of strong cultural insurance, we can assume that this label is not useful.

Some conditions of a sustainable development

How to export?

Out of the production area, no more consumers are really experts, with the exception of some specialist or some people knowing well the type. In this case of less cultural and geographical proximity, the risk of fuzzy representations is high. We can suppose that there is a border between archetype and stereotype. An archetype in the origin area become progressively a stereotype when decreasing the proximity with consumers' competencies.

To succeed in enlarging the market, the only possibility is to comply with a crossbreeding process (as for example Parma Ham), a hybridisation to enforce the acceptance outside of the origin area.

How to connect generations?

Cultural proximity is also a problem among the generations. As we learned a lot from the ancient people on typical products, we must stimulate the learning process to reach the capacity of typicality judgement in the youngest people: a work with children and students.

Conclusions on meat products

In our study, originality of dry cured hams seems to be closely linked to raw material and local breeds as IB and CO, being archetypes and requiring a strong cultural capacity to be appreciated. At the contrary, PA (and by the way, BA and IT) must be considered as a stereotype with a large market and weak consumers competencies. But, all the local products with a strong identity are they doomed to become stereotypes?

References

Casabianca F. & Coutron Cl., 1998. Relations entre caractérisation et définition d'un produit typique non industrialisé: le Prisuttu ou jambon sec corse, in EAAP publication no. 90, 359-364.

Giraud G. 2005. Typicality, a marketing approach of consumers' attitude. Working Paper.

Monin G., Virgili R., Cornet M., Gandemer G. & Grasso F., 1996. Composition chimique et caractéristiques physiques de 6 types de jambons d'Europe latine, in Produzione Animale Vol IX, III Serie, 219 - 230

Rousset-Akrim S., Martin J.F., Bayle M.C. & Touraille C., 1996. Etude des préférences des consommateurs de l'Europe du Sud pour le jambon sec, in Viandes Prod. Carnés, 17 (6) 361-364

Traceability of typical products: possibilities and difficulties to trace out the specificities of typical products

D. Chaisemartin

World Organisation for Animal Health, 12 rue de Prony 75017 Paris, France

Summary

The traceability of typical products should help to promote their specific characteristics. Up to now, standard-setting work at the international level has been concerned with only a very general definition of traceability. Traceability of typical products, regardless of the methods used to implement it, requires a means of identifying animals and animal products, farms and production sites and recording the required information, and must take into account the characteristics to be traced. As an adjunct to existing tracing methods, DNA analysis can help to confirm documentary traceability but is not in itself sufficient to ensure full traceability. There must also be analytical methods to verify the type of feed used in a typical production system. Traceability of typical products firstly requires a precise organisational analysis of the breeding and the way the products are produced, so as to determine how traceability can best be achieved without being too complicated to put in practice or too reliant on the registered information. Traceability will be much easier to achieve if the animals from which the typical products are derived are not mixed with other animals, if there are not too many movements to be taken into account, if the processing of the products takes place in the same establishment and if there is a clear separation from any other products being processed in the same establishment. Proper implementation of traceability requires concerted action by all stakeholders, the acceptance of certification by an external body and official checks.

Keywords: traceability, origin, DNA.

Introduction

The term traceability appeared during the bovine spongiform encephalopathy crisis in 1996. The primary aim in implementing traceability was to restore consumer confidence in meat products. Yet traceability is also useful to manage and monitor animal health, public health and to promote products with recognised specific characteristics, such as 'protected designation of origin', 'protected geographical indication' or 'traditional speciality guaranteed' products. These specific products existed before the start of the bovine spongiform encephalopathy crisis, but the crisis brought with it the need to identify and trace these products more effectively, and clearly distinguish them from other products. This traceability and distinction from other products enable consumers to choose products with potential advantages, with the aim of rebuilding confidence in the consumption of products of animal origin. Moreover, in order to promote the typical products of a given region one must now be able to guarantee the specific characteristics of the product. In which case traceability can be used as a tool to maintain and trace information from the origin of the animal and the specific characteristics, whether these relate to the production of the animals or to the processing of the product.

After a reminder of the definition of traceability and the present situation regarding international standards on the subject, we will briefly examine the various methods used to trace animals and products and then analyse the elements of typical products that can be traced. Difficulties that exist in tracing these various elements will be emphasised.

Current standards on traceability

In order to guarantee the traceability of animals and products not just within a country but within the framework of international trade, it is important to use standards that are recognised at the international level within the framework of the Sanitary and Phytosanitary Agreements adopted by the member countries of the World Trade Organisation. For the animals and animal products, two organisations are recognised as the reference bodies for international regulations:

- the Codex Alimentarius Commission, for international trade in products of animal origin;
- the World Organisation for Animal Health (OIE), for international trade in animals and animal products up to primary processing, the latter part of the regulations being developed in collaboration with the Codex Alimentarius Commission so as to avoid duplication and fill any gaps in the existing regulations in this field.

Until July 2004, only very general international definitions on traceability were provided by the International Standards Organization (ISO). Standard ISO 9000:2000 defines traceability as the ability to trace the history, application or location of that which is under consideration. In the case of a product, it may be concerned with the origin of materials and parts, the processing history, the distribution and location of the products after delivery. Standard ISO 8402:1994 defines traceability as the ability to trace the history, application or location of an entity, by means of recorded identifications.

In July 2004, after several years of discussions, the Codex Alimentarius Commission agreed to insert the following definition in its procedural manual: traceability is "the ability to follow the movement of a food through specified stage(s) of production, processing and distribution". During its twenty-seventh Session, the Codex Alimentarius Commission indicated that this general definition should be completed by additional work on the subject, notably within the framework of certification of international trade.

The OIE, within the framework of its General Session in May 2004, presented the results of a questionnaire on identification and traceability that it had sent to its 167 member countries. The ensuing discussions led to a resolution that included a request for the OIE, in close collaboration with the Codex Alimentarius Commission, to develop a joint definition for animal traceability and propose guidelines for the development of identification and traceability systems commensurate with the sanitary risks involved to attain the desired outcomes. A first meeting on the subject was due to take place in June 2005 within the framework of the OIE Ad hoc Group set up in response to this request.

Traceability of animals and products

Traceability of an animal and their products presupposes the existence of an identifier attached to the animal and its products and an information system that keeps a record of all the entries relating to the animal and its products.

The system of traceability must allow for tracing back and tracing forward. This means that the traceability of an animal and their products must enable:

- from the finished product, to trace the different stages in the processing of the product and as far back as the farm of origin, and thus to be able to trace all the raw materials involved in the processing of the product;
- from the animals' farm of origin, to trace the different production and processing stages through which the animals pass before being slaughtered and processed, and thus to be able to trace all the finished products manufactured from these raw materials.

To be able to trace the different places where an animal was raised and a product was processed presupposes the identification of all the various sites where animals are raised, pass through or are held as well as all the various slaughter sites and processing sites to the final distribution point.

To enable a product to be recognised, taking into account specificities of animal production or specificities of product processing, and to guarantee that its specificities are monitored and controlled, it is important to register information on these specificities and to indicate at what step from the animal production chain to the final product the information will be recovered and accessible. To enable this information to be traced, it will have to be linked to the animal or to the products derived from the animal.

To ensure that the traceability system is reliable, it is important to analyse the entire information circuit and the various stages where information is changed, notably information that is essential for the traceability, with the constraints arising from the transfer of this information subject to change and to verify that it matches the physical and spatio-temporal characteristics of the item being traced. All the changes relating to an animal or product must be recorded and analysed so as to identify the different operational actions which, if not properly performed, might result in a loss of traceability. Solutions aimed at minimising these risks must be proposed.

Traceability can either be ensured at the level of the individual or at the level of the batch. The choice of identification by individual or by batch will depend on an economic analysis of the potential risks of having to withdraw animal products or even having to destroy animals.

To ensure the traceability of products derived from an animal, it is essential to attach an identifier to the carcass of the animal and establish a link with the identifier of the animal, and then at each stage in the processing of the carcass and the products derived from it create a link between the product to be processed and the resulting products. Throughout the chain of slaughter, cutting of the carcass, and processing of the products, traceability will only be achieved if all the establishments involved have taken the necessary steps to ensure continuity in the monitoring of information on a product within their premises and if they have complemented their material accounts with the minimum required information.

Nowadays, traceability of animals and products is achieved by means of documentary records kept at each link in the chain:

- at the level of the animal holder, the identification is entered in the farm records with any additional information needed to ensure the traceability of a product's specific characteristics. The origin of animals entering the holding and the destination of animals leaving must also be taken into account;
- at the level of the slaughter house, meat-cutting and processing halls, the identifiers of the items entering their establishment must be recorded in registers along with a link to the items produced and sent either to another establishment or for distribution. The origin of products entering the establishment and the destination of products leaving it must also be taken into account;
- the distributor keeps the information supplied by the processing establishment and ensures the continuity of traceability by recording the products received.

The introduction of electronic recording equipment for traceability will help to make the system more reliable – especially if information can be automatically retrieved, without having to be re-entered, between the different steps in the movement of an animal and the processing of their products – and help to develop tools for rapid analysis and research.

Methods used in traceability

For animals, traceability involves marking them with an identifier, using one of a number of methods.

Depending on the type of animal and the husbandry method, identification of the animal can be achieved either:

- by directly marking the animal by ear-tattooing or by branding (hot-iron branding, liquid nitrogen branding);
- by attaching a metal ear tag;
- by attaching a plastic ear tag, with either a number, a number and a bar code, or an electronic tag;
- by electronic implant;
- or by ruminal bolus.

Taking into account the criteria of visibility and cost of the identifier, the most economic solution is the ear tag. However, ear tags come in many different shapes and materials and this can affect the way the tag is attached to the animal and its durability. To minimise criticism relating to the loss of a plastic ear tag the possibility of falsification, it is important to clearly define its technical characteristics and the technique for attaching it, to strictly manage the way tags are issued and applied, and to clearly define rules in the event of the loss of an ear tag. The use of handwritten ear tags that can easily be obtained from shops should be prohibited for traceability purposes.

Marking the body of an animal is more difficult to carry out, depending on the size of the mark to be applied, skill in application, notably the amount of pressure required to mark the animal correctly, and taking into account the movement of the animal during marking (not all movements can be controlled), which can reduce the visibility of the mark. In the case of tattooing, it is important to remember the risk of bleeding during marking and therefore the risk of transmitting animal diseases (e.g. enzootic bovine leukosis) from animal to animal while they are being tattooed.

Electronic identification has the advantage of facilitating recording of the identifier during the various movements of the animal and when animal production specificities have to be recorded in association with this identification. The analysis of this method should not be based only on the cost of the marking device but should also take into account the potential gains from using this type of material. The following points need to be taken into account when deciding which form of electronic identification to choose:

- an electronic tag is easy to apply and is visible, demonstrating to everyone that the animal is identified;
- an electronic implant is a little more difficult to insert, frequently migrates and does not make it easy to show that the animal is identified. If the implant does not work properly, it may result in a new implant being inserted without one necessarily being aware of the existing implant or at least not being able to recover it;
- a ruminal bolus is more difficult to insert, generally has to be done by a technician and is not always retained by an animal weighing less than 25 kg. In some regions, livestock producers deposit magnets in animals to avoid the risk of injury from ferrous material that the animals might ingest. Furthermore, as the ruminal bolus is not visible, if it starts to dysfunction it may lead to a second bolus being inserted. Depending on how these boluses function, they may both work together or alternately, leading to complications for monitoring and identification.

The microchip used in the electronic tag, electronic implant and ruminal bolus is generally a low frequency read only chip, with a reading distance ranging from 10 cm to a maximum of 80 cm. Given the very many different types of chip and types of reading device, it is preferable to use products that comply with standards ISO 11784: 1996 and 11785: 1996 for read only chips for use in animals. Standards are also being developed for read/write chips that enable information to be written in addition to the country code and the national identification number (Standard ISO 14223-1: 2003).

For animal products, traceability also requires them to be marked with an identifier. In most cases, and especially when these products have resulted from cutting and processing, the identifier is for a batch of products.

Identification of products is achieved either:

- by applying a number directly to the product. This is notably the case with whole carcasses, half-carcasses and even primal cuts. Edible ink must be used and it is important to ensure that humidity and refrigeration do not damage the marking;
- by printing on the boxes;
- by applying a label to a tray, a plastic overwrap on the product ready for sale, a box of products or a pallet. The label may include a bar code; the type of code will depend on the type and quantity of information that the manufacturer wishes to include. Generally speaking, the type of bar code should comply with standard EAN 128;
- within firms, by using hooks identified with a label (electronic or otherwise) or identified trays, especially for small pieces;
- or by using RFID (radio frequency identification). The microchips used for products are high-frequency chips and the existing standards are ISO 18000-1: 2004, 18000-2:2004, 18000-3:2004, 18000-4:2004, 18000-6:2004, 18000-7:2004. However, many of the chips are proprietary products and are not necessarily compatible.

As a complement to the definition and application of the identifier for an animal, DNA provides a guarantee of the animal's identifier throughout the chain and can be used to trace all the products derived from it. DNA analysis complements or confirms documentary traceability by providing irrefutable biological proof of the origin of the animal. It does, however, require a tissue sample to be taken at the same time as the animal's identifier is applied.

Animal identification can be performed using a small number of regions of the genome for which typing is easy and can be automated. These regions (loci) must present variant forms or alleles (polymorphism) in the animal population of interest and must be sufficient in number to ensure that each individual is uniquely characterised within the entire population of the species. This can be used to define the genetic markers to be looked for in a DNA test. Samples that are suitable for DNA analysis are skin, hair, blood and meat, though hair samples have proved to be more difficult to deal with (Sancristobal-Gaudy *et al.*, 2000). Furthermore, to minimise the risk of error right from the start of primary processing of the carcass of an animal, it is recommended that a control sample is should be taken when the animal is slaughtered, to ensure that the animal's identifier is indeed the right one to provide the link with the identified carcass (Vasquez *et al.*, 2004).

With the various methods that can be used in the study of genetic markers, it is also possible to detect and trace either an individual or a group of individuals with the same parent, this having been used in the pig sector to trace animals from the same mother using SNP (single nucleotide polymorphism) technology (China *et al.*, 2001). It is not possible to trace directly a batch of animals with no common genetic criterion, in other words a batch of animals purchased from different sources or born to different individuals within the same farm. For such groups of animals, DNA tracing involves genotyping each animal and indicating in a database that these animals belong to a given batch in the farm of origin and that they have a batch reference that is either the identification number of the holding or of a specific batch within the holding.

The genetic markers that can be used are microsatellites and SNPs. The use of SNPs as genetic markers is still the subject of debate since the information they contain is relatively slight compared with microsatellite markers (China *et al.*, 2001). It has been estimated that it takes at least five SNP markers to provide an equivalent amount of information to that contained in a microsatellite (Chakraborty *et al.*, 1999). Microsatellite markers are particularly useful because they present a high degree of polymorphism (Sancristobal-Gaudy *et al.*, 2000).

The implementation of traceability for animals and meat that makes use of DNA no longer requires mastery of molecular genetic techniques, and is more a question of organising the sector to sample animals, and extracting, storing and genotyping DNA. With regard to the sampling of DNA at the time the animal is marked, it is worth mentioning that there are plastic ear tags with a small individually identified tube that can be used to recover the small piece of ear that is perforated when the ear tag is attached. This means that any livestock holder can mark the animal and send the sample at the same time as the notification of birth and marking of the animal.

Tracing of typical products

The traceability of typical products is important for consumer confidence in such products, for the health of the animals from which the products are derived and for food safety. The traceability of typical products is also important for the commercial identity of typical products and to ensure that this is not usurped, and that the specifications for the typical product, where these exist, are properly respected.

The components that may characterise typical products and their production systems are as follows:

- an extensive animal production system;
- production within a defined geographical zone;
- farming of typical breeds;
- typical feed (which may be specific: e.g. acorns for typical ham with a protected designation of origin in Spain as "*Jamón de Huelva*");
- specific methods for preparing products.

These typical products are often produced with limited resources, particularly financial resources. A careful analysis is therefore needed of ways of achieving traceability at a reasonable cost.

Tracing the origin of typical products

The first problem is the definition of the term 'origin', notably for the labelling of a product from a typical production system. It would be logical to consider that when referring to the origin of a product, it should imply that the animals from which the product was derived were born, raised and slaughtered in the same country, and that all the steps in processing the product were also carried out in that country. European Union legislation on labelling of beef (Regulation (EC) No 1760/2000 of the European Parliament and of the Council of July 17th 2000) limits the definition of the term 'origin' to cover only where the animals were born, fattened and slaughtered. Moreover, at the international level, the latest discussions of the Codex Alimentarius Commission, during its twenty-seventh Session, have raised the problem of the name of the country of origin on the label being changed when the product is processed in a different country (rule of the Codex General Standard for the labelling of prepackaged foods).

Regardless of whether the definition of origin only takes into account the path taken by the animal from birth to slaughter or also includes the various processing stages to achieve the finished product, traceability of origin will require a record of the animal's identification and place of birth and a record of all its movements as well as the link with the carcass derived from the animal and a record of the different stages of processing leading to the product supplied to the consumer. It is therefore necessary to have the appropriate recording tools and to carry out data recording methodically. It is also important to control the allocation of identification marks (to avoid any duplication or errors) and any loss of the identification mark.

Another risk of loss of traceability is when information relating to a link in the chain has not been registered. The risk is increased when there are animal movements, when product processing involves movement through different establishments, or when there is no organisational system in place to clearly differentiate between the work being carried out on the product in question and other products being processed in the same establishment.

The use of DNA tests cannot guarantee the origin of the product. It simply provides a means of guaranteeing, if samples were taken when the animal was marked and when it was slaughtered, that the identifier of the animal and that of the product correspond, or of recovering the identification of an animal or product where this has been lost, though this would of course mean having to keep a complete repository of DNA samples, carrying out DNA analyses and maintaining the link between the result of the DNA analysis and the initial identification of the animal. A DNA analysis system of this kind is very expensive and this is a major limiting factor for typical production systems, which do not necessarily have a sufficient economic margin to meet the costs of DNA analysis.

The same remarks apply to the traceability of a product to a given geographical area as those relating to the traceability of origin.

Tracing the characteristics of animals in typical production systems

The simplest solution appears to be to register the characteristics of the animals to be traced when the animals are initially identified and to make these data accessible throughout the different stages in production of the animals and the processing of products derived from the animals. In this case, the risks of loss of traceability are non-registration of information, failure to provide the information and, lastly, loss of identification of the marked animals.

Some of the animals' characteristics can be defined by DNA analysis, with samples being taken as indicated above. The sex of the animal can easily be determined by testing for DNA markers of the Y chromosome, and an analysis of the frequency of different alleles of microsatellite DNA markers can be used to characterise breeds, as already demonstrated for numerous bovine breeds (Moazami-Goudarzi *et al.*, 1997). Genetic markers can be used to reveal the genetic diversity of domestic ungulate breeds.

The identification of species present in products can be used to ensure that a product has been made from a given animal species and does not contain material from other animal species.

Tracing feed given to animals in typical production systems

For traceability of this type, details of the feed given to animals must be recorded on a regular basis, linking it with their initial identification and providing access to this information throughout the animals' production chain and processing of products derived from them. In this case, the risks of loss of traceability are non-registration of information, failure to make the information available and, lastly, loss of identification of the marked animals.

Nevertheless, some feed components, once ingested, induce the production of specific elements within the animal which can be detected by analyses. In this case, if the specificity of the feed component is clearly detectable in a product derived from the animal, chemical methods can be used to verify whether or not the component has been ingested. One can therefore tell whether the animal's rations included the component in question or certify that the animal did not ingest the component. These methods cannot however be used to certify that the quantities to be ingested have been respected or determine the period when these feedstuffs were ingested.

The analytical tools currently being developed to trace animal feed use tracers of vegetable origin such as terpenes and carotenoids (Coulon and Priolo, 2002). The preliminary results indicate the need to use several markers in conjunction and to cross the measurements for the product (meat, milk, cheese) with the measurements for the animal's tissues and fluids. These preliminary results show a great variability in response, making it essential to obtain confirmation on a larger scale, identify the sources of variation and quantify their effects. These techniques have already been used to discriminate milk products and meat products obtained with highly contrasted types of feed, for example grass compared to corn silage in cattle, or grazed grass compared to animal feed concentrate and hay in lambs. However, intermediate feed types with a mixture of components will undoubtedly be more difficult to characterise. However, this approach is rather more a form of statistical verification than individual traceability.

Tracing the manufacturing characteristics of a product

For the manufacturing characteristics of a product, two options can be considered: firstly to indicate either that the product does not contain a given substance or that the substance has been incorporated in the product and secondly to guarantee the key stages in the preparation of the product, such as the period of curing for hams.

If the specified characteristics relating to substances either incorporated or excluded during manufacture have a characteristic DNA fingerprint, it is possible to verify the presence or absence of these substances. However, this amounts to verification rather than traceability of the product and its manufacture.

Regarding the parameters for product preparation such as, for example, the duration of the curing process for a product, an organoleptic analysis of the product may indicate that the product has indeed undergone curing but will not be able to indicate the length of time. The traceability of these characteristics therefore requires that at each stage of production the characteristics to be traced are registered.

Here, too, the risk of loss of traceability is the non-recording of information for a link in the chain. This risk is greatly increased if the product in question needs a lot of steps to be produced or if the product has to pass through several establishments or if there is no organisation in place to enable work on this product to be clearly distinguished from other products being processed in the same establishment.

How to improve the traceability of a typical product

It would be useful to undertake an organisational analysis of the animal production and processing system used to obtain a typical product and to examine all the factors that could undermine the achievement of low-cost traceability.

It is important to determine whether there are any animal movements between the different animal production centres and if there is any mixing of animals that do not correspond to the same typical production system.

A typical production system generally involves only one animal species raised extensively. If, therefore, there is no mixing of animals of different species and animals of the same species are being used for different purposes, it would be more simple to consider the whole zone as forming a single unit, which would greatly facilitate management for traceability. However, such a choice would have a major impact in the event of a sanitary problem: the entire zone would have to be placed under surveillance, regardless of its size.

Likewise for the stages of processing from the animal carcass to the finished product, if the slaughter, cutting and processing operations take place in the same establishment, or even in a small number of establishments with a clear distinction between the materials used to achieve this typical product from materials used for other products, it will greatly facilitate traceability by simplifying the organisation and facilitating the recording of information required throughout the production chain from the animal to the product.

The implementation of traceability will be improved if the elements of a typical production system that are to be promoted are clearly defined, if there is certification by an external body and if professionals are willing to allow their sites of animal production, collection, slaughter, cutting and processing to be officially controlled to verify that traceability is being properly implemented. Prior consultation with all the stakeholders in the production sector will therefore be required before traceability can be implemented.

Conclusion

As with any form of production, the traceability of a typical product requires the identification of the animals and products to be traced and the recording of this information, as well as the specificities that must be monitored throughout the chain. DNA analysis can confirm the identification of the animal and certain of its characteristics, but does not in itself ensure traceability. Likewise, some chemical techniques can help to detect certain feed substances but this is merely a way of verifying the presence or absence of these substances.

The traceability of typical products can be facilitated by analysing the organisation of the production system and proposing solutions to minimise the risk of these products being confused with others or of information being lost. The introduction of consultation between all the stakeholders in a given sector, drawing up a list of specifications for approval by a competent authority and the implementation of official controls will facilitate the management of traceability.

References

Chakraborty, R., Dn. Stivers, B. Su, Y. Zhong & B. Budwole, 1999. The utility of short tandem repeat loci beyond human identification: implications for development of new DNA typing systems. Electrophoresis, 20, 1682-1696.

China B., V. Evrard, R. Noirfalise, A. Clinquart & G. Daube, 2001. La traçabilité dans la filière viande. II Les marqueurs génétiques. Annales Médecine Vétérinaire, vol. 145, 15-24.

Coulon J.B. & A. Priolo, 2002. La qualité sensorielle des produits laitiers et de la viande dépend des fourrages consommés par les animaux. INRA Productions animales, Vol. 15, 333-342.

Moazami-Goudarzi K., D. Laloë, J.P. Furet & F. Grosclaude, 1997. Analysis of genetic relationships between 10 cattle breeds with 17 microsatellites. Anim. Genetics, 28, 338-345.

Sancristobal-Gaudy M., G. Renand, Y. Amigues, M.-Y. Boscher, H. Leveziel & B. Bibe, 2000. Traçabilité individuelle des viandes bovines à l'aide de marqueurs génétiques INRA Productions Animales, 13, 269-276.

Vasquez J. F., T. Perez, F. Ureña, E. Gudin, J. Albornoz & A. Dominiguez, 2004. Practical application of DNA fingerprinting to trace beef. Journal of Food Protection, Vol. 67, no. 5, 972-979.

Discrimination of production system and origin of animal products using chemical markers

R.J.B. Bessa[1], S.P. Alves[1], R. Figueiredo[2], A. Teixeira[2], A. Rodrigues[2], A. Janeiro[2], M.C. Costa[2], J. Santos-Silva[1] & J.A.M. Prates[3]

[1]*Estação Zootécnica Nacional, Fonte Boa, 2005-048 Vale de Santarém, Portugal*
[2]*INETI, Estrada do Paço do Lumiar, 1649-038 Lisboa, Portugal*
[3]*Faculdade de Medicina Veterinária - CIISA, Avenida da Universidade Técnica, Pólo Universitário do Alto da Ajuda, 1300-477, Lisboa, Portugal*

Summary

The geographical origin and production system of meat and dairy products are particularly important for those legally protected in UE with designation of origin (PDO) and geographical indications (PGI). The development of analytic tools that allow the quality control and authentication of these products are needed and are an important research topic. Some chemical markers and techniques are briefly reviewed. Particular emphasis is given to the potential of fatty acids as markers of production system. In addition, the utilization of plant secondary metabolites as markers of geographical origin and production system are discussed. The potential of stable isotopes ratios methods and some spectroscopic methods are also reviewed. The recent research produced interesting and promising results indicating that effective discrimination of geographical origin and production system of animal foods may be achievable.

Keywords: geographical origin, production system, fatty acids, terpenoids, stable isotopes.

Introduction

Most of the diversity of typical animal products in Europe is linked to the Mediterranean culture. In fact, more than 80% of total products protected by European legislation (CEE 2081/92) are from Mediterranean countries (including Portugal). The implementation of this EC regulation in order to protect and promote the edible products with Protected Designation of Origin (PDO) and Protected Geographical Indication (PGI) is an important contribution to the sustainability of the production chains.

The PDO and PGI products are obtained according to defined specifications in a well delimited region. Moreover, at least for PDO products, specifications can be very precise considering the genotype of animals, feeding and management and technological practices. In fact, the determinant factors of the sensorial "identity" of an animal product, particularly cheese, have been attributed not only to a technological process but also to upstream factors linked to the animals (genetics, physiological stage and health status) and animal feeding (type of diet, forage preservation method and grass botanical composition) (Coulon *et al.*, 2004).

The added value promoted by legal protection of PDO products may introduce a strong pressure to the intensification of production systems particularly those where the specifications are not too strict. Although the technological factors have a much greater influence than production factors, the effects of intensification of the production system on sensory characteristics of Cantal cheese have been recently demonstrated in a plant scale study (Agrabriel *et al.*, 2004). When product specifications allowed, as in Serpa-PDO cheese, an increase in the introduction of exotic breeds can be observed, as well as improved pasture management and higher levels of supplementation with concentrates. The development and implementation of methods that contribute to define objectively the typicity of animal products will be important either to ensure the authenticity and origin or as a tool to define the limits of intensification.

Moreover, recent health concerns, mostly linked to BSE crisis, increased the awareness of consumers to the geographic origin and the production system of beef (and other foods), pressing the food industry for accurate labeling of products. So, the development of analytical techniques that can identify the geographical origin and main features of production systems are at the moment a very important research topic.

We will review, non-exhaustively, some of the chemical markers and techniques that have been employed to discriminate the production systems and the geographical origin of animal foods.

Fatty acids and related volatile metabolites

Fatty acids in tissues and milk can either be synthesized endogenously or derived directly or indirectly from feed. The major polyunsaturated fatty acids are mostly derived from metabolization of dietary linoleic (18:2 n-6) and linolenic (18:3 n-3) acids. In ruminants, greater complexity is added by ruminal metabolization of dietary fatty acids and by microbial lipid synthesis. Both 18:2 n-6 and 18:3 n-3 are extensively hydrogenated in the rumen, leading to the formation and consequent presence in milk and tissues, of a complex pattern of octadecenoates (18:1), octadecadienoates (18:2) and octadecatrienoates (18:3) isomers (Bessa *et al.*, 2000). A high number of studies indicate that dietary conditions are determinant of the isomeric pattern found in the meat and milk/cheese. Diets rich in concentrate induce high levels of 18:1 *trans*-10 (Bessa *et al.*, 2005) and in conjugated octadecadienoic isomers (CLA) with double bonds in position 10 and 8 (Bessa *et al.*, unpublished) whereas forage based diets are characterized by the predominance of 18:1 *trans*-11 and 18:2 *cis*-9, *trans*-11.

Diets rich in linolenic acid, including pastures and forage based diets supplemented with linseed oil, led to a much more CLA isomer diversification with high proportions of *trans*-11, *trans*-13 and *trans*-11, *cis*-13 isomers as well as in the non conjugated 18:2 *trans*-11, *cis*-15 isomer (Collomb *et al.*, 2004). The identification of feeding systems, seasonal variations and altitude of pasture has been achieved using the fatty acids patterns in products. Martin *et al.* (2005) review the data relating the type of forage and its conservation method (silage vs hay) and the milk fatty acid pattern, confirming its interest as indicators of cow feeding. Santos-Silva *et al.* (2002) were able to discriminate lambs produced in three production systems using the fatty acid profile (Figure 1).

Besides the biohydrogenation derived fatty acids, also microbial structural fatty acids have some potential as indicators of the feeding practices. Odd chain and *iso*- and *anteiso*-branched chain fatty acids have been recently explored as indicators of the ruminal function (Cabrita *et al.*, 2003). These authors found a positive association between *anteiso*-15:0 and soluble sugar and a negative association between 17:0 and protein in the diet of dairy cows.

One practical application of fatty acids as markers of production system is found in the grading system implemented in Spain for Iberian pig finished in a range regime (mainly acorn and grass) in savannah-like pastures (called "montado" in Portugal and "dehesa" in Spain) where oaks (*Quercus ilex* and *Q. suber*) are the predominant trees. Every year, the Spanish Ministry of Agriculture approves a contract type for sale of Iberian pig carcasses that includes the weight percentages ranges for

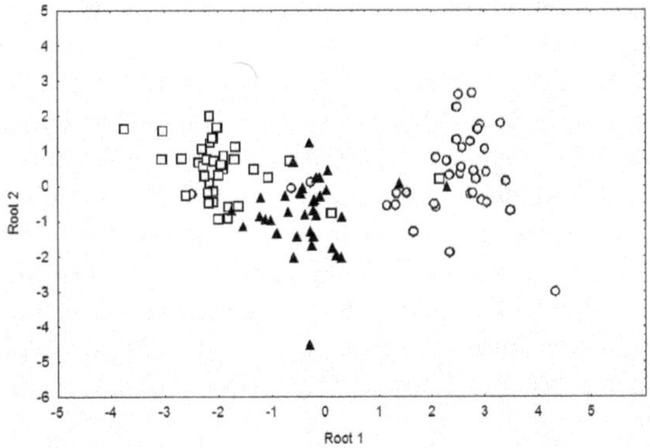

Figure 1. Lamb meat discrimination according to the feeding system (○ – indoors, concentrate, ▲-pasture supplemented with concentrate, □ -raised on pasture with their dams). Canonical discriminating analysis with 18:3 n-3, 18:2 cis-9,trans-11, 18:2 n-6, 20:4 n-6, 15:0, 14:0 and 16:1 as predictor variables (Santos-Silva et al., 2002).

palmitic, stearic, oleic and linoleic acids, measured in subcutaneous fat, that allows the classification in three commercial categories ("bellota", "recebo" or "pienso") (Garcia-Olmo *et al.*, 2002).

There are indications that some metabolites derived from fatty acids can be useful as chemical markers of the feeding system. Urbach (1990) reviewed the dietary effects on ³-lactone pattern in milk. Diets that reduce the ratio between acetic and propionic acids produced in the rumen (high grain diets or lipid supplements) promote the ruminal hydration of unsaturated fatty acids, and the endogenous conversion of the hydroxi-fatty acids in lactones that can be found in milk and meat (Urbach, 1990; Joblin & Hudson, 1997). The complexity of lactone pattern and its relation to dietary factors like basal diet and type and level of unsaturated fatty acids, resembles the generation of the isomeric pattern that occurs in rumen biohydrogenation and could produce similar results in terms of animal feeding discrimination.

A linoleic acid oxidation product, the 2,3-octanedione, found in sheep fat is highly specific to grass feeding (Young *et al.*, 1997). Priolo *et al.* (2004) confirmed the value of 2,3-octanedione as marker to distinguish lambs raised and finished in pasture from lambs raised and finished in concentrate. Moreover, they found that 2,3-octanedione concentrations decrease linearly ($r = -0.88$) with the time of concentrate feeding, considering as well lambs raised in pasture but finished on concentrate feeds. Young *et al.* (1997) hypothesized that the formation of 2,3-octanedione is related to vegetal lipoxygenase, absent in maize and abundant in leafy plants, which should explain the association with grass feeding.

Isoprenoids and other secondary metabolites of plants

Isoprenoids (or terpenoids) are a vast class of compounds composed of characteristics C5 units, including the monoterpenes (C10), sesquiterpenes (C15), diterpenes (C20) and tetraprenes (C40).

Monoterpenes and sesquiterpenes

Monoterpenes and sesquiterpenes are highly volatile compounds and synthesized mostly by plants. The accumulation and active air emission of monoterpenes is highly variable between plants, even within the same genus and depends on season, temperature, light conditions, farming condition and several stress factors (Kesselmeier & Staudt, 1999). The presence of monoterpenes and sesquiterpenes in milk/cheese (Dumont & Adda, 1978; Moio *et al.*, 1996) and ruminant meat (Larick *et al.*, 1987; Priolo *et al.*, 2004) has been reported and related to the animal diets. The type and concentration of mono- and sesquiterpenes in forages are dependent on botanical composition, geographic origin and forage conservation method, and these differences are reflected on the terperne pattern of milk/cheese as reviewed by Martin *et al.* (2005). The transfer of terpenes from diet to milk has been studied (Viallon *et al.*, 2000), and it is not clear if it is the digestive or pulmonary route the main source of entry of terpenes in ruminants. Viallon *et al.* (2000) found that the maximum secretion of terpenes was 4 days after the introduction of an aromatic dried plant in diet, decreasing thereafter, which suggests a metabolic adaptation of rumen microbial ecosystem and/or hepatic and urinary clearance and detoxification routes.

Application of terpenes in the discrimination of cheese origin has been attempted. It seems particularly useful for identification of mountain cheeses (Bosset *et al.*, 1999). In the Alps, higher altitude enhances the botanic diversity of pastures decreasing the *Graminacae* species, poorer in terpenes, and increasing the dicotyledonous species, particularly *Compositae*, *Rosaceae*, and *Plantaginaceae*, much richer in terpenes (Mariaca *et al.*, 1997, Jeangros *et al.*, 1999). The usefulness of terpenes to discriminate milk from cows fed common dairy cows diets (barley based concentrates, maize silage, rye-grass hay or silage, and mountain grassland hay) were evaluated (Martin *et al.*, 2005). These authors concluded that the terpene fraction of milk was not sufficient to accurately discriminate between all diets, but the discrimination was excellent when combined with milk fatty acid profile and adequate chemometrics techniques.

The application of terpene profile analysis in meat and body fat is much less explored than in milk/cheese. Priolo *et al.* (2004) studied the terpenes desorbed from fat of sheep raised and finished on grass, raised and finished on concentrate or raised on grass and finished with concentrates. The terpenes were found in trace concentrations. The only monoterpene significantly affected by treatments was p-cymene but several sesquiterpenes were affected by treatments. One of them, the β-caryophyllene, was highly represented in the fat of lambs raised and finished on pasture and virtually absent in lambs finished with concentrate. The distinction between grass-raised/concentrate finished and concentrate-raised/concentrate finished lambs was not achieved.

Diterpenes and tetraprenes

Some diterpenes (tocopherols and tocotrienols) and tetraprenes (caretonoids) of vegetal origin also are potential markers of the production system. The quantity and forms of tocopherols and tocotrienols in plants vary among species (Hess, 1993) and, at least for *Brassicaceae* family, have great chemotaxonomic value (Goffman *et al.*, 1999).

Beef meat has poor diterpene pattern, with the predominance of α-tocopherol, followed by γ-tocopherol (Prates *et al.*, 2005). In meat from Portuguese PDO beef, γ-tocopherol is much more affected by slaughter season (reflecting the availability of grass and level of concentrate supplementation) than α-tocopherol (Quaresma *et al.*, unpublished). Pastures are rich in α-tocopherol but have low concentration of γ-tocopherol while and concentrate (depending of ingredients) are usually richer in γ-tocopherol than pastures (Pontes *et al.* , unpublished).

The diterpene pattern of poultry meat is much more complex with α-, β-, γ-, δ-tocopherols, and corresponding tocotrienols, present. However, these profiles do not allow the discrimination between poultry raised with or without access to pasture (Pontes *et al.*, unpublished). Pigs finished in mediterranean ranges have higher γ-tocopherol content than concentrate fed pigs, reflecting the high concentration of γ-tocopherol found in acorn (Daza *et al.*, 2005; Rey *et al.*, 1998).

The carotenoids are among the most widespread pigment classes in living organisms but only plants and protists are able to perform their synthesis (Torrissen, 2000). Green leaves of vegetables are the richer sources of carotenoids in animal feeds, but hay-making results in great losses. The major carotenoids of grass pastures are lutein, β-carotene, 13 cis-β-carotene, zeaxantin (Prache *et al.*, 2003a) however, only lutein (ovine, caprine) or lutein and β-carotene (bovine) can be found in adipose tissues of ruminants (Yang *et al.*, 1992; Prache *et al.*, 2003b). Prache & Theriez (1999) demonstrated that these carotenoids can be used as biomarkers of grass feeding in ruminants and proposed a non-invasive method based on the reflectance spectrum by carcass fat in the range of 450 to 510 nm (corresponding to the light absorption by carotenoids). Application of this method to the discrimination between grass-raised-grass-finished, grass-raised-concentrate-finished and concentrate-raised-concentrate-finished lambs has been explored (Prache *et al.*, 2003a, b). However, the results are only satisfactory if coupled with plasma carotenoids content. Martin *et al.* (2005) measured colour (b*) and β-carotene in milk from cows fed diets based on distinct forages and were able to partially distinguish them. High variation in colour of milks from cows fed diets based in hay were observed, preventing its distinction from all other diets in trial.

Phenolics compounds

Phenolics compounds are a vast group of vegetable compounds and some have potential positive effects on human health (Scalbert *et al.*, 2005). The structural diversity of polyphenols is huge (more than 4000 flavonoids are identified) and a phenolic molecule is often characteristic of a plant species or even of a particular tissue of that plant.

Scalbert and Williamson (2000) reviewed the bioavailability of dietary polyphenols and found that they are extensively metabolised both in gut and tissues resulting in mostly unknown metabolites that can be extensively conjugated in the liver. Some phenolic compounds have been found in milk (Lopez & Lindsay, 1993; Besle *et al.*, 2005; O´Connell & Fox, 2001) and most of them seem to be rather unespecific and resulting from rumen catabolism of cell wall phenolics (lignin and hydroxycinnamic acids). However, others were found to be specific of a particular diet. The potential of phenolic compounds and its metabolites as chemical markers of animal feeding exist, although detailed information is still very scarce.

Stable isotopes

Recently, an increasing number of reports on the use of stable isotopic ratios to identify the origin and production system of meat (Piasentier *et al.*, 2003; Boner & Forstel, 2004; Renou *et al.*, 2004a; Schmidt *et al.*, 2005) and milk and dairy products (Rossman *et al.*, 2000; Pillonel *et al.*, 2003a; Renou *et al.*, 2004b; Brescia *et al.*, 2005) have been published. The (multi) element isotopic ratios, usually expressed as d values defined as "(sample ratio – reference ratio)/reference ratio x 1000", can function as natural fingerprints of vegetable compounds containing relatively complex environmental signatures. These methods have been extensively applied in the authentication of wines, fruit juices and other vegetable products (Rossman, 2001).

The δ^2H and $\delta^{18}O$ are clearly affected by climatic factors and are valuable in the determination of the geographic origin, as review by Martin & Martin (2003). The $\delta^{15}N$ is influenced mostly by soil conditions, intensity of agricultural practices, sources of N fertilizers and botanical composition of pastures (Piasentier *et al.*, 2003). The $\delta^{15}N$ in soil, plants and animal products increases with the utilization of organic fertilizers (Kreitler & Jones, 1975) and decreases in the presence of leguminous plants (Yoneyama, 1995). The $\delta^{87}Sr$ can also be useful for origin assignments as it is dependent only on the types of rocks and soils and not on human activity, climate or season of production (Pillonel *et al.*, 2003a; Rossmann *et al.*, 2000).

The $\delta^{13}C$ is mainly dependent on different isotopic fractionation in metabolic pathways in plants. In photosynthetic carbon fixation, C3 plants (which assimilate carbon via the Calvin cycle) have lower $\delta^{13}C$ than C4 plants (which assimilate carbon via the Hatch-Slack cycle) (O'Leary, 1981). The $\delta^{13}C$ in animal tissues reflects the $\delta^{13}C$ in the diet although the lipid synthesis induces an isotopic fractionation, during the oxidation of pyruvate to acetil coenzyme A, resulting in lower $\delta^{13}C$ in lipid rich tissues than in other tissues (DeNiro & Epstein, 1977). Most of the concentrate feeds include ingredients derived for C4 plants (maize, maize by-products, sorghum or molasses) and foods from animals fed concentrates have higher $\delta^{13}C$ than those fed with temperate pastures or forages (Figure 2). So, feeding maize silage and concentrates will result in animal products with higher $\delta^{13}C$ (De Smet *et al.*, 2004; Janeiro *et al.* 2005 (Figure 2). Moreover, the $\delta^{13}C$ of animal tissues have potential to trace back animal feeding and can be useful to produce information about the duration of concentrate supplementation. As pointed out by De Smet *et al.* (2004), distinct tissues have different metabolic turnover rates and may need different equilibration times for $\delta^{13}C$. So, the measurements of $\delta^{13}C$ in different biological samples could eventually be used to define the duration of concentrate supplementation. However, more research is needed in order to account for numerous factors that might affect the $\delta^{13}C$ in animal tissues, including the rate of growth rate, extent of fat deposition and the proportion of fat derived for *de novo* synthesis (subjected to isotopic fractionation) or transferred from dietary sources. Data presented in figure 2 reflects some of this complexity. The differences observed in $\delta^{13}C$ on muscle from lambs fed with lucerne or lucern supplemented with soybean oil, might be explained by prevailing *de novo* lipid synthesis on lucern fed lambs whereas dietary transfer of fatty acids is expected to predominate in lambs supplemented with oil.

Other chemical markers and spectroscopic techniques

There are other chemical markers that have been explored and that will not be reviewed in this paper. However, a brief mention on the usefulness of major and trace elements in the geographical

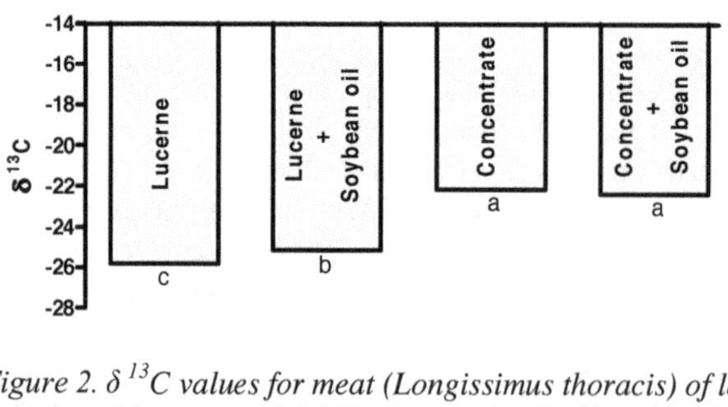

Figure 2. $\delta^{13}C$ values for meat (Longissimus thoracis) of lambs fed dehydrated lucerne, dehydrated lucerne with 10 % of soybean oil, commercial concentrate and commercial concentrate with 10% of soybean oil.
Columns with different indices – a, b or c, are significantly different – P<0,05. (Janeiro et al., 2005; Details on production trial in Bessa et al., 2005).

origin assignment of cheese (Pillonel *et al.*, 2003a) and of skatol (3-methylindol) as marker for ruminants finished in pasture (Priolo *et al.*, 2001) is made.

Other methodological approaches based on spectral patterns reflecting the chemistry of complex food matter have been applied to the geographical origin assignment of foods. These include non-destructive instrumental techniques such as infrared spectroscopy and front-face fluorescence. Chemometrics combined with infrared spectroscopy was used to discriminate between Emmental cheeses originated from different European countries (Pillonel *et al.*, 2003b; Karoui *et al.*, 2005a). The discrimination between cheeses according to the geographic origin of the milk using front-face fluorescence spectroscopy has also been reported (Karoui *et al.*, 2005a; Karoui *et al.*, 2005b). Those results showed that fluorescence emission spectra recorded directly on cheese samples are fingerprints that allow the identification of cheeses as a function of their region of manufacture. These analytical techniques are rapid, relatively inexpensive and can be applied in fundamental research, in control laboratories, and on-line in the factory to control dairy products (Karoui *et al.*, 2004).

Conclusions

Recent research demonstrates the potential of several techniques to discriminate geographical origin and animal production systems of dairy foods and meat. Both geographical origin and production system may be accurately determined by the combination of several chemical markers and adequate chemometrics. As pointed out by Martin *et al.* (2005), most of the techniques can be rather cumbersome and costly for routine application, although their mere existence and divulgation may have a dissuasive effect on frauds.

The product specifications for some PDO-meats, like Alentejana-PDO and Mertolenga-PDO beef, include a defined period range for concentrate finishing. So the development of an effective method for the determination of the time on concentrate is needed. The discrimination between meat (or carcasses) from ruminants finished with concentrate feeds or pasture is easily achieved with several chemical markers. However, the distinction between animals raised on pasture and finished on concentrate and animal raised and finished on concentrate feeds seems to be rather difficult. The potential of $\delta^{13}C$ in tissues with different equilibration times and the use of fatty acids combined with other chemical markers should be explored.

References

Agabriel, C., B. Martin, C. Sibra, J.C. Bonnefoy, M.-C. Montel, R. Didienne, & S. Hulin, 2004. Effect of dairy production systems on sensory characteristics of Cantal cheeses: A plant-scale study. Anim. Res. 53: 221-234.

Besle, J.M., J.L. Lamaison, B. Dujol, P. Pradel, D. Fraisse, D. Viala & B. Martin, 2005. Flavonoids and other phenolics in milk as a putative tool for traceability of dairy production systems. In: Indicators of milk and beef quality. Ed. J.F. Hocquette & S. Gigli., EAAP publication no. 112. Wageningen Academic Publishers, Wageningen, The Netherlands, 345-348.

Bessa, R.J.B., J. Santos-Silva, J.M.R. Ribeiro & A.V. Portugal, 2000. Reticulo-rumen biohydrogenation and the enrichment of ruminant edible products with linoleic acid conjugated isomers. Livest. Prod. Sci. 63, 201-211.

Bessa, R.J.B., P.V. Portugal, I.A. Mendes & J. Santos-Silva, 2005. Effect of lipid supplementation on growth performance, carcass and meat quality and fatty acid composition of intramuscular lipids of lambs fed dehydrated lucerne or concentrate. Livest. Prod. Sci. 96: 185-194.

Boner, M. & H. Forstel, 2004. Stable isotope variation as a tool to trace the authenticity of beef. Anal. Bioanal. Chem. 378: 301-310.

Bosset, J.O., B. Jeangros, T. Berger, U. Butokofer, M. Collomb, R. Gauch, P. Lavanchy, J. Scehovic, J. Troxler & R. Sieber, 1999. Comparaison de fromages à pâte dure de type Gruyére produits en régions de montagne et de plaine. Rev. Suuise Agric. 31: 17-22.

Brescia, M.A, M. Monfreda, A. Buccolieri & C. Carrino, 2005. Characterization of the geographical origin of buffalo milk and mozzarella cheese by means of analytical and spectroscopic determinations. Food Chem. 89: 139-147.

Cabrita, A. R. J., A. J. M. Fonseca, R. J. Dewhurst & E. Gomes, 2003. Nitrogen supplementation of corn silages. 2. Assessing rumen function using fatty acid profiles of bovine milk. J. Dairy Sci. 86: 4020-4032.

Collomb,M., R. Sieber & U. Butikofer, 2004. CLA isomers in milk fat from cows fed diets with high levels of unsaturated fatty acids. Lipids 39: 355-364.

Coulon, J-B., A. Delacroix-Buchet, B. Martin & A. Pirisi, 2004. Relationships between ruminant management and sensory characteristics of cheese: a review. Lait 84: 221-241.

Daza, A., A.I. Rey, J. Ruiz & C.J. Lopez-Bote, 2005. Effects of feeding in free-range conditions or in confinement with different dietary MUFA/PUFA ratios and α-tocopheryl acetate, on antioxidants accumulation and oxidative stability in Iberians pigs. Meat Sci. 69: 151-163.

De Niro, M. J. & S. Epstein, 1977. Mechanism of carbon isotope fractionation associated with lipid synthesis. Science, 197 (4300): 261-263.

De Smet, S., A. Balcaen, E. Claeys, P. Boeckx, & O. Van Cleemput, 2004. Stable carbon isotope analysis of different tissues of beef animals in relation to their diets. Rapid Commun. Mass Spectrom. 18: 1227-1232.

Dumont, J.P. & J. Adda, 1978. Occurrence of sesquiterpenes in mountain cheese volatiles. J. Agric. Food Chem. 26: 364-367.

García-Olmo, J., E. De Pedro, A. Garrido, A. Paredes, C. Sanabria, M. Santolalla, J. Salas, J.R. García-Hierro, I. Gonzalez, M.D. Gracía-Cachan & J. Guirao, 2002. Determination of the precision of the fatty acid analysis of Iberian pig fat by gas chromatography. Results of a mini collaborative study. Meat Sci. 60: 103-109.

Goffman, F., W. Thies & L. Velasco, 1999. Chemotaxonomic value of tocopherols in Brassicaceae. Phytochemistry 50:.793-798.

Hess, J.L., 1993. Vitamin E, alpha-Tocopherol. In *Antioxidants in Higher Plants*. Alsher, R.G & Hess, J.L. (Eds.) CRC Press, Boca Raton, FL.

Janeiro, A.I., L. Ramalho, B. Henriques, A.J.R. Teixeira, M.C. Costa, M.J. Marcelo-Curto, J. Santos-Silva & R.J.B. Bessa, 2005. Stable isotope ratio analysis for authentification of Portuguese sheep meat. A preliminary study. Proceedings of 7° "Encontro de Química dos Alimentos (Viseu, 13 a 16 Abril 2005), pp. 40.

Jeangros, B., J. Scehovic, J. Troxler, H.J. Bachmann & J.O. Bosset, 1999. Comparaison de caractéristiques botaniques et chimiques d'herbages pâtures en plaine et en montagne, Fourrages 159: 277-292.

Joblin K.N. & J.A. Hudson, 1997. Management of milk flavour through the manipulation of rumen microorganisms. In: Milk Composition, production and biotechnology, R.A.S. Welch, D.J.W. Burns, S.R. Davis, A.I. Popay & C.G. Prosser (Editors) CAB International, pp. 455-463.

Karoui, R., E. Dufour, L. Pillonel, D. Picque, T. Cattenoz & J. O. Bosset, 2004. Determining the geographic origin of Emmental cheeses produced during winter and summer using a technique based on the concatenation of MIR and fluorescence spectroscopic data. Eur. Food Res. Technol. 219: 184-189.

Karoui, R., E. Dufour, L. Pillonel, E. Schaller, D. Picque, T. Cattenoz & J. O. Bosset, 2005a. The potential of combined infrared and fluorescence spectroscopies as a method of determination of the geographic origin of Emmental cheeses. Int. Dairy J. 15: 287-298.

Karoui, R., J. O. Bosset, G. Mazerolles, A. Kulmyrzaev & E. Dufour, 2005b. Monitoring the geographic origin of both experimental French Jura hard cheeses and Swiss Gruyère and L'Etivaz PDO cheeses using mid-infrared and fluorescence spectroscopies: a preliminary investigation. Int. Dairy J. 15: 275-286.

Kesselmeier, J. & M. Staudt, 1999. Biogenic volatile organic compounds (VOC): An overview on emission, physiology and ecology. J. Atmospheric Chem. 33: 23-88.

Kreitler, C.W., & D.C. Jones, 1975. Natural soil nitrate: the cause of the nitrate contamination of groundwater in Runnels County, Texas. Ground Water, 13, 53-62.

Larick D.K, H.B. Hedrick, M.E. Bailey, J.E. Williams, D.L. Hancock, G.B. Garner & R.E. Morrow, 1987. Flavor constituents of beef as influenced by forage and grain feeding. J. Food Sci. 52: 245-251.

Lopez, V., & R.C. Lindsay, 1993. Metabolic conjugates as precursors for characterizing flavor compounds in ruminant milks. J. Agric. Food Chem. 41: 446-454.

Mariaca, R.G., T.F.H. Berger, R. Gauch, M.I. Imhof, B. Jeangros & J.O. Bosset, 1997. Occurrence of volatile mono- and sesquiterpenoids in highland and lowland plant species as possible precursors for flavour compounds in milk and dairy products. J. Agric. Food Chem. 45: 4423-4434.

Martin, B., Cornu, A., Kondjoyan, N., Ferlay, A., Verdier-Metz, I., Pradel, P., Rock, E., Chilliard, Y., Coulon, J.B. & Berdagué, J.L., 2005. Milk indicators for recognizing the types of forages eaten by dairy cows. In: Indicators of Milk and Beef Quality, J.F. Hocquette & S. Gigli. (Editors), EAAP publication no. 112. Wageningen Academic Publishers, Wageningen , The Netherlands, 127-136.

Martin, G.J. & M. Martin, 2003. Climatic significance of isopope ratio. Phytochem. Rev. 2: 179-190.

Moio, L., L. Rillo, A. Ledda & F. Addeo, 1996. Odorous constituents of ovine milk in relationship to diet. J. Dairy Sci. 79: 1322-1331.

O'Connell, J.E. & P.F. Fox, 2001. Significance and applications of phenolic compounds in the production and quality of milk and dairy products: a review. Int. Dairy J. 11: 103-120.

O'Leary M. H. 1981. Carbon isotope fractionation in plants. Phytochemistry. 20: 553-567.

Piasentier, E., R. Valusso, F. Camin & G. Versini, 2003. Stable isotope ratio analysis for authentication of lamb meat. Meat Sci. 64: 239-247

Pillonel, L., R. Badertscher, P. Froidevaux, G. Haberhauer, S. Holzl, P. Horn, A. Jakob, Pfammater, E., U. Piantini, A. Rossmann, R. Tabacchi & J.O. Bosset, 2003a. Stable isotopes ratios, major, trace and radioactive elements in emmental cheeses of different origins. Lebensm.-Wiss. u.-Technol. 36: 615-623.

Pillonel, L., W. Luginbuhl, D. Picque, E. Schaller, R. Tabacchi & J. O. Bosset, 2003b. Analytical methods for the determination of the geographic origin of Emmental cheese: mid- and near-infrared spectroscopy. Eur. Food Res. Technol. 216: 174-178.

Prache, S. & M. Theriez, 1999. Traceability of lamb production systems: Carotenoids in plasma and adipose tissue. Anim. Sci. 69: 29-36.

Prache, S., A. Priolo & P. Grolier, 2003a. Persistence of carotenoid pigments in the blood of concentrate-finishing grazing sheep: Its significance for the traceability of grass-feeding. J. Anim. Sci. 81: 360-367.

Prache, S., A. Priolo & P. Grolier, 2003b. Effect of concentrate finishing on the carotenoid content of perirenal fat in grazing sheep: its significance for discriminating grass-fed, concentrate-fed and concentrate-finished grazing lambs. Anim. Sci. 77: 225-233.

Prates, J.A.M., M.A.G. Quaresma, R.J.B. Bessa, C.A. Fontes & C.M.M. Alfaia, 2005. Simultaneous HPLC quantification of total cholesterol, tocopherols and β-carotene in Barrosã-PDO veal. Food Chem. In Press.

Priolo, A., D. Micol & J. Agabriel, 2001. Effects of grass feeding systems on ruminant meat colour and flavour. A review. Anim. Res. 50: 185-200

Priolo, A., A. Cornu, S. Prache, M. Krogmann, N. Kondjoyan, D. Micol & J. L. Berdague. 2004. Fat volatiles tracers of grass feeding in sheep. Meat Sci. 66: 475-481.

Renou, J-P., G. Bielicki, C. Deponge, P. Gachon, D. Micol & P. Ritz, 2004a. Characterization of animal products according to geographic origin and feeding diet using nuclear magnetic resonance and isotope ratio mass spectrometry. Part II: Beef meat. Food Chem. 86: 251-256.

Renou, J-P., C. Deponge, P. Gachon, J-C. Bonnefoy, J-P. Coulon, J-P. Garel, R. Verité & P. Ritz, 2004b. Characterization of animal products according to geographic origin and feeding diet using nuclear magnetic resonance and isotope ratio mass spectrometry. Cow milk. Food Chem. 85: 63-66.

Rey, A.I., B. Isabel, R. Cava & C.J. Lopez-Bote, 1998. Dietary acorns provide a source of gamma-tocopherol to pigs raised extensively. Can. J. Anim. Sci. 78: 441-443.

Rossmann, A., 2001. Determination of stable isotope ratios in food analysis. Food Rev. Intern. 17: 347-381.

Rossmann, A., G. Haberhauer, S. Holzl, P. Horn, F. Pichlmayer & S. Voerkelius, 2000. The potential of multielement stable isotope analysis for regional origin assignment of butter. Eur. Food Res. Technol. 211: 32-40.

Santos-Silva, J., R.J.B. Bessa & F. Santos-Silva, 2002. Effect of genotype, feeding system and slaughter weight on the quality of light lambs II. Fatty acid composition of meat. Livest. Prod. Sci. 77: 187-194.

Scalbert, A. & G. Williamson, 2000. Dietary intake and bioavailability of polyphenols. J.Nutr. 130: 2073S-2085S.

Scalbert, A., C. Manach, C. Morand & C. Rémésy, 2005. Dietary polyphenols and the prevention of diseases. Critical Rev. Food Sci. Nutr. 45: 287-306.

Schmidt, O., J.M. Quilter, B. Bahar, A.P. Moloney, C.M. Scrimgeour, I.S. Begley & F.J. Monohan, 2005. Inferring the origin and dietary history of beef from C, N, and S stable isotope ratio analysis. Food Chem. 91: 545-549.

Torrissen, O.J., 2000. Dietary delivery of carotenoids. In: Antioxidants in Muscle Foods., Ed. E. Dekker, C. Faustman, C.J. Lopez-Bote. John Wiley & Sons, Inc. Publ., New York, pp.289-313.

Urbach, G., 1990. Effect of feed on flavor in dairy foods. J. Dairy Sci. 73: 3639-3650.

Viallon, C., B. Martin, I. Verdier-Metz, P. Pradel, J.P. Garel, J-B. Coulon & J-L. Berdagué, 2000. Transfer of monoterpenes and sesquiterpenes from forages into milk. Lait 80: 635-641.

Yang, A., T.W. Larsen & R.K. Tume, 1992. Carotenoid and retinal concentrations in serum, adipose tissue and liver and carotenoids transport in sheep, goats and cattle. Aust. J. Agric. Res. 43: 1809-1817.

Yoneyama, T., 1995. Nitrogen metabolism and fractionation of nitrogen isotopes in plants. In: Stable isotopes in the Biosphere. Ed. E. Wada, T. Yoneyama, M. Minagawa, T. Ando, & B.D. Fry, Kyoto, Kyoto University Press, pp. 92-102.

Young, O.A., J-L. Berdagué, C. Viallon, S. Rousset-Akrim & M. Theriez, 1997. Fat-borne volatiles and sheepmeat odour. Meat Sci. 45: 183-200.

A new SNPs panel for cattle traceability

E. Genzini[1], A. Lanza[1], A. Sassano[2] & M. Blasi[1]

[1]L.G.S. Laboratorio di Genetica e Servizi Società Cooperativa, Via Bergamo, 292 Cremona/Potenza, Italy
[2]Parco Scientifico e Tecnologico del Molise, Moliseinnovazione, Via dell'edilizia, Potenza, Italy

Summary

L.G.S. Laboratorio di Genetica e Servizi, since 1980 in service of the italian livestock industry, is developing a new protocol, based on a new type of genetic marker, Single Nucleotide Polymorphisms (SNPs), for identity and meat traceability.

In order to define the SNP panel two different biological tissues from 100 Podolica Breed animals were collected: hair roots from alive animals and muscle fragments after slaughter.

The comparison between the two DNA samples was made using 20 SNPs selected from those of public domain.

Allelic frequencies, heterozygosity (He) and PI have been calculated analysing 50 Holstein Friesian and 50 Brown Swiss DNA samples.

SNPs can become the standard marker for identity and meat traceability because of the ease of scoring, low cost assay development and high-throughput capability.

Keywords: traceability, meat, DNA, SNPs.

Introduction

Many molecular biology laboratories working in the animal production field are involved in the gene mapping programme. The aim of this work was to find a good number of genetic markers, highly polymorphic, distributed among chromosomes, to increase the rate of genetic improvement in cattle.

Such a high number of markers gives an extremely powerful instrument for the correct genetic identification of animals, parentage testing and genetic traceability of animal products.

To achieve this aim till today the microsatellites are used. Molecular biology laboratories all over the world are studying a new class of genetic markers, Single Nucleotide Polymorphisms, spread all over the genome.

These markers are usually biallelic and can be analyzed by completely automatic systems, high throughput, in a binary way (Vignal *et al.*, 2002).

The aim of this study was to develop a new protocol for the individual identification and meat traceability through the analysis of SNPs selected from those of public domain (Heaton *et al.*2002; Plancon *et al.* 2004). These markers present the following peculiarity:

- Frequency of each allele close to 50%.
- Homozygo genotype frenquency close to 10%.
- Different chromosome localization.
- Not to be in linkage disequilibrium.
- Follow mendelian inheritance.

Materials and methods

Two different biological tissues were taken from 100 Podolica animals from farms located in Potenza and Matera: hair from the alive animals and a fragment of muscular tissue after slaughter.

Unrelated and random selected animals from the LGS data-base (50 Holstein Friesian e 50 Brown Swiss) were tested to calculate allelic frequencies, heterozygosity (He) and PI.

DNA was extracted from samples using 2 surface–active agents: Tween 20 and NP40.

The 19 analyzed markers, the related SNP and the size of the amplified fragment are shown in table 1.

Amplification was carried out in a 25 μl reaction volume containing 100ng DNA, 1x PCR buffer, 0.75mM $MgCl_2$, 0.2 mM dNTPs, 40 μM primers and 0.5 U Taq DNA polymerase, using a DNA Thermal Cycler 9700 (Applied Biosystems) with a denaturation of 30 seconds at 94°C, annealing at 55°C and extension of 1 minute at 72°C, for 35 cycles, final extension of 10 minutes at 72°C.

DNA fragments were loaded on 3% agarose gel containing ethidium bromide and visualized on a long wavelength UV transiluminator.

The technique of ddNTPs single base extension, to determine SNPs genotype, was used.

The 19 analyzed SNPs, related probe primers and the base present in the mutation point (alleles) are described in table 2.

A mini sequencing composed by the single nucleotide located in the mutation site, was carried out.

The SnapShot ddNTPs Primer Extension kit (Applied Biosystems) was used. The reaction was carried out in a 10 μl volume containing 0.15 pmol/μl of amplified DNA , 0.15 pmol/μl of primer and 5μl of Snapshot Ready Reaction Mixture.

Table 1. Locus name, primers forward/reverse and fragment size (bp).

Locus	Forward	Reverse	Fragment size (bp)
BTA3250	tgcgtttcaaaatggaacag	gggacttccaagatgcaatg	283
BTA5343	accgcagaggtgattaatgg	caagggttgagctggtttgt	226
BTA6343	catcgaccagaaccgtgac	ctcactgggtgggggaagt	221
BTA2019	gaagaggggcaagggagag	tcaggagtgaggcaggaaat	151
BTA5597	ggttgccctcctacacctg	ttccctctcaggggttacct	100
BTA4575	cagaagggcaggactgacac	tggttaagaacccagaccaga	159
BTA4083	aagacagggacaccaaccac	agttccagcttgttcctctg	207
BTA5033	ttccggagacagctcctcta	atggtgggagccctcttaat	235
BTA5089	tgtgcttgccgtctacacat	tctgcacttgatggcaggta	170
BTA2121	gctcgctgcttctcctgtag	aggcaatgttggcaaaacaa	201
BTA4081	atctccattggggagatggt	agagagctcaggggagacg	209
BTA3144	tctactgccttgctccgatt	tctccttgtctgggaagagc	161
BTA2889	cctactcccagtccaagcag	agaagtagcctgggggaggt	130
BTA5067	ctctggtgggtgtggaatg	ttgtcccacctctcagatcc	110
BTA6899	ttctgacagctgctcactgc	tcttgtggtcctgatttcca	150
BTA3463	caagcagaggaggtccagag	acctcctgggaaagaaaagg	114
BTA5279	acttggcattgttctaggttctg	tgtatattgtctgtctccccattc	100
BTA3635	gccctcactcccagtgtaaa	agcgtgatcaaggcagaact	98
BTA5915	ctttgacctggagcagcac	atcctcagggtcacctggtt	150

Table 2. Analyzed locus, related probe primers and base present in the mutation point (alleles).

Locus	probe primers	Alleles
BTA3250	18(c)ttc ccc agt ctc agg tct	G,A
BTA5343	ctt tga acc ggg ctg caa	A,G
BTA6343	6(c)cag ccc ctc tca cac cag	G,C
BTA2019	18(c)gtg gct cca tcc tcc cct	C,A
BTA5597	18(c)cta gct aga aca aag gct	G,T
BTA4575	42(c)gct cca tca cct ctc atc	G,A
BTA4083	36(c)aaa aga ctg tag aga gca	T,C
BTA5033	30(c) tgt ggg caa tcc ctt agc	T,C
BTA5089	30(c)gtc ttt aca gta atg ctc	G,T
BTA2121	ttt aag gtt att tat tta aa	G,T
BTA4081	30(c)cct agc ctc agg gct tcc	G,A
BTA3144	24(c)tca gat tgc ctg att tcc	T,G
BTA2889	24(c)ttc gag ctg att gcc acc	T,C
BTA5067	12(c)aat ggg aga ccc gcg g	C,A
BTA6899	12(c)cta ccc tac ctc agg cct	C,G
BTA3463	24(c) tca aaa tct gag act tgt a	C,T
BTA5279	6(c)ttt gct ctg ttt tgt tca	A,G
BTA3635	6(c)act tcc tct gtg aca aca	G,A
BTA5915	36(c)cgc ggg tcg gaa ggt ccc	T,C

The reactions were performed in a DNA Thermal Cycler with 25 cycles of 10" at 96°C, 5" at 50°C and 30" at 60°C.

One μl of reaction was incubated with 12 μl of formamide and 1 μl of Genescan LIZ120 size standard, and analyzed with a ABI 3100 Avant Genetic Analyzer sequencer (Applied Biosystem) and analyzed using Genescan Analysis Software.

In order to verify the accuracy of the results, obtained with the SNPs traceability protocol, the results obtained using microsatellites, were used for comparison.

Nine microsatellites (TGLA227, BM2113, BM1824, SPS115, TGLA126, TGLA122, INRA023, ETH225 and ETH10) were amplified in a single multiple PCR.

The PCR reaction was carried out on 1 μl of DNA solution in a 10 μl final volume containing 1 μl 10X PCR Buffer (Applied Biosystems), 0.6 μl of $MgCl_2$ (25 mM, Applied Biosystems), 0.25 μl of dNTPs (10 mM each), 0.06-0.48 μl of each primer (10 μM and 0.25 μl of AmpliTaqGold™ (5U/μl Applied Biosystems).

The thermal cycling profile was: denaturation for 12' at 95°C, 30 cycles of 30" at 94°C, 30" at 55°C and 1' at 72°C and final extension for 10' at 72°C.

PCR products were mixed with GeneScan 350 Tamra internal size standard before loading in a single gel lane for electrophoresis. Gel electrophoresis and genotype determination were performed on an ABI Prism 3100 DNA Sequencer equipped with GeneScan and Genotyper softwares (Blasi *et al.*, 2001).

Allelic frequencies and heterozygosity (He) have been calculated using GENEPOP ver. 3.4 (Raymond & Rousset, 1995).

Results and discussion

Hair and meat genotypes were compared to verify that samples were taken from the same animal. This comparison was done using a software generated in LGS, in collaboration with Italservice.

The comparison between two identical genotypes and two different genotypes, is shown in table 3. Using this protocol, comparing the genotype from hair and the genotype from muscular tissue were found 15 incompatibilities.

This data indicates the importance to carry out the genetic traceability in order to focus the critical steps of the productive chain. Therefore, it is important in marginal areas where the slaugthering is carried out in slaughter houses of limited ability and control systems are not comparable to slaughter houses of industrial ability. Furthemore, the breed considerated in this work, the Podolica, is an autochtonous genetic type, reared in the wild or half-wild state, in a mountainous area of Southern Italy, where the productive chain is quite difficult to follow.

Besides this difficult environment, genetic traceability is useful as well as to ensure consumers about food safety.

This protocol can be used to verify cases of cattle-stealing, false identification and wrong parentage.

Parentage control is very important for a correct assessment of the animal productive and reproductive characteristics and to improve the herdbook managment.

The use of SNPs is a good alternative instead of microsatellites: results are easily evaluated in a binary way by completely automatic systems and high throughput. The allelic and genotype frequencies, heterozygosity (He) and PI are reported in table 4.

Table 3. Comparison between two identical and two different genotypes (Hair – Meat).

Identical genotypes			Different genotypes		
	Sample n. 1	Sample n. 2		Sample n. 3	Sample n. 4
Locus	Hair roots	Meat	Locus	Hair roots	Meat
BTA3250	G/A	G/A	MBS042-1	G/A	A
BTA5343	A	A	MBS029-1	A	A
BTA6343	G	G	MBS048-1	G	G
BTA2019	C/A	C/A	MBS007-1	A	A
BTA5597	G	G	MBS043-1	G	G/T
BTA4575	G/A	G/A	MBS044-1	A	G/A
BTA4083	C	C	MBS014-1	T/C	T
BTA5033	C	C	MBS046-1	T	T/C
BTA5089	T	T	MBS047-1	G	G/T
BTA2121	G	G	MBS018-1	G/T	G
BTA4081	G/A	G/A	MBS020-1	G/A	G
BTA3144	T/G	T/G	MBS021-1	T/G	T/G
BTA2889	C	C	MBS034-1	T	T/C
BTA5067	C/A	C/A	MBS051-1	C	C/A
BTA6899	C/G	C/G	MBS049-1	C	G
BTA3463	C	C	MBS025-1	T	T
BTA5279	G	G	MBS035-1	G	A
BTA3635	G/A	G/A	MBS028-1	G/A	A
BTA5915	T/C	T/C	MBS051-1	T/C	T/C

Table 4. Allelic frequencies and heterozygosity (He).

	Brown Swiss			Holstein Friesian		
	Allele 1	Allele 2	He	Allele 1	Allele 2	He
BTA 3635	0.48	0.52	0.45	0.58	0.42	0.53
BTA 5067	0.87	0.13	0.25	0.97	0.03	0.05
BTA 2121	0.80	0.20	0.70	0.60	0.40	0.40
BTA 6343	0.33	0.67	0.45	0.56	0.44	0.41
BTA 3250	0.27	0.73	0.25	0.45	0.55	0.60
BTA 4081	0.13	0.87	0.25	1.00	0.00	0.00
BTA 5597	0.50	0.50	0.37	0.47	0.53	0.53
BTA 2144	0.45	0.55	0.60	0.83	0.17	0.20
BTA 3463	0.29	0.71	0.37	0.33	0.67	0.45
BTA 2019	0.70	0.30	0.70	0.70	0.30	0.41
BTA 5343	0.73	0.27	0.45	0.33	0.67	0.35
BTA 6889	0.30	0.70	0.00	0.69	0.31	0.17
BTA 2889	0.87	0.13	0.15	0.90	0.10	0.00
BTA 5089	0.92	0.08	0.15	1.00	0.00	0.00
BTA 4083	0.45	0.55	0.40	0.80	0.2	0.25
BTA 4575	0.92	0.08	0.05	1.00	0.00	0.00
BTA 5033	0.63	0.37	0.35	0.65	0.35	0.45
BTA 5279	0.52	0.48	0.40	0.40	0.60	0.45
BTA 5915	0.30	0.70	0.40	0.33	0.67	0.36

References

Blasi M., Lanza A., Genzini E. & Dall'Olio S. 2001. Evaluation of the variability of nine microsatellite markers in seven Italian cattle breeds for parentage control. Proceeding XIV Cong. Naz. ASPA Firenze 97-99.

Heaton M.P., Harhay G. P., Bennet G.L., Stone R. T., W. Grosse M., Casas E., Keele J. W., Smith T.P.L., Chitko-McKown C. G. & Laegreid W. W. 2002. Selection and use SNP markers for animal identification and paternity analysis in U.S. beef cattle - Mammalian Genome 13, 272-281.

Plançon C., Duva D., Henry E., Kucharzak R., Lechner D, Meghen C., Angiolillo A., Sanchez A. & Gut I. 2001. Evaluation of panel of SNP markers for the traceability of cattle. Proc. of the 29th International Conference on Animal Genetics - Tokio 2004.

Raymond M. & Rousset F.. 1995. GENEPOP: population genetics softwares for exact tests and ecumenicism. Journal of Heredity. 86: 248-249 (version 3.4).

Vignal A., Milan D., SanCristobal M. & Eggen A. 2002. A review on SNP and other types of molecular markers and their use in animal genetics. Genet. Sel. Evol 34: 275-305.

Session 5. Free communications and Posters

Sardinian goat breeding and utilisation of environmental resources

V. Carcangiu, M.C. Mura, G.M. Vacca & P.P. Bini

Dipartimento di Biologia Animale, Università di Sassari, via Vienna 2, 07100, Sassari, Italia

Summary

Sardinian goat population is the largest in Italy and one of the most important in the Mediterranean area. In the last few decades a small decrease in the number of these animals has been observed whereas in other European countries goat population has been increasing. The aim of this study was to assess the real situation of this breed and investigating problems that slow down the development of the breed, with a view to preserving environmental integrity. Data on the characteristics of 50 farms with a major vocation for this goat breed situated in the area were recorded. Data recorded in every farm were altitude, construction of the farm, pasture composition, morphological traits of the animals and their number, age and sex. Moreover the number of persons working on the farm has been recorded. Milk sample and milk yield records were taken on 25 does from each herd. Milk sample were analysed for fat, lactose and protein content. Majority of animals had white (39.1%) coat, middle size ear (58.4%), horns (72.1%) and globose udder (56.3%). The highest number of kidding (60%) took place in the period between late October and early December. The greater part of the farms had shelters constructed with simple materials (stones for walls and shrub branches for roofs) and only a small number had milking machines and modern constructions for storing cereals and hay. The very low milk yield results in having a high percentage of fat and proteins. The genetic improvement of animals and structure modernization could lead to the improvement of family livelihood and to the betterment of farmer net income thus reducing off-farm migration. Breeding development is essential for the exploitation of the natural resources in the marginal areas but safeguarding the actual environmental conditions.

Keywords: Sarda goat, breeding system, morphological traits.

Introduction

In the last 20 years the world breeding of the small ruminants has had a remarkable change in their population and management. In fact the number of the goats has had a great increase in opposition to the sheep number decrease. The same has also been generally observed in Europe despite goats are considered animals of the zones in course of development (Haenlein, 2001). Especially this increase has been registered in the countries of North Europe, where the goats had been almost eliminated during the last century (Morand-Fehr & Boyazoglu, 1999). This positive trend is linked with the milk quota system established in the European Union and with the "Slow Food" movement which considers goat cheeses as a genuine food. On the other hand in the European countries that have a long tradition of goat breeding goats number tended to remain stable during these last years or decrease slightly. In Italy goat population has had only a decrease in the number of heads in Sardinia, where there are present still about 200 000 heads equal to one sixth of the national herd. Such observation could be linked to the little productivity of the Sardinian goat population and to the low price of goat milk, although Sarda goat milk has a higher protein and fat content as compared to other breeds (Vacca *et al.*, 2002). The low productivity of these animals has led to the introduction of other more productive breeds leading to an uncontrolled crossbreeding (Brandano & Piras, 1978). Crossbreeding surely conduced to an increase in milk yield but it came to the detriment of milk quality. Moreover, the Sarda goat is very rustic and it allows the exploitation of marginal areas otherwise destined to neglect. Therefore, crossbreeding could undetermine the tolerance of Sarda goats to the adverse environments and it could limit their ability to utilise rough forages, with the loss of the Sardinian goat characteristic rusticity (Morand-Fehr & Boyazoglu, 1999).

The main objective of this study was to assess the present state of the sardinian goat breed and investigate problems that slow down the development of this sector, in order to put in action a plain for the breeding progress with a view to preserve the environmental integrity.

Material and methods

In the period between January 2002 and December 2004 data were recorded on the characteristics of 50 farms situated in the areas with major vocation for the goat breeding. This type of rearing in Sardinia is located mainly in the mountainous regions of south-west (Sulcis, Guspinese), centre (Barbagia), east (Baronia, Ogliastra) and south-east (Sarrabus) of the island (Figure 1). In every herd the number, the age, the sex of the animals, their morphological traits and the date of parturition were recorded. Moreover, in each farm records were taken on pasture composition, extension and altitude of the grazing area and construction of the farm (goats pen, milking machine, storehouses for cereals and hay) as well as the number of persons working and if they are wage-earner or member

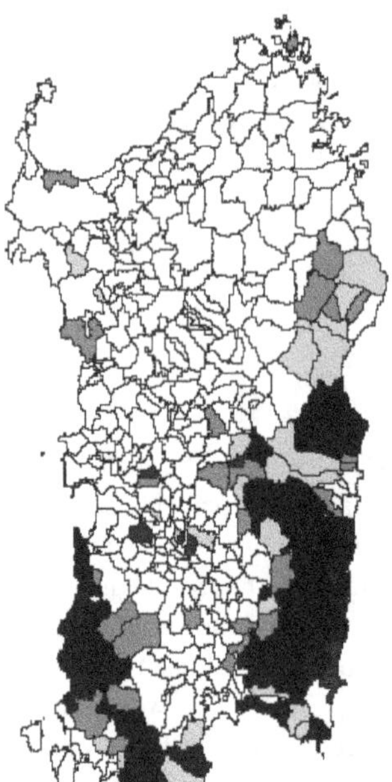

Figure 1. The greater intensity of colour corresponds to the highest number of raised heads.

of the same familiar nucleus. Twenty five does from each herd were milk-recorded and milk samples taken for fat, lactose and proteins contents using Milko-scan spectrophotometer (Foss Electric); sample also, somatic cell count using Fossomatic machine (Foss Electric); total microbial count (TMC) using flow cytofluorimeter Bacto-Scan (Foss Electric) were measured. We have also registered if the milk is transformed in cheese in the farm or sent to a cheese dairy.

Results

Farms differed in the size of the herd: small herd presented a range number from about 50 to 200 heads (27 farms); medium herds from about 200 to 500 heads (16 farms); large herds from 500 to over 800 heads (7 farms). Age of animals showed remarkable variation with goats that can exceed the ten years but maintaining good level productions. Number of young replacements are low (23%) and the percentage of bucks is high. The highest number of kidding (60%) has been registered in the period between late October and early December, the remainder between late February and early March. The majority of does kidded for the first time at about 16 month of age between March and April. Even morphological characters presented a noteworthy variability. White (39.1%) and spotted (33.8%) coats outnumbered red, grey and black ones. The majority of the animals showed a middle size ear (58.4%), instead the small size and large size ear were present in 13.2% and 28.4% of the animals, respectively. The horns were present in the 72.1% of the animals and in the 68.7% of them backwards inclination. Udder was globose in 56.3%, intermediate in 19.0% and pear shaped in 24.7%.

Pasture was 80% of Mediterranean scrub and was in the majority of commounal property. The remaining 20% of pasture was mostly by *Leguminoseae* spp. and *Graminaceae* spp. During the pregnancy and in the periods of the year in which natural feed resources running out, the animals received an integrated diet composed of broad beans, barley, maize and hay. The greater part of the farms possessed shelters constructed with simple materials (stones for walls and shrub branches for roofs) typically used in the traditional Sardinian husbandry. Only few farms (18%) had milking machines and modern constructions for recovering animals and storing cereals and hay. The very low milk yield resulted in high percentage of fat and protein and an acceptable number of SCC and TMC (Table 2).

Milk was transported to the collection centre where it was processed into cheese. Only few number of breeders made cheese on the farm and possessed refrigeration tanks. The majority of the person that worked in the farms belonged to the familiar nucleus and only few were wage-earner.

Table 1. Mean value ± SD of breeding goat in the present observation.

Number of farms	Number of heads per farm	Farm area, ha	Altitude, m asl	Age, year	Prolificacy %[1]	Buck %
50	230±160	600±200	460±240	6.3±3.5	1.3	10 %

[1]Number of kids born per doe kidding.

Table 2. Mean value ± SD of milk composition.

Number of goats	Milk yield (g/day)	Fat %	Protein %	Lactose %	SCC (n/ml)	TMC (CFU/ml)
1250	928.3±41.6	5.4±0.1	4.4±0.1	5.0±0.1	$13x10^5$±$14x10^5$	50 000±211 000

Discussion

Result show that goat breeding in Sardinia is run in traditional manner without modern structures. In fact this breeding type occupies mountain marginal areas where the land presents an impervious surface for agricultural exploitation (Macciotta *et al.*, 2002). These areas in most cases are communal property where many herds graze and for this reason the farmers do not take the trouble to improve the pasture (Wilson, 1988). The most productive areas, instead, are utilized for other types of animal production such as sheep which are more profitable than goats. In many cases the areas where goats graze are places of naturalistic interests and so they cannot be utilized for other uses and must not have too high animal density. This fact has prevented investment in new technology and modernisation of farm structures by the breeders. In fact in our observation only few farms were served by accessible roads provided with milking machine, goat pen or storehouse for cereals and hay. However, this traditional management has concurred to attain environmental protection but it has constrained the development of this breed. The failed development of the goat breeding techniques has lead to the abandonment by young people of this job and also this has discouraged the breeders not to undertake financial investments in the improvement of their farms. However, the presence of goats in mountain areas guarantees the control of the expansion of the Mediterranean scrub and permits to keep the fire breaks free from shrubs and pasture by grazing. Moreover, the presence of the breeders in these uninhabited areas allows the vigilance for the protection of environment and from fire (Glimp, 1995; Warren *et al.*, 1984). This production system and the poor pasture are factors that cause low production. In fact grazing principally composed of Mediterranean scrub gives limited nutrients for adequate lactation. Moreover, the advent of the summer season determines the withering of the grass and this causes quick termination of lactation with the loss of body weight as we have observed in previous studies (Carcangiu *et al.*, 1998; Vacca *et al.*, 2004a). The diet supplement in periods of lack of feed only, is not sufficient anyhow for the lactation demands. Another reason of the low production for Sardinian goat is linked to the insufficient genetic selection. In Sardinia the goat breeding has been ever considered of a secondary level compared to sheep and so it has not aroused interest. The smaller importance of this breeding is also derived from a law instituted in the past century that taxed goat raising thus discouraging whatever structural and genetic improvement. This led to the introduction of other breeds such as Maltese, Saanen, Alpine breed, that have been used to produce more productive crossbreds although with an inferior milk quality (Boyazoglu & Morand-Fehr, 2001).

Also in this study it has been observed that goats pure Sarda breed gave the best milk quality of high fat and protein content and acceptable hygienic level despite manual milking, according to what observed by Vacca *et al.* (2003). Considering also the minor resistance of these crossbreds to adverse environments and their limited abilities to utilise rough forages, the farmers have rediscovered the importance of Sarda goats. In fact in the last few years the regional farmers associations have established the standards of this breed for inclusion in the genealogical book of Sarda goat breed even if the animals registered are still few in number, about 15 000 heads (www.assonapa.com).

Therefore the best course to follow is genetic selection of this autochthonous breed, because some subjects have exhibited an elevated yield and maintained the high milk quality typical of this breed. Furthermore, polymorphism of α-s1 casein has been evidenced in Sarda goats as also demonstrated in some French breeds (Grosclaude *et al.*, 1987; Vacca *et al.* 2004b). It permits a different utilization of milk produced. On the basis of this polymorphism animals are organised in four classes according to the amount of milk protein synthesis (Martin, 1993). Consequently, we can suppose specific genetic selection programs based on the quality of milk produced: the animals with high content of α-s1 casein (strong and intermediate alleles) produce a milk suitable for cheese making; those with low content of this protein (weak alleles) produce a milk appropriate for human consumption; goats without α-s1 casein (null alleles) produce a milk suitable for many people with food allergies and gastrointestinal disorders and are a valid alternative to cow milk.

We registered the effect of introducing bucks of other breed to improve milk production; it also caused a remarkable morphological variability. This variability, like long ears and black heads, derives from Maltese breed crosses; short hair and brown coat, from Alpine breed crosses. The introduction of these breeds has also led to body size and prolificacy increase, as found in other countries of the Mediterranean (Shrestha & Fahmy, 2005). From our data we see that 60% of the goats drop in Autumn; lactation continues from October to May: the period of the year in which natural green grass and buds are present, giving an optimal nutritional support to ensure lactation. On the other hand the animals that drop after February have a shorter four month lactation period. Environmental temperature increase from July lowers pasture quality and quantity, leading to a drastic decrease in the milk yield. Considering that about 40% of the animals drop after February, we can see how great the economic loss is for the Sardinian farmer. This reproductive behaviour is linked to pasture availability: a condition of inadequate feeding at the moment of buck introduction in the flock can determine, in lean animals, a drop in reproductive activity.

Conclusions

From the data obtained we observe that there is a large margin for improvement in sardinian goat breeding. The modernization of structure such as the construction of goat pens and milking machines could improve economic yield of farms. Shelter provides a protection from climate changes and better control of the animal's condition. Milking machines would lead to a reduction in personnel and time spent in animal management.

The application of a genetic selection scheme is essential for this breeding system, that must lead to an improvement of production, considering milk use. Furthermore this selection program safeguards this genetic resource which risks extinction in favour of more productive goat breeds.

Good marketing methods promoting the originality and purity of typical products from Sarda goat milk (by systems of appellation of origin and geographic indication), are necessary to justify their higher prices than sheep or cow dairy products. This higher cost must be accepted by consumers as an essential step to improve farming families' livelihood and net income, as well as reducing land abandonment. The development of goat breeding is essential to exploit the natural resources of the marginal areas, so that these areas are not abandoned, while safeguarding the environment.

References

Boyazoglu, J. & P. Morand-Fehr, 2001. Mediterranean dairy sheep and goat products and their quality. A critical review. Small Rum. Res. 40: 1-11.

Brandano, P. & B. Piras, 1978. La capra Sarda. Ann. Fac. Agric. Sassari XXVI: 1-36.

Carcangiu, V., G.M. Vacca, P.P. Bini, A. Ghibellini, A. Ortu & A. Caddeo, 1998. Goat breeding in future park in Sulcis: Metabolic-nutritional condition of the animals. Proceeding of the Biodiversità, germoplasma locale e sua valorizzazione IV: 1099-1102.

Glimp, H.A., 1995. Meat goat production and marketing. J. Anim. Sci. 73: 291-295.

Grosclaude, F., M.F. Mahe, G. Brignon, L. Di Stasio & A. Jeunet, 1987. A Mendelian polymorphism underlying quantitative variation of goat alpha-s_1 casein. Génét. Sél. Evol. 19: 399-412.

Haenlein, G.F.W., 2001. Past, present, and future perspectives of small ruminant dairy research. J. Dairy Sci. 84: 2097-2115.

Macciotta, N.P.P., A. Cappio-Borlino, R. Steri, G. Pulina & P. Brandano, 2002. Somatic variability of Sarda goat breed analysed by multivariate methods. Livest. Prod. Sci. 75: 51-58.

Martin, P., 1993. Genetic polymorphism of caprine lactoproteins. Le Lait 73: 511-522.

Morand-Fehr, P. & J. Boyazoglu, 1999. Present state and future outlook of the small ruminant sector. Small Rum. Res. 34:175-188.

Shrestha, J.N.B. & M.H. Fahmy, 2005. Breeding goats for meat production: a review: 1. Genetic resources, management and breed evaluation. Small Rum. Res. 58: 93-106.

Vacca, G.M., V. Carcangiu, M.L. Dettori & P.P. Bini, 2004a. Relationships between body condition score, milk yield and milk composition of Sarda goat. J. Anim. Feed Sci. 13 suppl 1: 705-708.

Vacca, G.M., V. Carcangiu, M.L. Dettori, L. Chianese, G. Longu & P.P. Bini, 2004b. Effect of αs_1-casein variants on yield and physicochemical properties of Sarda goat milk. Proceedings of the 55th EAAP. 55: 252.

Vacca, G.M., V. Carcangiu, A. Ghibellini, M. Galioto, M. Cancedda & P.P. Bini, 2002. La qualità del latte della capra censire Sarda. Proceeding of the Fe.Me.S.P.Rum. X: in press

Vacca G.M., L. Chianese, A. Ghibellini, V. Carcangiu, R. Mauriello & P.P. Bini, 2003. α_{s1}-casein genetic variants in sarda breed goat. Italian Journal of Animal Science. Procedings of ASPA 15: 55-57.

Warren, L.E., D.N. Ueckert, M. Shelton & D. Chamrad, 1984. Spanish goat diets on mixed brush rangelands in South Texas. J. Range Manage 37: 340-348.

Wilson, R.T., 1988. Small ruminant production systems in tropical Africa. Small Rum. Res. 1: 305-325.

Milk yield and composition in Sarda goats and their relations across lactations

G.M. Vacca, V. Carcangiu, M. Porqueddu, M.L. Dettori & P.P. Bini

Università di Sassari, Dipartimento di Biologia Animale, via Vienna 2, 07100 Sassari, Italy

Summary

The research was carried out on 42 Sarda goats raised on open pasture. Milk yields and chemical composition evolution in the first three lactations were investigated. The aim was to describe the milk production in Sarda goat and verify how parameter of the first lactation are correlated with the next two. Chemical composition showed the physiological fluctuations linked to lactation stage. From the second lactation milk yield and lactose content increased, while fat content decreased. Protein content and cryoscopic index were higher in third lactation. SCC registered progressively higher values as parity increased. TBC evidenced a good hygienic quality of the product throughout the three lactations. Correlation analysis has pointed up many relationships among parameters in all lactations, but the higher r coefficient values for yields, fat, protein and somatic cell content were registered between third and second lactation. These results show that the second lactation is the most useful to assess the productive value of the subjects.

Keywords: goat, parity, milk composition.

Introduction

Sarda goat breed, with a population of more than 200 000 producing about 15 000 tons of milk, is the most important Italian goat breed (www.istat.it). Thanks to its rusticity, it is able to productively exploit vast territories, mostly in hills and mountains, since long time representing for the local populations one of the few sources of income. Its breeding is characterized by traditional way of management. Structural facilities are poor; feeding is based on the extensive use of native pasture with rare supplementations. Selection is almost exclusively based on dam's production. Oestrus synchronization, artificial insemination and controlled mating are rarely practised. Individual milk yields are not very high on average, but milk is characterised by high fat and protein content. The great part of milk yield is used for cheese making. This breed population is characterised by considerable genetic (Ajmone-Marsan *et al.*, 2001), morphologic and productive variability (Vacca *et al.*, 2003). It could be possible, in a relatively short time, to considerably improve the quantitative aspects of production, trying to maintain high the qualitative value, just beginning from this huge reservoir of biodiversity, and designing an efficient breeding scheme (Moioli *et al.*, 1995). However, not interfering with the standards of selection on a large scale, it is possible to operate a selection program in the individual herds, taking also into account quality traits of different lactations. These decisions have to be taken as far in advance as possible. Hence, it could be useful to know how to estimate, from the first lactation, productive value of the animals. The present study was carried out with a double aim: i) to provide a description of yield and composition of Sarda goat milk, with regard to stage of lactation and parity; and ii) to estimate correlations between parameter values of the first lactation, with those of the next two.

Material and methods

The first three lactations of 42 Sarda breed goats were monitored. The animals, with good clinical and feeding conditions, belonging to a group of 100 does whose quantitative and qualitative lactation traits were studied. At the beginning of the study all the goats were two years old, primiparous and gave birth in the first ten days of February. All the goats in the following lactations that did not give birth in the same period of the year or were affected by any problems were excluded from statistical analysis. Goats were reared in a farm located in South-West Sardinia (latitude: 39°10' N; longitude: 8°45' E) and grazed on open-pasture, predominantly composed of Mediterranean scrub (400-600 m asl), without receiving diet supplement. Milking was manual once-a-day. Data were taken monthly for each lactation, from March (lactation day 45) to July (lactation day 165). Milk samples were collected in proper containers that were maintained at constant temperature (4°C) and arrived to the laboratory in two hours. Fat, protein and lactose contents were determined using I.R. Milko-Scan spectrophotometer (Foss Electric, DK-3400 Hillerød, Denmark) according to the standard method (FIL-IDF 2000:141C). Cryoscopic index (C.I.) was determined using Cryoscope (FIL-IDF, 1982:108). Somatic cell count (SCC) was determined using Fossomatic machine (Foss Electric), with FIL-IDF method 1995:148A and total bacterial count (TBC) using flow cytofluorimeter Bacto-Scan (Foss Electric), according to FIL-IDF method 2000:358. Milk production was estimated by functional data analysis proposed by the Genealogy Book of Sarda breed goat (www.assonapa.com). In order to asses the effects of stage of lactation and parity, data were analysed, according to GLM procedure, using MINITAB™ statistical software. Preliminarily, SCC (cells/ml) and TBC (colony forming units, CFU/ml) values were transformed into logarithm form ($\log_{10}$) to normalize their frequency distribution. In order to estimate relationships among the first three lactations all parameter values were analysed using the Pearson's correlation matrix (r) and Bonferroni significance test.

Results

Milk yield during the first three lactations are shown in figure 1.

Milk yield showed the typical lactation curve reported for this breed by Vacca *et al.* (1995). The second and the third lactations produced significantly more milk (P<0.01). The total milk yield in 180 days in the three lactations was 120, 173 and 158 kg, respectively.

The mean values of fat, protein and lactose content, cryoscopic index, SCC and TBC, in relationship with stage of lactation and parity are shown in table 1.

In all lactations fat, protein and lactose content showed the fluctuations that are usually found with respect to stage of lactation and that are, in part, related to the amount of milk produced. As observed by Dulin *et al.* (1983), SCC showed an increase as the lactation progressed and with respect to parity (P<0.01). Primiparous goats had significantly higher values of fat and lower of lactose (P<0.01). The third lactation showed higher values of protein content (P<0.05) and cryoscopic index (P<0.01). The interaction effect between stage of lactation and parity, for protein, lactose and cryoscopic index was significant (P<0.01).

Table 2 shows correlation coefficients (r) of the different parameters in the three lactations.

Table 1. Least squares means of milk composition in parities 1 (P1), 2 (P2) and 3 (P3).

n = 42		Parity P1	P2	P3	SE	Level of significance M^i	P^j	M^i x P^j
Fat content (%)	March	5.56^{BC}	4.38^{A}	4.89^{AB}				
	April	5.48^{BC}	5.03^{AB}	4.65^{A}				
	May	5.57^{C}	5.38^{B}	5.28^{B}				
	June	5.54^{C}	5.19^{B}	5.19^{B}				
	July	5.82^{D}	5.61^{C}	6.00^{D}				
	Mean	5.59^{B}	5.12^{A}	5.20^{A}	0.11	**	**	NS
Protein content (%)	March	4.40^{B}	4.22^{A}	4.41^{B}				
	April	4.28^{A}	4.33^{AB}	4.52^{B}				
	May	4.42^{B}	4.07^{A}	4.30^{AB}				
	June	4.04^{A}	4.02^{A}	4.41^{B}				
	July	4.21^{A}	4.93^{C}	4.84^{C}				
	Mean	4.27^{a}	4.31^{a}	4.50^{b}	0.06	**	*	**
Lactose content (%)	March	5.02^{D}	5.51^{E}	5.51^{E}				
	April	4.77^{C}	5.31^{E}	5.31^{E}				
	May	4.54^{B}	5.04^{D}	5.04^{D}				
	June	4.44^{B}	4.84^{CD}	4.84^{CD}				
	July	4.26^{A}	4.41^{B}	4.41^{B}				
	Mean	4.61^{A}	5.02^{B}	5.02^{B}	0.04	**	**	**
CI (°Hortvet)	March	-0.564^{AB}	-0.567^{B}	-0.563^{A}				
	April	-0.562^{A}	-0.567^{B}	-0.564^{AB}				
	May	-0.566^{B}	-0.570^{C}	-0.570^{C}				
	June	-0.570^{C}	-0.561^{A}	-0.572^{C}				
	July	-0.566^{B}	-0.568^{BC}	-0.572^{C}				
	Mean	-0.566^{A}	-0.567^{A}	-0.569^{B}	0.001	**	**	**
SCC ($\log_{10}$ cells/ml)	March	5.40^{A}	5.44^{A}	5.60^{B}				
	April	5.55^{A}	5.74^{B}	5.86^{B}				
	May	5.94^{B}	5.83^{B}	6.16^{C}				
	June	5.85^{B}	5.97^{B}	6.32^{CD}				
	July	5.91^{B}	6.34^{CD}	6.45^{D}				
	Mean	5.73^{A}	5.86^{B}	6.08^{C}	0.05	**	**	NS
TBC ($\log_{10}$ CFU/ml)	March	4.21	4.17	4.03				
	April	4.24	4.19	4.16				
	May	4.22	4.31	4.04				
	June	4.02	4.20	4.36				
	July	4.07	4.33	4.23				
	Mean	4.15	4.24	4.16	0.06	NS	NS	NS

Capital letters and ** indicate significant differences at $P<0.01$; lower case and * at $P<0.05$.
NS = not significant. M^i = Month; P^j = Parity.

The results show many significant relationships between the considered parameters and there are strong links between corresponding parameters in the three lactations (in bold). TBC has not been considered in correlation analysis, because it mostly depends on external factors. As regards fat and protein content, yields and SCC the highest values of r were between P3 and P2. Lactose content displayed the highest significant correlations between P2 and P1 ($P<0.001$). Cryoscopic index showed a stronger correlation between P3 and P1. Fat content had highly significant correlations ($P<0.001$)

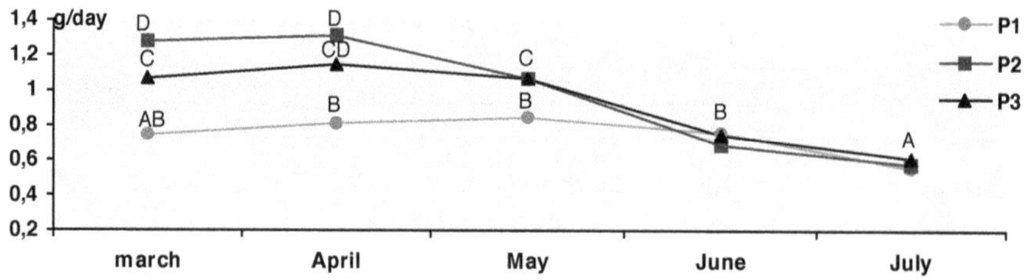

Figure 1. Milk yield in the first three lactations (means with different letters differ significantly at P<0.01).

Table 2. Correlation coefficients (r) between yield and milk composition in lactations 1, 2 and 3.

	Yield (1)	Fat (1)	Protein (1)	Lactose (1)	CI (1)	SCC (1)
Lactation 1						
Yield (2)	0.407***	-0.307***	0.025	0.580***	-0.088	-0.254**
Fat (2)	-0.248**	0.548***	0.314***	-0.270**	0.256**	0.076
Protein (2)	0.204*	0.333***	0.257**	-0.103	0.144	0.021
Lactose (2)	0.132	-0.044	0.163*	0.568***	-0.223**	-0.334***
CI (2)	0.061	0.099	0.059	0.037	0.286***	0.045
SCC (2)	-0.092	-0.009	0.143	-0.388***	0.176*	0.417***
Lactation 2						
Yield (3)	0.385***	-0.306***	0.092	0.493***	0.026	-0.246**
Fat (3)	-0.269***	0.538***	0.300***	-0.223**	0.315***	0.027
Protein (3)	-0.167*	0.387***	0.330***	-0.048	0.017	-0.148
Lactose (3)	0.091	0.047	0.118	0.437***	-0.223**	-0.334***
CI (3)	0.041	0.018	0.011	0.086	0.464***	0.043
SCC (3)	-0.030	0.007	-0.152	-0.381***	0.194*	0.405***
Lactation 3						
Yield (3)	0.752***	-0.300***	-0.185*	0.420***	0.076	-0.343***
Fat (3)	-0.376***	0.651***	0.390***	-0.283***	0.249**	0.081
Protein (3)	-0.257**	0.440***	0.567***	-0.243**	0.069	0.110
Lactose (3)	0.123	-0.081	0.141*	0.443***	-0.202*	-0.342***
CI (3)	-0.024	-0.006	0.078	0.034	0.358***	0.216**
SCC (3)	-0.116	0.110	0.137	-0.473***	0.069	0.524***

****P*< 0.001; ** *P*< 0.01; * *P*< 0.05.

with milk yield and protein content. High correlations between milk yield and lactose content, and between lactose and SCC were found out too.

Discussion

First lactation yields were similar to those reported for primiparous Sarda goats by AssoNaPa. On the other hand, P2 and P3 values were slightly lower in regard to those described for pluriparous goats. This is due to the fact that also in these parities lactation was shorter. Indeed, considering the three consecutive lactations, only goats that gave birth in February (that is in the final period of the reproductive season), were considered for the trial. Analysis of chemical composition showed the high fat and protein content of Sarda goat milk. On the whole, the first lactation showed lower (P<0.01) fat and protein amount than in the second and third lactations 12.1 kg *vs* 16.6 kg and 15.4 kg respectively (data not shown). This is due to the fact that in the 1st lactation physiological maturity is not reached and the animal cannot completely express its potentiality. SCC values may appear high (range $2.5x10^5$ to $2.5x10^6$ cells/ml) if compared to other species. This is probably due to the fact that milk secretion in goat occurs through an apocrine process. This leads to the loss of the apical region of the mammary gland cell and with commonly used methods also these particles are counted as cells (Haenlein, 2002). Our results are similar to those reported by Morgan *et al.* (2003) for French and Greek farms and by Delgado *et al.* (2003) for different management of the flock. TBC values (on average<$23x10^3$ CFU/ml) show the very good hygiene-sanitary quality of milk, despite it was obtained by manual milking. The low TBC value, together with the strong interrelations of SCC with milk yield and lactose, suggest the need for further investigations about the significance to ascribe to SCC values in goat milk (Haenlein, 2002). The higher correlation coefficients found in almost all parameters between P3 and P2, point out how the second lactation is more appropriate than the first one to evaluate the productive performance of the animals, usually assessed in 3rd lactation. Furthermore, observing individual values, it is clear how some subjects, although highly productive, maintain high fat and protein content of milk. These results also show how it is possible, in the herd, to act a selection that takes into account quantitative and qualitative milk traits. The improvement of production and the inclusion of the product in specific commercial contexts, could make Sardinian goat breeding more profitable. This may lead to promote conservation of autochthonous genetic patrimony, as well as maintenance of a socio-economic activity that has been transmitted for centuries and that takes place in the full respect for environment and the correct use of territory.

References

Ajmone-Marsan, P., R. Negrini, P. Crepaldi, E. Milanesi, C. Gorni, A. Valentini & M. Cicogna, 2001. Assessing genetic diversity in Italian goat populations using AFLP® markers. Anim. Genet. 32: 281-288.

Delgado-Pertiñez, M., M.J. Alcalde, J.L. Guzmán-Guerrero, J.M. Castel, Y. Mena & F. Caravaca, 2003. Effect of hygiene-sanitary management on goat milk qualità in semi-extensive systems in Spain. Small Rum. Res. 47: 51-61.

Dulin, A.M., M.J. Paape, W.D. Schultze & B.T. Weinland, 1983. Effect of parity, stage of lactation, and intramammary infection on concentration of somatic cells and cytoplasmic particles in goat milk. J. Dairy Sci. 66: 2424-2433.

Haenlein, G.F.W., 2002. Relationship of somatic cell counts in goat milk to mastitis and productivity. Small Rum. Res. 45: 163-178.

Moioli, B.M., A.M. Pilla, A. Rosati, G. Cavillo & T. Fresi, 1995. Ereditabilità e ripedibilità della produzione di latte nella razza caprina Saanen e valutazione genetica dei riproduttori. Zoot. Nutr. Anim. 21: 231-236.

Morgan, F., T. Massouras, M. Barbosa, L. Roseiro, F. Ravasco, I. Kandarakis, V. Bonnin, M. Fistakoris, E. Anifantakis, G. Jaubert & K. Raynal-Ljutovac, 2003. Characteristics of goat milk collected from small and medium enterprises in Greece, Portugal and France. Small Rum. Res. 47: 39-49.

Vacca, G.M., L. Chianese, A. Ghibellini, V. Carcangiu, R. Mauriello & P.P. Bini, 2003. α_{s1}-casein genetic variants in Sarda breed goat. Italian J. Anim. Sci. 15: 55-57.
Vacca G.M., W. Pinna, P. Lai & A. Cappio-Borlino, 1995. Curva di lattazione di capre di razza Sarda. Proc. XLIX Congr. S.I.S.Vet.: 981-982.

Evaluation and improvement of a traditional practice in the salting of artisanal ham: salt and moisture diffusion

C. Diaferia[1], G. Madonia[2], S. Margiotta[2], S. Palmisano[3], V. Pruiti[4], S. Iori[1] & P. Baldini[1]

[1]Stazione Sperimentale per l'Industria delle Conserve Alimentari, Viale Tanara 31/A Parma,Italy
[2]Istituto Sperimentale Zootecnico per la Sicilia, via Roccazzo 85 Palermo, Italy
[3]Centro ricerche produzioni animali, Corso Garibaldi 42 Reggio Emilia, Italy
[4]Servizio Assistenza Tecnica Regione Sicilia, Castell'Umberto, Messina , Italy

Summary

The mechanism that regulate the mass transfer process in meat products is important for the final keeping quality of dry-cured ham. The salting period allows the ham to absorb the exact amount of salt required to guarantee the final microbial and chemical stability of the meat. Salt and water distribution in typical artisanal ham produced in the area of Nebrodi (Sicily) from native swine was investigated and compared to Parma and Veneto techniques. The results obtained have not evidenced differences between artisanal and Parma salting techniques.

Keywords: salting period, salt distribution, water distribution, AGT.

Introduction

Products made from meat from native pigs, which are reared under extensive bio-sustainable system are provoking an ever increasing and renewed interest in customers. In recent years the demand for traditionally crafted "artisanal" products has grown in answer to increased health and quality requirements. The study of the traditional techniques and their improvement constitute an important objective for the success of these products (Diaferia *et al.*, 2004; Andrés *et al.*, 2004).

In the production of the entire product (ham, coppa) the salting phase assumes an important role in how much salt (an amount that must be between a determined minimum: motivated from requirements of technological character, and a certain maximum: dictated by sensory and nutritional requirements) is favoured in the absorption. Hence, the acknowledgment of the laws that regulate the mass transfer mechanism in meat products is very important on the preservability of the product and its final sensory characteristics (Diaferia & Miccio, 2001; Ruiz-Cabrera *et al.*, 1998; Gou *et al.*, 2004). The scope of the work, detailing in depth the artisanal technique of producing ham in the zone of Nebrodi (Sicily) from native swine and standardization of the final characteristics, has been done to compare the modified traditional technique with two other referenced techniques (Parma and Veneto technique) applied in the production of the protected denomination of ham. Salt and moisture distribution is studied at the end of salting period.

Material and methods

After the completion of the salting period three hams salted with differing techniques: a) standard Veneto type; b) standard Parma type; and c) modified traditional process in use at some artisanal laboratories located in the area of Nebrodi (Sicily), were analyzed for salt and water content in the lean portion and to obtain the concentration profiles of chloride and water in a cross section of the ham. The difference between the two techniques of standard processing is that in the Veneto type ripening, the hams have been subject to a "saturation" salting, that foresees an excess of dry salt on the surface of the product for the entire duration of the phase (10-14 days according to the weight of the thigh), while in the Parma type ripening, there's a reduced amount of salt distributed on the surface, which in some areas depletes quickly, (often before the end of the period), being compensated with an extension of the salting phase (3-4 weeks). In the traditionally crafted process executed at an artisanal laboratory of Castle Umberto (Messina), the amount of salt is approximately equal to 5% of the weight of the thigh, being distributed over the entire muscular surface for a duration of 10 days. At the end of this term any residual salt is removed, and thighs are placed in a press for another 10 days.

To determine the concentration profile of chloride and water, a slice about 1 cm thick was cut from each ham at the point indicated by the indicated area in figure 1. From each slice 10 samples were obtained for concentration profiles of salt, using a 4 mm internal diameter cork borer, inserted in the *semimembranosus* muscle along two lines parallel and perpendicular to the upper part of the thigh. The amount of salt (NaCl) has been determined by means of extraction in distilled water and followed by titration with mercury nitrate.

The water amount has been determined according to the AOAC method on meat rectangles drawn from the interior *semimembranosus* muscle in direction parallel to the upper area of the slice.

The scheme for sampling used for the salted hams with the standard Veneto technique and the standard Parma technique have then been extended to native autoctonous pig thighs salted by the artisanal technique in a processing laboratory located in the zone of Nebrodi. The analytical results presented refer to a single ham.

The values of salt absorbtion were estimated on the same hams stripped of their fat, rind and bones. In accordance with sectioning outlines employed at SSICA laboratories (Figure 1) 8 portions were obtained on which the salt and water content were determined according to the official AOAC methods. The dotted area shows the position of the slices for salt and water determination.

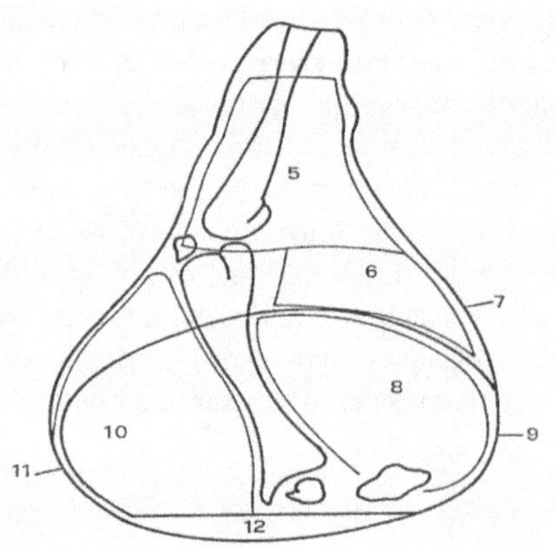

Figure 1. Ham trimming outline.

Results and discussion

The course of the salt penetration indicated in both referenced techniques a distribution of salt inside of the *semimembranosus* muscle in relation of the distance from the surface: parabolic in that Veneto type (Figure 2), linear in the Parma technique (Figure 3). The penetration of salt mostly through the uncovered muscular part, and if it is assumed that the dry salt in excess covers the surface of the product for all the duration of the salting (as is the case of the Veneto technique), then it's possible to consider the external salt solution as semi-infinite phase containing one uniform concentration initially of NaCl (Palmia *et al.*, 1992). In the artisanal process (Figure 4) the distribution of salt is similar to that of the Parma technique, with a downward slope of slightly less, probably attributable to the greater amount of salt absorption also in reference to elevated amounts of fat present in thighs subject to the traditional technique (advanced values of the relationship salt/lean part).

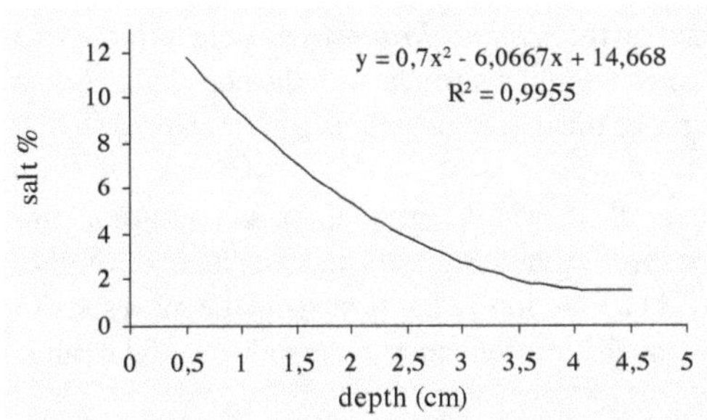

Figure 2. Veneto salting (10 days): plot of salt concentration.

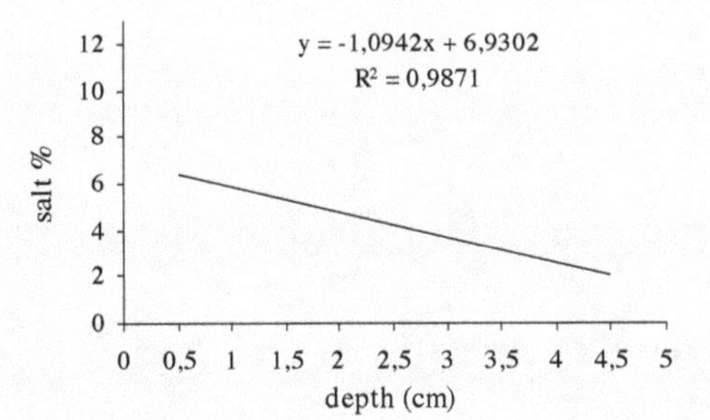

Figure 3. Parma salting (28 days): plot of salt concentration.

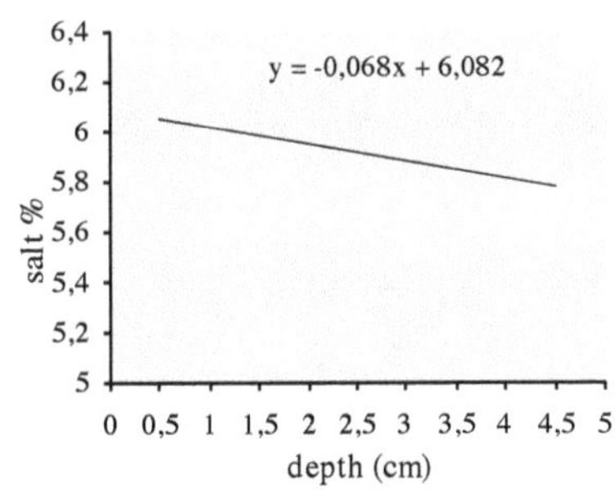

Figure 4. Artisanal salting (20 days): plot of salt concentration.

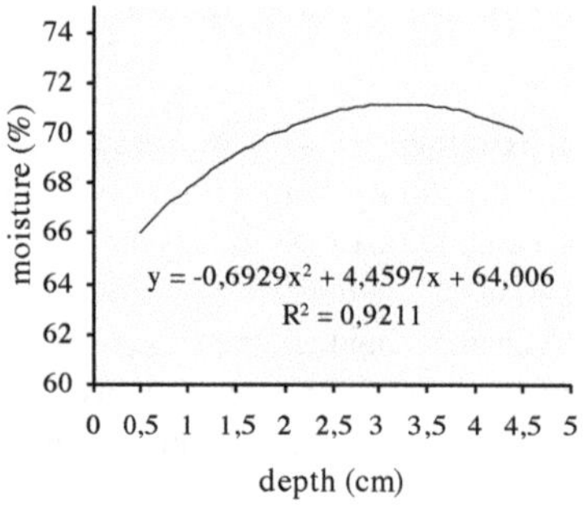

Figure 5. Artisanal salting (20 days): plot of water concentration.

The profile of the moisture content determined on the *semimenbranosus* muscle introduces a gradient of concentration in relation to the distance from the surface with the parabolic course (Figures 5 to 7). In the artisanal process the course of the moisture content is more similar to the Parma test rather than that of the Veneto test.

In accordance with a previous job (Diaferia *et al.*, 1992) the amounts of salt and moisture determined on the 8 fractions (Table 1) indicate respective values greater and smaller in the higher fractions (6, 8, 10) regarding those inferiors (7, 9, 11). Fraction 12 has turned out the most salty in the Veneto and in the Artisanal techniques in how much more tendency to the slide out of the salt.

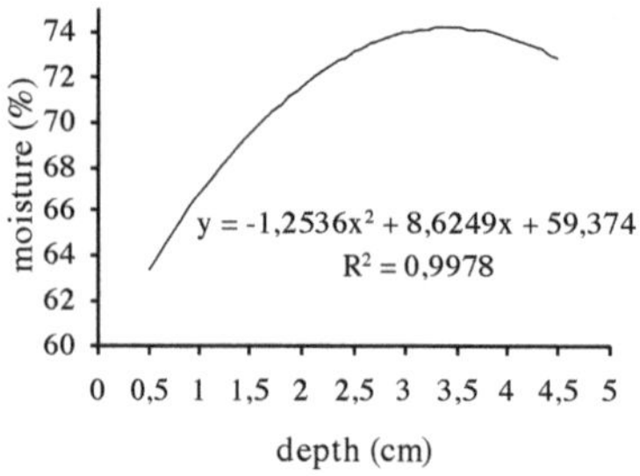

Figure 6. Veneto salting (10 days): plot of water concentration.

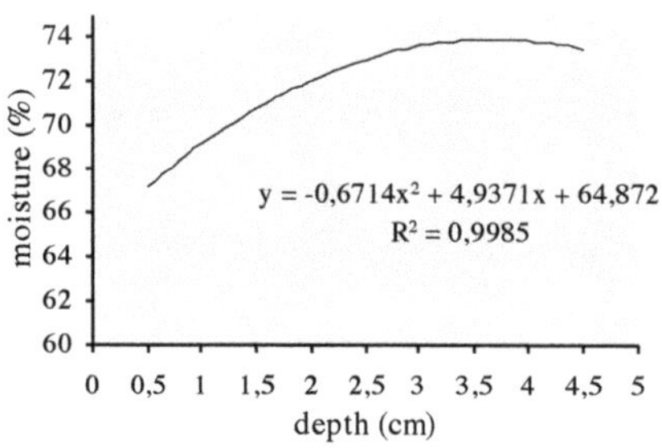

Figure 7. Parma salting (28 days): plot of water concentration.

Table 1. Salt and water (mean and standard deviation) of the hams at the end of salting[1].

Portion	Parma salting		Veneto salting		Artisanal salting[2]	
	Salt	Water	Salt	Water	Salt	Water
5	1.75 (0.20)	72.06 (0.26)	1.87 (0.27)	71.55(0.22)	1.84	69.17
6	4.18 (0.03)	68.91 (0.60)	2.85 (0.40)	70.58(1.86)	2.75	69.78
7	0.50 (0.08)	73.15 (0.71)	1.03 (0.58)	71.90(1.03)	0.70	72.30
8	2.24 (0.12)	68.05(1.12)	4.28 (0.19)	67.24(0.59)	4.91	65.86
9	1.01 (0.26)	72.74 (0.46)	0.71 (0.28)	73.30(0.47)	1.78	70.96
10	4.09 (0.17)	68.44 (0.87)	3.63 (0.32)	69.14(0.34)	5.65	65.87
11	1.44 (0.05)	73.62(1.25)	0.80 (0.08)	74.21(0.53)	2.20	71.02
12	3.53 (0.15)	65.58 (0.95)	5.97 (0.62)	65.87(1.60)	8.59	62.70
(Salt/l.p)[3] *100	2.34		2.69		3.06	

1All data are expressed as % (w/w).
2The values are calculated on a single ham.
3The values are calculated on the basis of the raw weight of lean part (l.p.).

In conclusion, the technique adopted has guaranteed a correct distribution of salt and moisture and seems not to remove itself considerably from the two comparative techniques of reference, (with a greater similarity for the Parma type process), also keeping in mind the longer duration of the salting phase.

References

Andrés A.J., R. Cava, J. Ventanas, V. Thovar & J. Ruiz., 2004. Sensory characteristics of Iberian ham: Influence of salt content and processing conditions. Meat Science 68: 45-51.

AOAC. 1980. Official methods of Analysis. 13th Edition of the Association of Official Analytical Chemists, Washington, DC.

Gou P., J. Comaposada & J. Arnau, 2004. Moisture diffusivity in the lean tissue of dry-cured ham at different process times. Meat Science 67: 203-209.

Diaferia C., F. Palmia & P. Baldini , 1992. Distribuzione del sale e dell'umidità nei prosciutti crudi tipo Veneto preparati con cosce di suini leggeri. Industria Conserve, 67: 3-9.

Diaferia C., & M. Miccio, 2001. Mass transfer mechanism in meat products: diffusion coefficient and surface exchange coefficient of water in salami. In proceedings of the fifth Italian Conference on Chemical and Process Engineering, Florence, Italy, 20-23 May.

Diaferia C., A. D'Amico, G. Madonia, S. Margiotta, D. Cartabellotta, V. Pruiti, 2004. In proceedings of the fifth International Symposium on the Mediterranean Pig, Tarbes , France, 16-19 November.

Palmia F., C. Mazoyer, C. Diaferia, P. Baldini & A. Porretta, 1992. Salt water distribution in typical Italian hams. Rev. Esp. Cienc. Tecnol. Aliment. 32(1), 71-83.

Ruiz-Cabrera M.A., P. Gou, J. Comaposada, J.D. Daudin, L. Foucat, J.P.Renou, 1998. Determination of the water diffusivity coefficient in pork meat. In Proceedings of 44th International Congress of Meat Science and technology, Barcelona, Spain.

Characterization of Zerasca ovine population: breeding management and zoometric data

J. Goracci, L. Giuliotti, N. Benvenuti & P. Verità

Department of Animal Production, Pisa University, Viale delle Piagge, 2, 56124, Pisa, Italy

Summary

Zerasca ovine breed is an important example of the close connection between Italian husbandry and the environment. The scientific aim of this research was to characterize breeding techniques and zoometrical traits of that population, which is now beginning to be recognized and supported. All the 37 farms studied in the Lunigiana district were inspected directly, in order to collect livestock data (2143 subjects), while biometrical records were created on a sample of 83 purebred animals belonging to four farms. Our results revealed interesting characteristics as well as the adaptability of sheep and lambs from Zeri, especially in relation to the long grazing periods and adverse climatic conditions. In addition, the evolution of morphological traits and the influence of past crosses made to increase productivity underline the need to safeguard the Zerasca population's strong bond with the ecosystem and its role as a significant economic resource of livestock biodiversity.

Keywords: ovine, Zerasca breed, zoometric data.

Introduction

Zerasca sheep is a native Italian breed with endangered status, whose meat reveals unique organoleptic characteristics regarding tenderness, tastiness and aroma. As a result, it has earned the hallmark of Slow Food *Presidia*, a record of hundreds of traditional products from all over the world, listed for documentation, guarantee and advertising purpose. Today, Zeri lamb has a considerable economic impact on the territory, ensuring increasing profits for shepherds.

The Zeri region is located in northwest Tuscany at an altitude of 600-1 200 m in the Lunigiana countryside, where natural pastures and woods of chestnut, beech, alder, hazel and acacia are the most important sources for animal nutrition. This region has an area of 7344 ha (Verità *et al.*, 2001a); the three valleys contain 1314 inhabitants, grouped into 671 families. Zeri is far from industrial centres, guaranteeing unpolluted air and a clean environment. A total of 192 farms and nearly 300 agricultural workers (22.8% of total population) give an idea of the importance of animal husbandry and agriculture in that region.

Moreover, the climate is generally extremely variable, implying inconstant quantitative and qualitative grazing land productions. This situation has considerable effects on meat production, forcing breeders to provide nutrition for the animals in winter and in the summer, requiring them to send the flocks to high altitude territory, where there is greater availability of grass. Generally, mothers lead their lambs to pasture all year round, except for winter periods, when they are kept indoors, due to adverse climatic conditions: snow and hail are not unusual in the cold seasons. However, regular and frequent rainfall assure the good fertility of the land.

Nowadays trends show that Italian counties are too often characterized by the exclusion and neglect of outlying territories, which also suffer progressive depopulation, especially of young people. Instead, Zeri is an example of the younger generation's determination to keep sheep breeding closely connected to the territory. Thus, after a period of total abandonment and serious risk of extinction, the Zerasca breed has begun to recover, thanks to various public interventions (Pisa University, Mountain Communities, Local Administration and Agencies) and to the aforementioned determination of individual farmers - especially women - organized by a *Consortium* ("Consortium for the Valorisation and Safeguarding of Zerasca Sheep and Lambs") born in 2001. This organization has coordinated about twenty farms and has tried to restore ancient productions with traditional breeding techniques and land use. The youth and resolve of breeders had lead to the success of these small-farming practices, aimed at stabilizing production techniques and establishing standards, in order to guarantee a viable future and economic returns for production.

The Zeri lamb is a sturdy animal, medium-large in size and with a white fleece. Horns may be present. Zeri sheep have milk high in nutrients, showing 6.1-6.4% protein content (Benvenuti *et al.*, 1998; Verità, personal communication), superior to fat (5.5-5.8%) and lactose (5.1%). This milk is used to feed lambs and, in some cases, to make traditional cheese as well. In the near future, several shepherds intend to organize cheese factories, which will have further positive economic and marketing implications. Moreover, they are attempting to obtain authorization for the creation of a small, but necessary, on-site abattitor. This would allow them to avoid expensive and stressful trips to the present slaughterhouse, located more than 20 km away from the farms, over a winding and rutted road that is quite dangerous for animal's welfare and, consequently, for meat quality.

Lambs are usually slaughtered at the age of 60-90 days for the production of heavy lambs (Martini *et al.*, 1993).

Few studies have been published regarding carcass and meat quality in lambs: in a comparative trial between Zerasca, Ile de France and related crossbreed, Verità *et al.* (2001b) reported the highest weight at slaughter, the lowest dressing percentage and the longest carcasses and legs in the first. Particularly lean meats and the best adaptability to the territory were also noticed.

Moreover, it is important to underline that Zeri has another typical production associated with sheep breeding: the *"mezzalana"*, a mixture of wool and hemp used for the manufacture of socks, sweaters, carpets, mattresses, pillows and special long full skirts (the traditional women's clothing).

This work aimed to evaluate the morphometric and morphological traits of the Zerasca population, in the light of possible modifications over the years typified by enhancement and development strategies. Furthermore, breeding techniques and productive parameters have been defined.

Material and methods

The study was conducted on all the 37 Zerasca sheep farms indicated by the Massa e Carrara Provincial Breeders' Association (APA). One each farm, after a direct inspection in order to collect all useful informations about Zerasca livestock, a questionnaire was filled out by a trained observer, with pertinent the data regarding activities, land, structures, animals, nutrition and productive traits.

The previously described rearing techniques were similar on all farms.

Morphologic and morphometric traits were collected on a sample of 83 animals from four different flocks, evaluating head width, forehead-nose arc and chord bridge, ear length, withers height, rump height, chest height, width and girth, body length, back width, ischiatic width and shin girth front and back (Meregalli, 1980). Measurements were carried out with animals in standing position on flat land, using a measuring tape and a Lydtin stick. Colour of fleece, head, limbs and skin were noted.

Means, variances and ANOVA tests were estimated using JMP software (2002).

Results and discussion

Farms were placed at 843±133.4 m (550-1 330 m) above sea level; 22% were higher than 900 m.

The Zerasca population totalled 2143 subjects, divided among 37 farms (mean 58±53.0 animals per farm). High standard deviation indicates the heterogeneous consistence of flocks: 64.9% had less than 50 animals (mean 32±11.4), 24.3% between 50 and 100 (mean 74±14.0) and 10.8% more than 100 (mean 177±85.0).

Age of breeders was 49±13.5 years, 22% were younger than 40 years old; more than half were women (54%). The average number of persons taking care of animals was 1.5±0.68 *per* farm. However, the actual number of individuals performing this work on each farm was not related to the number of animals on the farm.

Purebred Zerasca were grazed on 92% of farms, in only four cases crosses were being raised. On 35% of farms other species were reared, mostly crossbreed cows for milk and meat production, but also pigs and horses.

The average area of the land on each farm was 9.0±7.82 ha, except for one 100 ha farm, while 22% of breeders owned no fields at all; they used 18.7±28.12 ha for pasture (range 2-140 ha), reflecting the frequent custom of utilizing state property areas for breeding or grazing purpose. Manure spread was the only care for pasture in 90% of cases.

In addition, sizes of the sheds available for each flock varied greatly: about 64 m^2 for less than 50 sheep, 136 m^2 between 50 and 100 animals and 220 m^2 for more.

Both pastures and sheepfold dimensions were influenced by flock size ($P \leq 0.001$ and $P \leq 0.001$, respectively).

Growth performances of lambs were quite unique in free range ovine rearing: they reached 20-22 kg of live weight in 60-62 days (Martini *et al.*, 1993; Verità *et al.*, 2001b), showing 230-375 g of daily gain, also thanks to the sheeps' special milk. Otherwise, our results indicated 23±2.8 kg of mean body weight: 44% of farms preferred to slaughter lambs between 20-24 kg, 25% between 16-20 kg and 31% required heavier weighs. Age of lambs at time of slaughter was 70±9.7 days: 57% between 55-74 days and 47% between 75-90 days. Farmers prolonged lamb growth in order to obtain bigger carcass ($P \leq 0.01$).

Twins births occurred in 32±22.3 % of cases, with rate of up to 75%: this could be considered a peculiarity of Zerasca sheep.

Ewes began their reproductive activity at different ages (mean 12.3±3.95 months): 19% under 8 months, 51% between 8-12 months and 30% at more than one year. Instead, rams began earlier (mean 9.6±2.78 months) and with a more homogeneous distribution: in 92% of cases rams were used for reproductive purpose before 12 months.

The ewes lived more than rams (10.2±1.57 vs. 4.2±2.15 years): normally (92%) females lived more than 8 years, while rams were bred generally (78%) less than 5 years.

Verità *et al.* (1993) retained that Zeri sheep came to maturity at the time of the second birth, even though the animals' age was difficult to determine, since it was extremely variable.

Compared to previous studies (Verità *et al.*, 1993; Verità *et al.*, 2001a), our results (Table 1) underlined some structural conformational differences: narrower heads, longer and more convex muzzles, larger shin girth, greater chest girth, lower chest width and longer body length. This reveals an evolution of the morphological traits of Zerasca sheep, even if this study has presented preliminary results on a small sample of purebred subjects.

Only 8.3% of farms used milk to produce cheese, while 16.2% utilized wool for *"mezzalana"* at the time of study.

Table 1. Zoometrical indexes of Zerasca breed (Mean ± Standard Deviation).

	Head width	Forehead-nose bridge (arc)	Forehead-nose bridge (chord)	Ear length	Withers height
Lambs	11.00±0.58	23.0±1.15	22.0±1.41	10.3±3.99	73.7±3.20
Primiparous	12.0±1.41	24.7±2.06	22.5±1.73	10.2±2.22	77.2±3.40
Adults	12.1±0.97	25.6±1.38	24.3±1.18	12.5±1.54	76.0±4.54
Rams	11.5±0.69	24.2±1.33	23.3±1.27	11.8±2.09	75.0±2.93
	Rump height	Chest height	Chest width	Chest girth	Body length
Lambs	76.1±2.85	33.6±3.31	20.9±3.02	90.3±15.04	79.1±7.45
Primiparous	79.0±2.71	34.2±5.74	23.2±5.31	104.0±11.34	83.7±7.41
Adults	78.6±3.49	35.7±1.45	23.2±2.60	102.2±5.71	85.4±4.72
Rams	77.5±2.07	33.5±1.13	22.0±2.12	100.0±8.05	82.9±3.05
	Back width	Ischiatic width	Shin girth (front)	Shin girth (back)	
Lambs	20.7±5.38	16.1±2.67	8.6±0.53	10.0±0.58	
Primiparous	19.5±0.58	17.5±1.00	9.2±1.26	11.0±1.41	
Adults	22.7±3.09	15.1±2.52	8.9±0.61	10.6±0.71	
Rams	21.6±4.08	17.0±3.19	8.4±050	10.1±0.30	

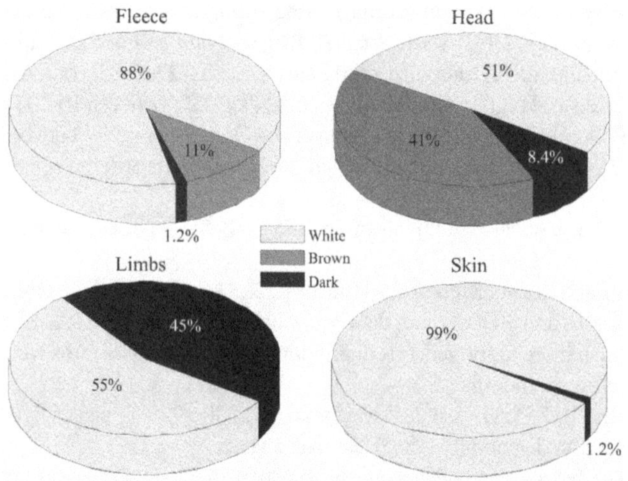

Figure 1. Frequency of colour characteristics.

Regarding morphologic traits (Figure 1), Zerasca skin was consistently fair (white or pink), but the colour of the fleece was more heterogeneous: 88% were the characteristic white, while the others were brown or black spotted. Heads were completely white in more than half the cases, but 41% displayed brown tints. Limbs presented many cases of dark colouring (brown or black). In spite of everything, compared to past studies (Verità *et al.*, 1993), this confirm a certain influence of past Zerasca x Massese crosses aimed at improving production, especially for milk.

Conclusions

The Zerasca is one of the most interesting sheep breed with endangered status and it now seems to be making a comeback, thanks to resolute young farmers and strong traditions. However, assistance and cooperation are still required to continue to maintain the Zeri special lambs. We confirm our interest in the development of this sheep, especially regarding the careful utilization of allopathic treatments (Benvenuti *et al.*, 2004; Giuliotti *et al.*, 2005) and pasture management (Verità *et al.*, 1990; Verità *et al.*, 2001a), in the name of a proficient and sustainable utilization of marginal territories.

Acknowledgements

The Authors thank Achille Guastalli, Valentina Merletti, Cinzia Angiolini, Patrizia Figaroli and Livio Quiligotti for their active cooperation, Olivia Paganelli and Brunella Antonioli for technical assistance.

Work supported by ARSIA-Tuscany Region founds.

References

Benvenuti N., Bottoni L. & Verità P., 1998. La pecora Zerasca. Una popolazione autoctona da conoscere, proteggere, valorizzare. L'Allevatore di Ovini e Caprini, XV (10): 9-11.

Benvenuti N., Goracci J., Giuliotti L. & Verità P., 2004. Resistance parameters to gastrointestinal parasitosis in Zerasca breed: a Tuscany indigenous line. Book of Abstract no.10 of the 55th Ann. Meet. EAAP, 5-9 Sept. 2004, Bled (Slovenia): 19.

Giuliotti L., Benvenuti M.N., Goracci J. & Verità P., 2005. Observation on gastrointestinal strongylosis resistance in Zerasca breed. Abstract XVI Congr. Naz. A.S.P.A., 28-30 June, Torino (Italy); (In press).

JMP, 2002. J.M.P. User's Guide ver. 5.0, S.A.S. Institute Inc., Ed. Cary (NC), USA.

Martini M., Verità P., Cecchi F., Ricci G., Giuliotti L. & Colombani B., 1993. Prove di accrescimento e rese alla macellazione della popolazione ovina Zerasca. Proc. XXVIII Simp. Int. Zoot., Ed. Dip. Sci. Zoot. Torino, 14 May 1993, Milan (Italy): 365-380.

Meregalli A., 1980. Conoscenza morfofunzionale degli animali domestici. Ed. Liviana Editrice, Padova (Italy), pp. 300.

Verità P., Corleto A., Martini M., Cazzato A. & Giuliotti L., 1990. Relation between environment and sheep farming system: food resources management and yield results. Proc. VI Meet. FAO, European Sub-Network on Mediterranean Pasture and Fodder Crops, Bari (Italy): 201-209.

Verità P., Martini M., Cecchi F. & Colombani B., 2001a. La popolazione ovina Zerasca: studi per una sua valorizzazione. Ed. Centro Tipografico Università di Pisa, pp. 91.

Verità P., Martini M., Leotta R., Cecchi F. & Colombani B., 1993. Studio biometrici della popolazione ovina Zerasca. Proc. XXVIII Simp. Int. Zoot., Ed. Dip. Sci. Zoot. Torino, 14 May 1993, Milan (Italy): 479-494.

Verità P., Russo C. & Preziuso G., 2001b. La produzione dell'agnello zerasco: indagine preliminare sulla qualità della carcassa e della carne. Proc. Conv. Naz. "Parliamo di...zootecnia e sviluppo sostenibile", Ed. Greppi G.F. & Ciceri A., 11-12 Oct. 2001, Fossano (Cuneo, Italy): 181-186.

Rediscovering the niche products of Sardinia: "Sartizza a lorika"

S. Porcu[1]*, C. Diaferia*[2]*, E. Daga*[1]*, M. Delrio*[1] *& S. Ligios*[1]

[1]*Istituto Zootecnico e Caseario per la Sardegna,Km 18,6 S.S. Sassari-Fertilia, 07040 Olmedo, Italy*
[2]*Stazione Sperimentale per l'Industria delle Conserve Alimentari, Viale Tanara 31/A, 43100 Parma, Italy*

Summary

Despite a thousand-year-old tradition of pork processing in Sardinia, in depth studies, aimed at the characterisation and finding the true value of the various typical delicatessen produce of the island, have never been carried out. This report aims to present a niche product which even today is produced in certain villages in the centre of Sardinia.Villages which have always kept up archaic culture and traditions. Here, we would like to present a preliminary and descriptive study on the characteristics of 'Sartizza a lorika' (ring-shaped, spiral sausage), comparing three different methods of production. In this trial, three types of processing were observed. For each process, the chemical-physical and microbiological parameters were recorded.

Keywords: sausage, typical products, ripening, tradition.

Introduction

In Sardinia, the vast oak and chestnut woods proved perfect terrain for the pig breeding (Cherchi Paba, 1974). The tradition of pig farming is age old (Porcu *et al.*, 2004), but an increase in the rearing was to be seen during the Roman domination, when Sardinians paid their taxes in fresh, cased and salted pork and lard (Cetti, 1774; Cherchi Paba, 1974; Meloni, 1990; Vera, 2004). Up until very recently it was usual, particularly in rural areas for families, to rear at least one pig for their own consumption. In spite of the extent of pig farming and sausage production, no study had ever highlighted the quality and tradition of the island's products. Recently, the Istituto Zootecnico e Caseario per la Sardegna (IZCS) has started research, that enabled them to 'discover' products which were not only unknown to most Sardinian people, but also to the farm workers in so much as, a list of typical products was never registered. Amongst these products, are shoulder ham with cheek (at times including the jaw), pork sausage produced with meat which is soaked in wine for more than four days (before casing) and sausages which are several metres in length. In Roman times, among the various cold cuts there were sausages known as 'Longaones'. These are long sausages (Vera, 2004) which are similar to ones produced today by some families in villages in the centre of Sardinia and some French regions. This report aims to present this niche product evaluating some features useful to obtain a designation like for the similar French sausage.

The name of the sausage changes from village to village: sartizza a lorika, saltizza a loriga, sardizza a loriga, sartithu a loriga (ring-shaped, spiral sausage) and sartizza longa (long sausage). This sausage is different in as it is cased in pig intestine which may be over three metres in length. Sometimes, the sausage is bound approximately every 40-50 cm with string making it more compact. For the drying, smoking and maturation phases the sausages (which become spiral shaped) are hung on long rods. Until fifty years ago this type of sausage was common in Sardinia. In some villages, where there was no casing machine, the process involved pushing the meat into the casing with a wooden spoon via a funnel. The binding of the ends was done by a red-hot spit which sealed the casing. Today, for convenience and preservation, smaller traditional shaped sausages are produced.

Material and methods

Three types of manufactures were observed between January-February 2005. A, a private producer and B a cured pork meat factory both in Villagrande Strisaili (Nuoro) and C a private producer in Littos – Sarule (Nuoro). The process in C followed a traditional recipe from Ollolai (Nuoro) where a 12 month old Sardinian sow was slaughtered. For A and B, two male cross-bred pigs of 24 months age (castrated at 14 months) were slaughtered. All the animals were fed acorns on wild pasture and were not on empty stomach before slaughter.

In process A and B meat was minced with a 16 mm grill and C with a 22 mm. The minced meat was put into pork sausage skins by hand. In A the drying and maturation processes were carried out in a loft, in C in a farmhouse kitchen whereas in B the first phase was in an air-conditioned cell but not the second phase. For each process the manufacturing methods and the length and weight of each sausage were noted. The thermo-igrometric parameters of the drying locations and the maturation of the products were taken by using a digital thermoigrometer (Oregon Scientific) twice a day, morning and evening for the whole maturation period.

The recordings of weight loss, physical-chemical and microbiological parameters were noted on day 0, 3, 7, 14, 21 and 28 of maturation. For A and B it was deemed necessary to take another sample on day 30 as the product did not seem fully mature.

Physico-chemical analyses

The pH was determined potentiometrically by mixing 10 g of sample with 90 ml of distilled water. Water activity was determined at 25°C using Novasina AW Sprint (Axair Ltd.). Proximate chemical composition was determined through the use of method AOAC and non protein nitrogen (NPN) as described by Careri *et al.* (1993), after precipitation with trichloroacetate acid. The iodine value and acidity were determined according to the Italian Norms for the Inspection of Fats and Derivatives. (NGD C 32 and NGD C 10- 1976). Fatty acid composition was examined after extracting the lipids as in the method modified by Folch (1989) and the subsequent transmethylation according to the method of Chin *et al.* (1992) and Stanton *et al.* (1997).

Microbiological analyses

The sampling of the sausages was performed starting by aseptically removing the casing. An aliquot of 20g of each sample was taken aseptically and homogenized with 180ml of sterile diluent containing peptone (1g/l), NaCl (8.5 g/l) and Tween 80 (1ml/l) in a Stomacher blender. Ten-fold dilutions were carried out and placed on MRS agar (30°C, 72h in anaerobic jars) to count the mesophilic lactic bacteria; Mannitol Salt Agar (30°C, 72 h) for *Micrococcaceae*; Slanets and Bartley agar (37°C, 24 h) for *Enterobacteriaceae*; Malt Extract Agar acidified at pH 3.5 with lactic acid 10% (25°C, 120 h) for moulds and yeasts. Results were expressed as $\log_{10}$ numbers colony forming units/g ($\log_{10}$ cfu/g).

Results and discussion

Sausage production technique

The technique used to prepare the batches required the cooling of the carcasses for 24 hours at room temperature before sectioning. In each process, the meat used was the remains of whole cuts of meat. Only in C was meat from the shoulder added. The manufacture parameters during the process are shown in table 1.

Physico-chemical analyses

The pH evolution during maturation was different in the three batches (Table 2). Batch A shows a decrease up to the 21st day. The evolution was more irregular in batch B. In the final phase the pH values alternated in both batches A and B. In batch C the pH values showed slight variations. The a_w (Table 2) of batch A had slight changes during the whole maturation period. It never reached values lower than 0.94, the high a_w even at the end of maturation was probably due to low amounts of salt used (Table 1). The a_w decreased regularly during maturation in batches B and C. Manufactures B and C show similar weight loss in the first seven days of the drying procedure. This differs in the second phase of maturation. A, on the other hand, initially presented an inferior weight loss reaching slightly lower values than B at the end of maturation. During maturation there was no evidence of an evolution of the acid profile (data not shown). However, a difference in products A and B compared to C was observed. A higher concentration of saturated fatty acids was present in product C. This data was confirmed by a lower iodine value (Table 3).

Microbiological analyses

The trend of mesophilic lactic bacteria and of *Micrococcaceae* during maturation is shown in figure 1. Mesophilic lactic bacteria increased rapidly between the 3rd and the 7th day of maturation in all three batches, although the increase was different in each case (3.79 log –2.8 log - 2 log). Figure 1 shows that batch C always had a lower mesophilic lactic bacteria count than batches A and B. In the latter batches the mesophilic lactic bacteria count reached 8.5 $\log_{10}$ cfu/g. At the beginning of maturation (3-7 days) *Micrococcaceae* outnumbered lactic bacteria in batches A and B but from the 7th day it was outnumbered by lactic bacteria, especially in batch A. In batch C the prevalence of *Micrococcaceae* was very evident in the first seven days of maturation with a maximum of 3.2 log on day three; from the 14th day the two microbial groups were equivalent in number. *Enterobacteriaceae* were absent from the 21st day of maturation in batch A, from the 7th day in batch B and batch C they were still present on the 28th day although their number was not high (1.68 $\log_{10}$ cfu/g) (data not shown). Yeasts were 1 log more numerous than *Micrococacceae* from the 14th day of maturation in batch A except on the 28th day. However, they were outnumbered by *Micrococcaceae* in batch B. In batch C *Micrococcaceae* and yeasts had a similar count. The number of *Enterococci* was generally higher in batch B (data not shown).

Table 1. Sausages manufacture parameters.

Batches	A	B	C
Processed meat (kg)	20	15	10
Salt (g/kg meat)	11	20	18
Pepper (g/kg meat)	2	1	2.5
Garlic (for kg meat)	a head	1 g (freeze-dried)	-
Wine (ml/kg meat)	12.5	-	85
Other	-	10[1]	-
Processing room T (°C)	4.3	4.5	8.6
Processing room U.R.	73	75	74
Mixture standing (h)	24	24	3
Sausage lenght (max) (cm)	180	320	423
Sausage weight (max) (kg)	2.0	3.14	4.3
Ripening T (min-max) (°C)	1.8 – 9.5	3.5 – 9.0	2.4 – 10.1
Ripening UR (min-max) (%)	62 - 88	64 - 85	54 - 80

[1]Commercial preparation: antioxidant E 300, preservative E 252 (3%), saccharose, salt, max 1%.

Table 2. pH, a_w and weight loss values of sausages during ripening.

	pH			a_w			Weight loss (%)		
Day	A	B	C	A	B	C	A	B	C
0	5.71	5.72	5.68	0.971	0.964	0.959			
3	5.82	5.84	5.85	0.971	0.943	0.952	7.47	15.40	15.30
7	5.63	5.80	5.79	0.956	0.929	0.948	14.34	25.83	26.60
14	5.49	5.47	5.86	0.956	0.891	0.892	23.17	33.21	39.70
21	5.37	5.77	5.77	0.950	0.875	0.874	28.95	34.82	44.20
28	5.52	5.21	5.86	0.941	0.865	0.830	32.28	35.94	46.90
35	5.30	5.70	n.d.[1]	0.941	0.869	n.d.[1]	34.48	36.62	n.d.[1]

[1]Not determined

Table 3. Physico-chemical parameters of sausages of the batches A, B, C at 0, 28 and 35 days of ripening.

	0			28			35		
	Batches			Batches			Batches		
Times (days)	A	B	C	A	B	C	A	B	C
Dry matter (%)	47.57	49.23	43.49	64.33	72.46	79.96	69.34	75.04	n.d.[1]
Iodine value	65.46	66.79	61.22	66.07	66.57	62.59	66.35	68.05	n.d.
Acidity (%C18:1)	0.74	0.76	0.54	1.99	1.49	1.26	1.78	1.49	n.d.
Saturated fatty acid[2]	33.51	30.47	40.45	33.08	31.31	40.25	33.39	30.54	n.d.
Unsaturated fatty acid[2]	66.49	69.53	59.55	66.92	68.69	59.75	66.61	69.46	n.d.

[1]Not determined.
[2]Fatty acid (wt.%).

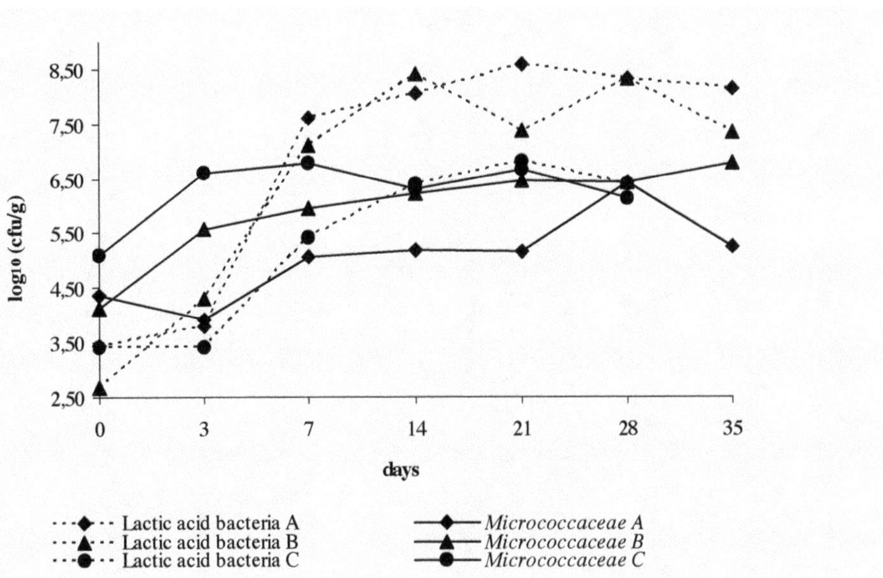

Figure 2. Evolution of lactic acid bacteria and Micrococcaceae during ripening of sausages.

Acknowledgements

This work was supported by a grant of European Union Interreg Project IIIA.

The authors would like to thank M. Barrili, V. Ecca and S. Lostia for their help during the manufacturing. N. Lei, S. Pintus, M.A. Sanna, M. Fiori, M. Addis and M.C. Fozzi for sampling and laboratory analyses.

References

Association of Official Analytical Chemists, 1995. Official methods of analysis, 16th Edition. Arlington.

Careri, M., Mangia, A., Barbieri, G., Bolzoni, L.,Virgili, R. & Parolari, G., 1993. Sensory property relationships to chemical data of Italian type dry-cured ham. J. Food Sci., 58: 968-972.

Cetti, F., 1774. I quadrupedi di Sardegna, Sassari, 87-92.

Cherchi Paba, F., 1974. Evoluzione storica dell'attività industriale, agricola, caccia e pesca in Sardegna. Regione Autonoma Sardegna (editors), Cagliari, voll. I – IV.

Chin S.F., Liu W., Storkson J.M., Ha Y.L. & Pariza M.W., 1992. J.Food Comp.Anal. 5: 185-197.

Meloni, P., 1990. La Sardegna romana. Chiarella (editors), Sassari, 205-206.

Porcu, S., Usai, G., Carta, A. & Ligios, S., 2004. L'elevage du porc en Sardaigne entre histoire et actualite. In Proceedings of V Symposium International sur le Porc Méditerranéen Tarbes 16-19 Novembre. In press.

Porcu, S., Usai, G., Cappai, P., Carta, A. & Ligios, S., 2005. Allevamento suino in Sardegna: storia; attualità; prospettive. Corte 11-10-04 Porcu e prisuttu: une affaire de famille. In press.

Stanton C., Lawless F., Kjellmer G., Harrington D., Devery R. & Conolly J.F., Murphy J.J., 1997. J Food Sci. 62: 1083-1086

Vera, D., 2004. Del suino e delle sue carni nella storia: dall'antichità all'alto medioevo. In http://www.museidelcibo.it/page.asp?IDCategoria=234&IDSezione=969&ID=32127.

Technical-economic characterization of dehesa farms using cluster analysis

P. Gaspar, F.J. Mesías, M. Martín, M. Escribano, A. Rodríguez & F. Pulido

Escuela de Ingenierías Agrarias. Universidad de Extremadura,
Ctra. de Cáceres, s/n, 06071 Badajoz, Spain

Summary

This paper deals with the characterization of dehesa farms in Extremadura using cluster analysis. Several groups have been created according to economic and livestock management criteria. Cluster analysis allowed to establish a farm typology, adjusting the dimensional and management levels to homogeneous groups of farms.

Keywords: cluster analysis, dehesa farms, economic management.

Introduction

This paper aims to get a typology of dehesa farms in Extremadura, identifying groups with common characteristics which will allow comparisons according to their behaviour. It will also allow to improve the knowledge about the situation of the livestock sector.

The data were obtained by surveying owners or managers from 69 farms with beef, sheep and/or Iberian pig throughout the year 2004. The farms were located all over Extremadura (SW Spain) following the selection criteria of the Interreg III project SP4.E13 "Development of an information system for the environmental and economic management of the dehesa/montado ecosystem in Extremadura and Alentejo".

Material and methodology

Cluster analysis is a descriptive and multivariate technique used to classify data from a with cases-variables, producing homogeneous groups (clusters) (Bisquerra, 1989, Carrasco & Hernán, 1993). In this paper, the variables obtained from the poll questionnaire were studied by using a K-means cluster, which is considered the correct approach when a big number of cases are included in the data base (SPSS, 1999). This technique requires to specify the number of clusters that are going to be considered. Thus, the first step is to determine the number of groups which provides significant relative group sizes (Malhotra, 2004). Afterwards, the analysis is repeated by using other procedures or with different number of clusters (Luque, 2000).

According to this procedure, several K-means cluster analyses were carried out with three and four groups. The variables were selected in order to allow the measurement of the different technical-economic management aspects in dehesa farms. Prior to the analysis, our variables were normalized in order to avoid problems with the measurement ranges. The resulting 21 variables are listed in table 1 with their corresponding measurement scale respectively.

Finally, four groups were established which provided a coherent explanation. The membership to a given cluster of each element was identified so as to subsequently interpret the clusters that were obtained.

Table 1. Variables and final cluster centres.

	Variable	Scale	Cluster 1	2	3	4	Total	Sig
V1=	Cattle Livestock Units (LU) in the farm/ha of Useful Farming Surface (UFS)	LU/ha	0.21	0.29	0.08	0.03	0.14	***
V2=	Sheep LU in the farm/ha UFS	LU/ha	0.24	0.12	0.19	0.56	0.24	***
V3=	Pig LU in the farm/ha UFS	LU/ha	0.07	0.02	0.01	0.00	0.02	***
V4=	Relationship Rented farm surface/Total farm surface	%	0.00	0.29	0.45	0.35	0.34	n.s
V5=	Relationship wooded farm surface/Total farm surface	%	0.99	0.57	0.67	0.43	0.64	*
V6=	Stocking rate	LU/ha	0.52	0.45	0.30	0.59	0.41	***
V7=	Total Cost Buying Raw Materials/ha UFS	€/ha	398.10	159.53	113.45	251.40	179.80	***
V8=	Cost External Services/ha UFS	€/ha	26.21	18.08	15.85	35.13	20.68	n.s
V9=	Labour Cost/ha UFS	€/ha	105.33	32.67	46.38	89.73	56.75	***
V10=	Total farm income selling animals/ha UFS	€/ha	631.53	248.08	139.64	294.57	248.09	***
V11=	Current Cost/ha UFS	€/ha	611.96	245.21	222.08	454.87	310.09	***
V12=	Final Output/ha UFS	€/ha	727.19	371.01	198.11	392.53	333.04	***
V13=	Gross fixed capital investment /ha UFS	€/ha	75.66	86.62	30.58	47.32	52.28	***
V14=	Intermediate Product/ha UFS	€/ha	163.62	77.03	46.70	76.53	72.49	***
V15=	Gross Product/ha UFS	€/ha	890.80	448.04	244.82	469.06	405.53	***
V16=	Farm net margin/ha UFS	€/ha	210.78	167.68	-25.99	-41.57	46.69	***
V17=	Net subsidies/ha UFS	€/ha	118.92	112.90	76.55	137.79	100.18	***
V18=	Net Surplus/ha UFS	€/ha	329.70	280.57	50.56	96.22	146.87	***
V19=	Farm profitability rate	%	4.18	4.74	0.93	1.74	2.38	***
V20=	Relationship Fixed land capital/Total Fixed capital	%	94.70	92.81	96.02	93.71	94.71	***
V21=	Relationship Livestock and machinery fixed capital/Total Fixed capital	%	5.30	7.19	3.98	6.29	5.29	***

*** $P<0.001$; ** $P<0.01$; * $P<0.05$; n.s.: non-significant.
Cattle LU = adult cows + adult bulls + (heifers + bullocks)*0.60.
Sheep LU = adult sheep (male+female)*0.15.
Pig LU = Adult pigs (male+female)*0.30.

Results and discussion

The size of the final clusters are listed in table 2, while table 1 shows the means corresponding to the different clusters, together with the level of significance for each variable of an analysis of variance

Table 2. Number of cases of the final clusters.

Cluster	Number of cases	%
1	8	11.59
2	17	24.64
3	33	47.83
4	11	15.94
Total	69	100.00

using the final clusters as groups, which allows to check the variability between clusters. These results can be used to detect those variables showing no differences between the clusters.

The characteristics which differentiate the clusters, according to the results in table 1, are described below.

Group 1. Mixed cattle, sheep and Iberian pig farms in tree-covered dehesas

This is the minor group and includes 11.6% of the farms. It is characterized by the high diversification of its productions because of the combination of cattle, sheep and Iberian pig in the farms. The stocking rate in these farms is high, and they are located mainly in a tree-covered pastures (*Q. ilex* and *Q. suber*). The high levels of stocking rate produce also high total cost Buying Raw Materials. Labour Costs are also important probably because of the different animal species present in the farms. Both costs make Current Cost the highest compared to the other groups. Iberian pig's breed is able to take profit of acorn (*Q. ilex* fruit) and this is probably the reason of the high level of Intermediate Product. In addition the total farm income from selling animals is high as well, both influence Final Output indicator making it highest. Concluding, this is a group of farms with high mean values of Farm net margin and Net subsidies which give as a result a mean value of Farm profitability of 4.1%.

Group 2. Beef farms

This group comprises 24.6% of the farms, that are specialized in beef production although sheep and Iberian pig are also found in the farms as a complement. The stocking rate is medium but the Total Cost Buying Raw Materials and especially Labour Cost are lower than the mean value of the whole sample. Both data explain the low level of Current Costs. However, Total Farm Income from selling animals and Gross Product are average values and this is why these farms keep a good Farm net margin.

The Farm net margin that these farms obtain, together with the subsidies (mainly from cattle and cereals) give as a result a good level of Net Surplus and Farm profitability rate. In conclusion, these are the best optimised farms.

Group 3. Mixed livestock and Low profitable farms

Enclosing 47.8% of the farms, the group represents nearly half of the sample. This group has low stocking rate and does not present preferences for any specie. Considering the low stocking rate the cluster shows high costs from labour and buying Raw Materials. These, together with Productions indicators lower than the average produces negative Farm net margins within this group.

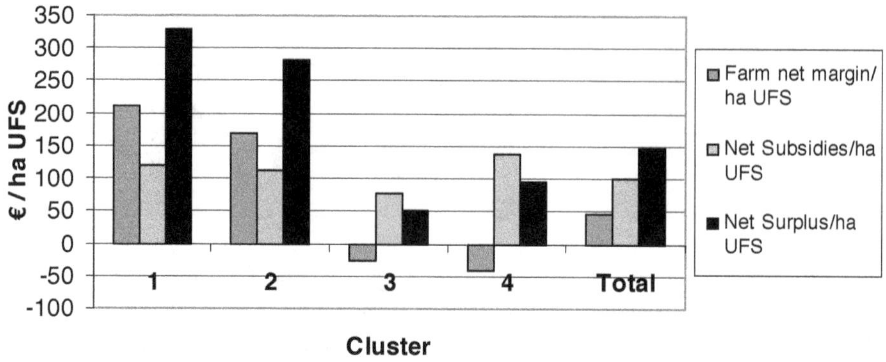

Figure 1. Farm Net Margin, Farm Net Surplus and Farm Net Subsidies for every cluster.

These negative values are slightly compensated by the subsidies received by the farms. Farm profitability rate is very low (0.92%), not just because of the low surplus, but mainly because in these farms fixed capital is biased towards land fixed capital. The high value of the land reduces Farm profitability rate. The above mentioned indicators make these farms the worst economic managed (Figure 1).

Group 4. Sheep farms

This segment contains 15.9% of the farms, and the usual livestock is sheep. These farms have the lowest rates of wooded surface, but high stocking rates. Due to those stocking rates, and also to the difficulties of sheep management, the farms of this cluster present higher feeding and labour costs. Although they show good intermediate and gross productions, the farm net margin is negative, but this figure is compensated by higher net subsidies, due to the stocking rates. They are productive farms, with a farm profitability rate of 1.7%.

From the study of cost, production and profit indicators, it has been observed that dehesa farms' economic results differs according to the ecosystem's characteristics, the livestock and the stocking rate they show. Although the subsidies linked to cattle and sheep breeding are important to keep these farms profitable, they are not a determinant factor. A more efficient management, as in group 2, allows the optimisation of the resources with a good mixed of livestock species and a correct destination for the productions.

References

Bisquerra, R., 1989. Introducción conceptual al análisis multivariable. Promociones y Publicaciones Universitarias, Barcelona, Spain, pp. 410.

Carrasco, J.L. & M.A. Hernán, 1993. Estadistica multivariante en las ciencias de la vida. Ediciones Ciencia 3, Madrid, Spain, pp. 625.

Luque, T., 2000. Técnicas de análisis de datos en investigación de mercados. Ediciones Pirámide, Madrid, pp. 557.

Malhotra, N., 2004. Investigación de mercados. Pearson Educación, Mexico, pp. 816.

SPSS Inc., 1999. SPSS Base 10.0 Aplications Guide. SPSS Inc., Chicago, USA.

Study of gastrointestinal parasite dynamics in Zerasca sheep aimed at reducing anthelmintic treatment

N. Benvenuti, L. Giuliotti, J. Goracci & P. Verità

Department of Animal Production, University of Pisa,
Viale delle Piagge, 2, 56124 Pisa, Italy

Summary

Counteracting parasite problems with anthelmintic treatment is only a short-term solution. More information is needed about the various factors that influence parasite epidemiology and the employment of integrated strategies in order to minimize the impact of pharmacological treatment. This study aimed at analyzing parasite burden in a flock of Zerasca sheep in Italy, taking into account individual, age, sampling data and physiological status as variability factors. Mean faecal egg count (FEC) was 340 (s.d. 355) and packed cell value (PCV) was 29 (s.d. 3.65). FEC was not significantly influenced by all factors studied, except age of animal alone. FEC and PCV appeared to be inversely correlated (r = -0.262; $P \leq 0.001$). Results regarding monthly FEC correlations showed good repeatability in certain periods of the year: October displayed a significant correlation with the months from April to December, characterised by low infestation. February, instead, was a strategic period for identifying the highest worm burden and, at the same time, appeared to be the best time to evaluate treatment opportunities.

Keywords: Zerasca, ovine, gastrointestinal parasites, sustainable control.

Introduction

Gastrointestinal parasite control is one of the most important aspects of health management in sheep breeding. In Italy today, sheep breeding is based on extensive production techniques, where feed is mainly provided from pasture. Gastrointestinal parasites carry out their biological cycle thanks to that substrate: grass helps preserve the larvae, which survive and re-infest grazing animals. For these reasons farmers rely on pharmacological treatment as a prophylactic strategy, even without any laboratory evidence of parasites. However, this practice can have detrimental effects on animal production and the environment, by increasing anthelmintic resistance and burdening the farm's budget with the high cost of treatments.

For a long time, the risks connected with such health programme have prompted researchers to investigate alternative methods of parasite control, suggesting the use of integrated and sustainable approaches.

Gastrointestinal parasite burden is closely linked to the environment, including climatic conditions, farm and pasture management (AA.VV., 2003), animal nutrition and individual response. Based on these considerations, it may be possible to employ integrated strategies by pharmacological treatment to control the host-parasite relationship, without parasites eradication. In fact, in balanced ecological systems, both the host and parasite populations are firmly controlled by a complex array of interacting factors. Domestication of livestock and related implications (increasing demand for productivity, restriction of movement, and so on) has shifted the equilibrium in favour of parasites (Waller, 2003).

Currently, it is possible to significantly reduce the number of anthelmilmintic treatments for grazing ruminants by employing some relatively simple management procedures, supported by knowledge of parasite epidemiology in relation to climate and production systems (Barger, 1997). Of available techniques, this approach suggested focusing on the suitable use of pasture including rotational grazing and agronomic care, proper animal nutrition, use of forage high in tannins, utilisation of soil organisms (i.e. *Duddinghtonia flagrans* fungus), breeding of resistant or resilient breeds and, finally, employment of homeopathic or phytotherapic treatments.

In any parasite control plan, the first step is to determine the effective worm burden on the farm and consider all the influencing factors (climatic, physiological): with this knowledge, it would be possible to clarify the need for effective anthelmintic treatment and, eventually, the best time of the year for the operation (AA.VV., 2003).

All these considerations have become increasingly important for local sheep farms, where environment and quality of production are closely linked. The availability of sustainable control of gastrointestinal parasites along with limited use of chemotherapy could prove to be of great value to farmers and beneficial to the environment.

This research aimed at aiding both producers and veterinary practitioners in gastrointestinal parasite control by studying of the gastrointestinal parasite burden in the Zerasca sheep. In fact, an important consideration in controlling nematode parasites in livestock, is an exact and timely diagnosis of helminthosis.

Materials and methods

From January 2004 to March 2005, faecal and blood samples were collected monthly from 48 naturally infested Zerasca ewes. The farm was located at an altitude of 800 m, typified by a climate midway between Mediterranean and Continental conditions, temperatures drop to -6°C in winter and average 23°C in summer. Yearly rainfall was estimated to be 2000 mm, being higher in November and lower between June and August (Verità *et al.*, 2001). The farm, a member of a "*Consortium*", contained 90 ewes and yielded heavy lamb (22-35 kg of live weight at 60-90 days of age). Zerasca farmers systematically use two or three conventional anthelmintic treatments, particularly against Strongyles.

Faeces were collected directly from the rectal ampoule and individually examined using a modified McMaster technique (Permin & Hansen, 1998) to estimate the FEC (expressed in EPG - Eggs Per Gram) of gastrointestinal nematodes. Blood samples were collected from the jugular vein and packed cell value (PCV) was estimated by capillary microhematocrit analysis in order to evaluate parasite damages.

Complete information regarding each animal was gathered during sampling, including age and physiological status (peri-parturient period, 30 days before and after birth).

Statistical analysis was performed by ANOVA. The tested factors in the model were date of sampling, age of animals and physiological status. Data were logarithmically transformed [y = log(EPG + 25)] to normalize error (Baker, 1997). Repeatability of monthly FECs and PCV were estimated by Pearson's correlations.

Statistical analysis were performed with JMP software (JMP, 2002).

Post-drench samples taken in May and June 2004 were excluded from statistical analysis.

Results and discussion

Mean FEC was 340 (s.d. 355); it was very similar to that in a study previously conducted in Massese sheep, where FEC reached 360 (Benvenuti *et al.*, 2004b).

Grouping FEC into four levels of infection (1=0; 2=1-300; 3=301-600; 4=more than 600), our samples showed an acceptable parasite burden (<300 EPG) in 56% of cases. Mean PCV was 29 with a standard deviation of 3.65; 64% of samples could be assessed as being in the normal range (28-35%) reported by Bertoni (1999), while 31% had values below the physiological limit. FEC and PCV were inversely correlated ($r = -0.262$; $P \leq 0.001$) and this may underline the parasite effect in altering haematocrit values, so that we could hypothesize that damage to the animals was a consequence of parasite burden and the probable presence of blood-suckling genera. Factors significantly influencing FEC were individual ($P \leq 0.001$), date of sampling ($P \leq 0.001$) and physiological status ($P \leq 0.001$), while age did not affect parasite count. Results showed a great variability among individuals.

Table 1 shows the mean value of the FEC in a group of ewes. A number of animals have very low FEC values (indicated by an asterisk), suggesting intrinsic properties of parasite resistance.

Date of sampling significantly influenced parasite burden. Data obtained in our study did not show the usual trend (Ambrosi, 1995; Urquhart, 1999); analysing the period after the second drench, the expected increase in autumn did not take place and in fact the gastrointestinal strongyle burden presented less than 230 FEC on average, allowing the farmer to skip the usual treatment of the flock.

We saw a steady increase in parasite levels only after January 2005, the highest level occurred in March of the same year.

The lambing period of the flock began at the end of November and births peaked during the months of January and February. Increased FEC may have been the result not only of climatic conditions, but also a consequence of lambing. This seems to agree with the peri-parturient influence and stress-related problems. The observations regarding FEC during the peri-parturition period were in agreement with other authors who reported a significant rise in faecal egg output occurring around lambing time and throughout the early lactation period (Benvenuti *et al.*, 2004a; Ng'ang'a *et al.* 2004; Ambrosi, 1995). In that period, immunity was relaxed in adult ewes (Abbott *et al.*, 2004).

Table 1. Mean EPG of some tested ewes .

Animal	EPG	St. Dev.	Animal	EPG	St. Dev.	Animal	EPG	St. Dev.
1	356	222	15	775	408	29*	148	209
2	257	270	16	466	486	30	460	857
3*	117	135	17	306	484	31	290	396
4	383	402	18	215	209	32	235	154
5	243	199	19	314	410	33	213	189
6*	97	133	20*	85	84	34	220	151
7*	104	170	21	429	449	35	274	204
8*	112	180	22	276	236	36*	89	113
9	280	225	23	307	304	37	418	379
10	486	406	24	270	277	38	215	151
11	629	591	25	260	238	39	453	265
12	280	285	26	382	339	40	583	392
13	750	509	27*	147	148	41	372	221
14	371	310	28	490	477	42	436	228

(* ≤ 150 EPG).

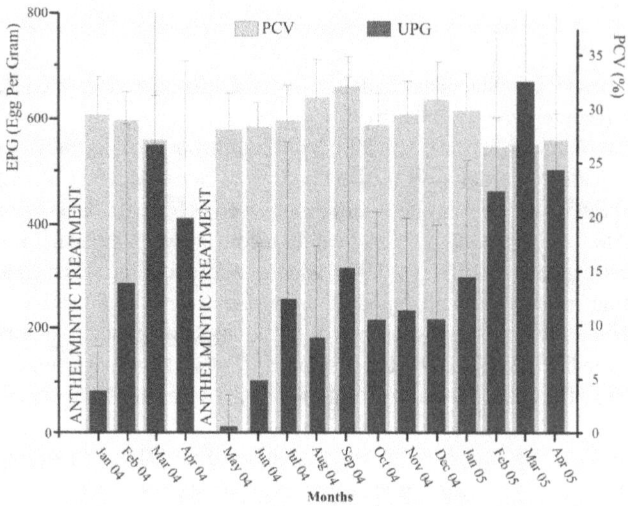

Figure 1. Trend of FEC and PCV with standard deviation.

Mean FEC of Zerasca ewes in the peri-parturient phase were significantly higher (429 s.d. 455) compared to non peri-parturient ewes (285 s.d. 326). Normally, lambs are more susceptible to parasitic disease, since innate and acquired resistance is poor (Sangster, 1998). However, in our study young animals did not show higher parasite burdens: FEC levels in lambs were consistently lower than those of adults but differences were not statistically significant (Figure 1).

Monthly correlations of EPG during the year were significant in many cases. The first 4 months of the two years cannot compared since in 2004 the animals had been treated with anthelmintics in the previous period, but this was avoided in the second year. The month that allowed the prediction of parasite burden was October, which showed a significant correlation with the months from April to December where a faecal sample taken in October could represent a long period characterised by low infestation. More interesting was identifying the proper period for evaluating the administration of anthelmintic treatment. Our data suggested that February would be the most suitable month to identify high parasite levels (spring rise) and consequently to evaluate the opportunity to treat the flock. In fact, this month was correlated with January ($r=0.521$; $P \leq 0.001$), March ($r=0.691$; $P \leq 0.001$) and April ($r=0.676$; $P \leq 0.001$).

Conclusions

In conclusion, this represent the first step to point out the effective lack of communication and back-up service to convince farmers not only to adopt, but also to persist with alternative schemes of parasites control (Waller, 1999). These approaches, possibly in combination, together with the minimal, strategic use of anthelmintics specifically selected for the specific parasite problem, could pave the way towards economically and environmentally sustainable parasite control systems for livestock. Besides, it is important to underline that worms cannot be eradicated from the environment completely (Scarfe, 2000), but infections can be limited by an adequate knowledge of parasite dynamics and prevalence.

Acknowledgements

The Authors thank Achille Guastalli (Breeders Association of Massa Carrara) and the breeder Valentina Merletti for active cooperation. Work was supported by ARSIA-Tuscany Region founds.

References

AA.VV., 2003. Code of recommendations for the welfare of livestock. Sheep. DEFRA Pub., London, pp. 34.

Abbott K.A., Taylor M. & Stubbings L. A., 2004. Sustainable worm control strategies for sheep. A technical manual for veterinary surgeons and advisers. Context Pub. (UK), pp. 43.

Ambrosi M., 1995. Parassitologia Zootecnica. Ed. Edagricole, Bologna (Italy), pp. 390.

Baker R.L., 1997. Resistance genetique des petits ruminants aux helminthes en Afrique. INRA Prod. Anim., 10: 99-110.

Barger I., 1997. Control by management. Vet. Paras., 72: 493-506.

Benvenuti M.N., Pisseri F., Cianci D. & Perrucci S., 2004a. The dynamics, prevalence and impact of strongyle (nematode) infections in a farm of massese sheep and evaluation of the efficacy of a homeopatic treatment. Atti XVI Congr. SIPAOC, Siena (Italy), 29/9-2/10: 131-132.

Benvenuti N., Giuliotti L., Goracci J. & Cianci D., 2004b. Le strongilosi gastrointestinali e la produzione del latte in ovini di razza Massese allevati con metodologie biologiche. Ob. e Doc. Vet., 9: 23-28.

Bertoni G. (Editor), 1999. Guida all'interpretazione dei profili metabolici. Ed. Perugina University, pp. 150.

JMP, 2002. J.M.P. User's Guide ver. 5.0, S.A.S. Institute Inc., Ed. Cary (NC), USA.

Ng'ang'a C.J., Munyua W.K., Maingi N. & Kanyari P.W., 2004. Occurrence of peri-parturient rise in trichostrongylid nematode egg output in Dorper ewes in a semi-arid area of Kajiado District of Kenya. Acta Trop. 2004 Nov-Dec; 92(3): 213-218.

Permin A. & Hansen J., 1998. Epidemiology, diagnosis and control of poultry parasites. FAO Animal Health Manual.

Scarfe A. D., 2000. Approaches to managing gastrointestinal nematode parasites in small ruminant. In: Meat Goat Production Handbook. (www.clemson.edu/agronomy/goats/handbook).

Sangster N., 1998. Internal parasites of sheep.(http://www.vetpath.usyd.edu.au/parasitology).

Urquhart G. M., 1999. Parassitologia Veterinaria. Ed. U.T.E.T., Torino (Italy), pp. 370.

Waller P.J., 1999. International approaches to the concept of integrated control of nematode parasites of livestock. Int. J. Paras., 29: 155-164.

Waller P.J., 2003. Global perspectives on nematode parasite control in ruminant livestock: the need to adopt alternatives to chemotherapy, with emphasis on biological control. Anim. Health Res. Rev., 4 (1): 35-43.

Verità P., Martini M., Cecchi F. & Colombani B., 2001. La popolazione ovina Zerasca: studi per una sua valorizzazione. Ed. Centro Tipografico Università di Pisa, pp. 91.

Effect of rearing system on meat quality and on fatty acid composition of subcutaneous fat in Cinta Senese pigs

C. Pugliese, F. Sirtori, L. Pianaccioli, O. Franci, A. Acciaioli, R. Bozzi & G. Campodoni

Dipartimento di Scienze Zootecniche, Via delle Cascine 5, 50144, Firenze, Italia

Summary

Thirty Cinta Senese pigs were employed in the trial: 10 pigs fed commercial mixture (3 kg/pig/d), 10 pigs fed chestnuts, 10 pigs fed acorn. The trial was carried out from November to February corresponding to the fattening period of pigs. Pigs were slaughtered at 148 kg of live weight on average. On *Longissimus lumborum* (LL) the following analysis were determined: colour (L*, a*, b*) water-holding capacity; shear force measurements; moisture; intramuscular fat; protein; ash. On backfat, colour and fatty acid profile were determined. Pig fed acorn showed, compared to pigs fed chestnut and feedstuff, the highest percentage of intramuscular fat, (4.5 vs. 4.2 vs. 3.3 % respectively). The diet did not affected the physical traits of muscle and fat. Pigs fed acorn and chestnut showed the highest values of oleic acid respect to pigs fed feedstuff (44.4 vs. 44.4 vs. 42.7 % respectively) and, consequently, the higher percentage of MUFA (47.3 vs. 47.1 vs. 45.5 % respectively). Outer layer showed, compared to inner layer, the highest MUFA (47.4 vs. 45.8 % on total fatty acids, respectively) and PUFA percentage (18.7 vs. 16.9% respectively).

Keywords: fatty acids, meat quality, acorn, chestnut, Cinta Senese pig.

Introduction

Cinta Senese pig breed represents one of the more interesting examples of a well preserved Italian autochthonous germplasm, realised with public grants, and it is a witness of the strong link between rustic breeds, extensive rearing and typical products.

Grazing on wood is variously used for the Cinta Senese breed; this type of grazing allows to exploit feed resources otherwise unused, to scatter the manures and to give a product that is recognised by the consumers as "natural" and "genuine".

Cinta Senese pigs are always reared outdoor with extremely different modalities of feeding ranging from total supply of concentrate, to exclusive employment of spontaneous resources of the wood (chestnuts and acorn), at least during the fattening phase. It is noteworthy that the use of the wood could be possibly fractioned along time with the aim to allow the feeding of the swine only for a certain period in order to obtain the product with the required organoleptic properties. The knowledge of the optimal range of this finishing phase could allow a correct utilisation of the vegetable ecosystem, with a full advantage of its sustainability. The rearing zone of Cinta Senese is more fragile and more complex in comparison with the Spanish Dehesa that already reached a stable equilibrium with the swine rearing (Lopez-Bote, 1998). Moreover, in the former situation, *Quercus* stands are flanked by chestnut stands; chestnuts productions, in excess for the human consumption, could be used by the swine with possible effects on the quality of the product (Coutron-Gambotti *et al.*, 1998; Pugliese *et al.*, 2004).

The aim of this trial was to asses the effect of rearing Cinta Senese on wood pasture on meat characteristics and fatty acid composition of subcutaneous fat.

Material and methods

Thirty Cinta Senese pigs were employed in the trial: 10 pigs were reared in paddock of 1 ha and fed commercial mixture (3 kg/pig/d), 10 pigs were reared in wood and fed chestnuts, 10 pigs were reared in wood and fed acorn. Each group reared in wood had 8 ha available for pasture and no alimentary supplement was provided.

The initial live weight of animals was 120 kg, on average. The trial was carried out from November to February corresponding to the fattening period of pigs. In table 1 the characteristics of diets are reported. Pigs were slaughtered at 148 kg of live weight, on average. On *Longissimus lumborum* (LL) the following analysis were carried out: colour measurements (L*, a*, b*) using a Minolta Chromameter; water-holding capacity (WHC) by drip loss and cooking loss in water-bath; shear force measurements (WB) on raw and cooked meat; moisture by lyophilising; intramuscular fat (IMF) by ether extraction; protein by Kjeldahl method; ash. On backfat, colour and fatty acid profile for inner and outer layer separately, were determined. Data were analysed by GLM procedure (SAS, 1996) following the model: $Y_{ijk} = \mu + D_i + b(X_{ijk}) + E_{ijk}$ where: μ = mean; D = diet; b = regression coefficient on slaughtered weight (X); E = error. For fatty acid the effect of layer was included in the model.

Results

In table 2 chemical-physical parameters of LL are reported. The only remarkable difference was recorded for ether extract: pig fed acorn showed the highest percentage of intramuscular fat, while pigs fed chestnut showed intermediate value.

In the literature there is lack of information regarding comparisons between acorn and chestnut, while many trials were carried out comparing acorn or chestnut with commercial feedstuff. Coutron-Gambotti *et al.* (1998) observed that high energy intake in Corsican pigs fed chestnut leads to deposition of large quantities of intramuscular lipids. The high percentage of intramuscular fat in

Table 1. Chemical analysis (% on weight basis) and fatty acid composition (%) of feeds.

	Chestnut	Acorn[1]	Concentrate
Moisture	61.04	43.0	10.60
Protein	3.08	3.12	16.10
Ether extract	0.84	1.69	3.58
Crude Fibre	1.57	1.96	3.9
C 16:0	13.9	15.0	16.7
C 18:1	37.5	46.8	19.7
C 18:2	42.3	30.2	56.2
C 18:3	3.9	4.0	4.5
MUFA	38.3	48.3	20.3
PUFA-n3	3.9	4.0	4.5
PUFA-n6	42.3	30.2	56.2

[1]without hull.

outdoor pigs must be also related to diet composition, characterized by the presence of acorns with high caloric value due to high contents of lipid and starch (Lopez-Bote, 1998). The higher intramuscular fat percentage in pigs fed wood products is in agreement with the observations made on other Mediterranean breeds reared in similar conditions, including Nero Siciliano (Pugliese *et al.*, 2004) and Iberian pigs (Tejeda *et al.*, 2002). The diet did not affect the physical traits of muscle and fat. This is probably due both to the same rearing condition of pigs (all three groups were reared outdoors) and to the same weight at slaughtering. It's well known that the main effects on physical traits of the outdoor system compared to indoor one, are linked to differences in growing rate and in tissular development. The effect of rearing system on physical parameters is reported in the literature with contradictory results, as discussed by Jonsäll *et al.* (2001). As regard fatty acid composition (Table 2) pigs fed acorn and chestnut showed the highest values of oleic acid and, consequently, of MUFA. For PUFA percentage, the only significant difference was found for PUFA n-3, highest in pigs fed chestnut. The first result can be explained by a instant incorporation into pork fat of dietary fatty acid (Fontanillas *et al.*, 1998). Indeed acorn and chestnut are characterized by a higher content of oleic acid compared to concentrate (Table 3). More difficult to explain is the higher value of PUFA n-3 in pig fed chestnut. Opposite results were found on adipose tissue of Corsican pigs fed chestnut in relation to pigs fed concentrate, showing no differences in MUFA content and higher value of total PUFA percentage (Coutron-Gambotti *et al.*, 1998). As it regards differences between layers, outer layer showed higher MUFA and PUFA percentage.

Conclusion

The pasture on chestnut and acorn did not affect physical traits of meat while significant effects were found on fat quality. This latter results can be exploited to a strong characterisation of products of

Table 2. Chemical-physical traits of Longissimus lumborum.

	Unit	Diet			RSD
		Chestnut	Acorn	Concentrate	
Moisture	%	70.80	71.21	71.57	2.18
Crude protein	%	22.87	22.41	23.09	1.77
Ether extract	%	4.22^{ab}	4.49^{a}	3.29^{b}	1.15
Ash	%	1.26	1.29	1.29	0.09
Water holding capacity					
cooking loss	%	23.31	20.39	20.61	4.44
free water	cm^2	11.53	9.39	10.37	2.17
Warner Bratzler					
on raw meat	kg	8.73	9.07	7.93	1.84
on cooked meat	kg	12.56	14.65	13.66	2.95
Color of LL					
L*		48.21	46.11	46.75	2.60
a*		12.27	11.81	13.12	1.44
b*		4.03	3.44	3.8	0.71
Color of subcutaneous fat					
L*		78.32	78.95	79.52	1.81
a*		1.64	1.42	2.36	1.52
b*		3.31	3.31	3.33	1.03

[a, b, c] Values within traits followed by different letters are significantly (P<0.05) different.

Table 3. Fatty acid composition of subcutaneous fat.

	Unit	Diet			Layer		RSD
		Chestnut	Acorn	Concentrate	Inner	Outer	
Total lipids	%	79.58	79.02	78.19	78.93	78.93	2.41
Fatty acid							
C14:0	%	1.22	1.22	1.21	1.23	1.20	0.1
C16:0	%	21.83	22.07	22.34	21.38[b]	22.79[a]	0.85
C16:1	%	1.71	1.74	1.75	1.82[b]	1.64[a]	0.20
C17:0	%	0.36[a]	0.36[a]	0.43[b]	0.4	0.37	0.07
C16:3	%	0.31[a]	0.310[a]	0.36[b]	0.37[b]	0.28[a]	0.06
C18:0	%	11.26[a]	11.65[ab]	12.10[b]	10.66[b]	12.68[a]	0.93
C18:1	%	44.39[a]	44.44[a]	42.66[b]	44.52[b]	43.14[a]	1.75
C18:2	%	15.59	14.93	15.82	16.17[b]	14.72[a]	1.65
C18:3	%	1.07[a]	0.91[b]	0.92[b]	1.01[b]	0.93[a]	0.1
C20:0	%	0.13[a]	0.16[b]	0.17[b]	0.14[b]	0.17[a]	0.04
C20:1	%	1.02	1.10	1.09	1.08	1.06	0.12
C20:2	%	0.76	0.75	0.83	0.83[b]	0.73[a]	0.15
C20:3	%	0.27	0.10	0.20	0.24	0.14	0.33
C20:4	%	0.11	0.12	0.10	0.12	0.09	0.06
SFA	%	34.88[a]	35.55[ab]	36.35[b]	33.89[b]	37.29[a]	1.58
MUFA	%	47.11[a]	47.30[a]	45.50[b]	47.44[b]	45.84[a]	1.79
PUFA	%	18.00	17.15	18.15	18.67[b]	16.87[a]	1.77
PUFA n-3	%	1.23[a]	1.05[b]	1.05[b]	1.18[b]	1.04[a]	0.11
PUFA n-6	%	16.45	15.79	16.75	17.12[b]	15.54[a]	1.71
PUFA:SFA	%	0.521	0.487	0.504	0.553[b]	0.454[a]	0.064

[a, b, c] Values within traits followed by different letters are significantly ($P<0.05$) different.

Cinta Senese pigs according to feeding system, but further investigations will be carried out in order to find other innovative techniques of traceability.

Acknowledgements

This study was supported by grant of ARSIA (Tuscany Region).

References

Coutron-Gambotti, C., Gandemer, G. & Casabianca, F. 1998. Effects of substituting a concentrated diet for chestnuts on the lipid traits of muscle and adipose tissues in Corsican an Corsican x Large White pigs reared in sylvo-pastoral system in Corsica. Meat Science. 50: 163-174.

Fontanillas, R., Barroeta, A., Baucells, M.D., Guardiola, F. 1998. Backfat fatty acid evolution in swine fed diets high in either Cis-monounsaturated, trans or (n-3) fats. Journal of Animal Science. 76: 1045-1055.

Jonsäll, A., Joansson, L. & Ludström, K. 2001. Sensory quality and cooking loss of ham muscle (M. *Biceps femoris*) from pigs reared indoors and outdoors. Meat Science. 57: 245-250.

Lopez-Bote, C.J. 1998. Sustained utilization of the Iberian pig breed. Meat Science. 49 (suppl. 1): S17-S27.

Pugliese, C., Calagna, G., Chiofalo, V., Moretti, V.M., Margiotta, S., Franci, O., Gandini, G. 2004. Comparison of the performances of Nero Siciliano pigs reared indoors and outdoors: 2. Joints composition, meat and fat traits. Meat Science. 68: 523-528.

SAS. 1996. SAS/STAT software, release 6.12. SAS Institute Inc., Cary, NC.

Tejeda, J.F., Gandemer, G., Antequera, T., Viau, M. & Garcia, C. 2002. Lipids traits of muscles as related to genotype and fattening diet in Iberian pigs: total intramuscular lipids and triacylglycerols. Meat Science. 60: 357-363.

Rumen-protected amino acids in diets for lactating buffaloes

A.M. Kholif[1], M.A. El-Ashry[2], H.M. El-Sayed[2], H.A. El-Alamy[1] & T.A. Aly[1]

[1]*Dairy Sci. Dept., National Research Center, Dokki, P.O. Box 12622, Cairo, Egypt*
[2]*Anim. Prod. Dept., Faculty of Agriculture, Ain Shams University, Cairo, Egypt*

Summary

Twelve lactating buffaloes were divided into four groups (three animals each) using 4 x 4 Latin square design. Animals were fed on a basic diet consisting of a concentrate feed mixture: berseem clover: rice straw (70: 15: 15, dry matter basis); control (C). Rumen protected amino acids were added to the basic diet at levels of 15 g Smartamine™ M (protected methionine) (Me), 40 g protected Lysine (Ly), and 15 g Smartamine + 40 g protected lysine (Me+Ly), and fed to buffaloes for 4 months of lactation. Milk yield, 4% FCM yield, Fat percent, total solids percent; total protein percent, casein percent and lactose percent significantly ($P<0.01$) increased with treated groups in comparison to the control. Also, milk NPN percent decreased ($P<0.01$) with treated groups.

In conclusion rumen protected amino acids supplementation to the ration of lactating buffaloes had beneficial effects on milk yield and compositions.

Keywords: protection, methionine, lysine, buffaloes, milk.

Introduction

The amino acid requirement is significantly affected by the rumen degradability of dietary crude protein. The differences in degradation of crude protein from forages and concentrates and the dynamics of microbial protein synthesis in the rumen have made it difficult to significantly alter the ratio and the quality of amino acids reaching the small intestine (Schwab *et al.*, 1992). Although it still is not common to consider individual amino acids when formulating diets for lactating animals, there is increasing evidence that the addition of certain amino acids can improve milk production and particularly milk protein content. Two amino acids, methionine and lysine, are of special interest for dairy cows as have been identified as the most limiting amino acids for the synthesis of milk protein (Schwab *et al.*, 1992; Rulquin & Verite, 1993; Guinard & Rulquin, 1995). Unfortunately, free crystalline methionine and lysine is easily and rapidly degraded by rumen bacteria and a variety of approaches have been used to achieve delivery of methionine or lysine to absorption sites.

In this study, the diet of early-lactation buffaloes was supplemented with rumen protected methionine (Smartamine™) and/or protected lysine in order to study the effects of these supplementations on milk yield and milk composition.

Materials and methods

Animals managements

A total number of 12 lactating buffaloes were divided into equal four groups, using 4 x 4 Latin square design. The experimental period was extended for four months (one month each period).

Dietary treatments were (1) control, (2) control + 15g Smartamine™ M (Me) (Protected Methionine, RHONE- POULENC product, France),(3) control + 40g protected lysine (Ly) (Lysine amino acid product of ADWIA company, Egypt), which was treated with formaldehyde 40% according to the method suggested by (Fergueson, 1975), and (4) control + 15 g Smartamine™ M + 40g protected lysine (Me+Ly). The control ration consisted of concentrate feed mixture (CFM): berseem (B): Rice straw (RS); (70: 15: 15) on dry matter basis. Chemical composition of feed ingredients is shown in table 1.

Amounts of daily feeds were assessed to cover the maintenance and the production requirements (Shehata, 1971). The CFM was individually weighed for each animal and offered twice daily during milking times at 5.00 and 16.00 h, while roughages were offered at 9.00 and 11.00 h after accessing the animals to fresh water. The daily supplemental rumen protected amino acids were mixed with CFM daily just before morning feeding to ensure that each animal had consumed its own supplement.

Analysis of feed and milk samples

Samples of CFM, RS and protected amino acids were analyzed for dry matter (DM), ash, crude protein (CP), crude fiber (CF), and ether extract (EE) was determined according to AOAC (1995). Nitrogen free extract (NFE) was calculated by difference.

The animals were hand-milked twice daily. Milk yield was recorded daily during the experimental period which extended for 120 days. During the last three days of each period, (30 days) samples of milk were collected from each animal at morning and evening milkings. Composite milk samples (relative to the quantity of milk produced) were taken from the two milkings to determine the total solids, fat, total protein (TP), and ash (Ling, 1963), lactose (Barnett & Abd El-Tawab, 1957), protein fraction (Aschaffenburg & Drewry, 1959).

The data of milk yield and milk composition were analyzed according to Statistical Analysis System (SAS, 1998) using Latin square design. Duncan multiple range test was carried out for separation among means.

Table 1. Chemical composition of concentrate feed mixture (CFM), berseem clover (B), rice straw (RS), Samartamine™ M (Me) and protected lysine (Ly) (% DM basis).

	Diet ingredients				
Items	CFM[1]	B	RS	Me	Ly
Dry matter	91.29	13.3	92.85	95.93	94.60
Organic matter	89.89	88.2	84.55	96.55	97.69
Ash	10.11	11.80	15.45	3.45	2.31
Crude protein	14.15	14.2	3.50	68.23[2]	63.61[3]
Ether extract	4.05	2.60	2.10	8.60	4.10
Crude fiber	15.33	27.50	33.90	0.00	3.30
Nitrogen-free-extract	56.36	43.9	45.05	19.72	26.68

[1]The CFM consisted of 25% undecorticated cotton seed meal, 35% wheat bran, 30% corn, 3% rice bran, 3% molasses, 2% limestone, 1% urea and 1% salt (NaCl),.
[2]Total nitrogen of methionine is multiplied by (11.1).
[3]Total nitrogen of lysine is multiplied by (5.26).

Results and discussion

Milk yield and composition

Effect of rumen protected methionine or/and lysine supplementations on milk yield and its composition of lactating buffaloes are shown in table 2.

Milk yield and 4% fat corrected milk was significantly (P<0.01) increased by including rumen protected amino acids in the ration. In other words, Me+Ly, Me and Ly treatments produced 25.9, 17.6 and 11.3% more milk, respectively, as compared to control. These results are in a good agreement with those obtained by Iwanska *et al.*, (1999), Younge *et al.*, (2001) and Noftsger and St-Pierre (2003) who observed that milk production increased with rumen protected methionine or/and lysine supplementation.

The increase in milk yield may be due to one or more of the following three reasons:

1. Higher nutrition digestibility and total volatile fatty acids (TVFA) content in rumen of animals feed on diets with rumen protected methionine or/and lysine (Aly *et al.*, 2005).
2. Apparent increase in the efficiency of nitrogen utilization as well as increased conversion and availability of nutrients significant for milk synthesis (Iwanska *et al.*, 1999).
3. Methionine and lysine appear to be most limiting for milk synthesis because they are both heavily utilized by the mammary gland.

Milk fat percent was significantly increased (P<0.01) with different rumen protected amino acid treatments compared with control. It is possible that similar mechanism of methionine and lysine action in the mammary gland may be responsible for the increase of milk fat content. The results of this study confirm the data of Xu *et al.* (1998) and Iwanska *et al.* (1999). While, Younge *et al.* (2001) and Kholif & Kholif (2003) showed that milk fat percent was not affected by rumen protected amino acids.

Table 2. Effect of rumen protected AA supplements on overall mean of milk yield and composition of lactating buffaloes.

	Treatments				
Items	Control	Me	Ly	Me+Ly	±S.E
Milk yield and composition:					
Milk yield (kg/d)	6.65^d	7.82^b	7.40^c	8.37^a	0.129
Fat-corrected-milk (kg/d)	9.30^d	11.41^b	10.52^c	12.34^a	0.179
Fat content (%)	6.68^d	7.01^b	6.85^c	7.18^a	0.035
Total solids content (%)	15.89^d	16.37^b	16.12^c	16.93^a	0.072
Solids – not – fat content (%)	9.21^b	9.35^b	9.31^b	9.75^a	0.064
Total proteins content (%)	3.72^d	4.00^b	3.91^c	4.17^a	0.018
Lactose content (%)	4.66^c	4.77^b	4.74^b	4.95^a	0.013
Ash content (%)	0.72	0.72	0.71	0.72	0.010
Milk protein fraction content:					
Casein content (%)	2.80^d	3.07^b	3.02^c	3.28^a	0.017
Whey protein content (%)	0.825	0.824	0.790	0.811	0.013
Non-protein-nitrogen content (%)	0.042^a	0.029^c	0.036^b	0.027^c	0.001

Each value of means obtained from 36 samples from 12 animals.
Means with different superscripts in the same row are differ significantly (P<0.01).
C= control diet,; Me = C + protected methionine; Ly = C+ protected lysine;
Me+Ly= C+ protected methionine + protected lysine.

Milk total solids (TS), total protein, and lactose contents were significantly increased (*P*<0.01) with different supplemented group compared with control; similar results of rumen protected protein supplementation, were obtained by Bharadwaj & Sengupta (1999). The increase in milk total proteins may be due to one or more to the following reasons:

1. The milk total proteins response to post-ruminal supply of limiting amino acids were much lower on low protein (≤ 14% CP) compared with high protein rations (Rulquin & Verite, 1993).
2. Methionine must contribute 5.3 to 5.6% of the total essential amino acids in duodenal digesta and 2.5-2.7% of the total amino acids, respectively for maximum content and yield of milk protein (Rulquin *et al.*, 1995).
3. Higher CP and OM digestibilities and higher ruminal true protein nitrogen when Baladi goats fed on diets with rumen protected methionine or/and lysine supplemented (Aly *et al.*, 2005). These results are in a good agreement with those obtained by Iwanska *et al.*, (1999), Younge *et al.*, (2001), Rulquin *et al.*, (2001), Kholif and Kholif (2003) and Noftsger *et al.*, (2005).

Milk solids not fat (SNF) present were significantly increased (*P*<0.01) with Me+Ly treatment, while Me and Ly treatments were insignificantly increased (*P*>0.05) compared with control. Similar results were obtained by Kholif & Kholif (2003), who found an insignificantly (*P*>0.05) increases in SNF with protected methionine. Milk ash percent was not significantly affected (P>0.05) by different experimental treatments. Similar trend of milk ash content was obtained by Kholif & Kholif (2003).

Milk protein fraction

The data in table 2 represents the effect of supplementing lactating buffaloes diets with rumen protected methionine or/and lysine on the milk protein fraction. Milk casein content was significantly (*P*<0.01) higher with Me+Ly, Me and Ly treatments by 17.1, 9.6 and 7.9%, respectively, than control diet. The increase in milk casein may be due to rumen protected methionine or lysine supplementations, which produce 100% of the casein response in early lactation and 47 to 58% of the casein response in peak to mid lactation (Schwab *et al.,* 1992). Similar results were obtained by Nichols *et al.,* (1998) and Younge *et al.,* (2001).

Milk whey protein percent was not significantly (*P*>0.05) affected by different treatments; similar results were obtained by Younge *et al.* (2001), who found that the milk whey protein percent was not affected by rumen protected methionine or/ and lysine supplemented to dairy animals' rations. While Nichols *et al.,* (1998) suggested an increase (*P*<0.05) in whey protein percent with rumen protected methionine plus lysine.

Milk non-protein-nitrogen (NPN) percent in all treated buffaloes groups decreased (*P*<0.01) by 35.7, 31.0 and 14.3% for Me+Ly, Me and Ly, respectively, compared with the control group. These results did not agree with the findings of Younge *et al.,* (2001) who reported that milk NPN percent was not affected by rumen protected methionine or lysine supplemented ration.

References

Aly, T.A, M.A. El- Ashry, A.M. Kholif, H.M El-Sayed, H.A. El- Alamy & M.M. Khorshed. 2005. Effect of rumen–protected methionine and /or lysine supplementation to the ration on nutrients digestibility and on some rumen parameters of male Balady goats. Egyptian J. Nutrition and feeds, 8: 41-51.

AOAC. 1995. Methods of Analysis. Vol. 1: Agricultural Chemicals, Contaminants, Drugs. 16th Edition, Washington, D.C., USA.

Aschaffenburg, R. & I. Drewry. 1959. New procedure for the routine determination of the various non-casein-proteins of milk. Int. Dairy Congr., 3, 1631.

Barnett, A.J.G. & G. Abd El-Tawab. 1957. Determination of lactose in milk and cheese. J. Sci. Food Agric., 8: 437-441.

Bharadwaj, A. & B.P. Sengupta. 1999. Effect of sources of dietary unprotected and protected protein on lactational performance of Murrah Buffaloes. Buffalo J., 3: 269-278.

Fergueson, K.A. 1975. Digestion and metabolism in the ruminant. I.W. McDonal & A.C.I. Warner, Univ. of New England. Publishing Unit, Armidale, Australia, pp. 448-464.

Guinard, J. & H. Rulquin. 1995. Effects of graded amounts of duodenal infusions of methionine on the mammary uptake of major milk precursors in dairy cows. J. Dairy Sci. 78, 2196-2207.

Iwanska, S., D. Strusinska & B. Pysera. 1999. Effect of rumen protected methionine supplementation on early lactation responses of dairy cows fed a grass silage and cereals diet. Acta Veterinaria Hungarica, 47: 191-206.

Kholif, S.M. & A.M. Kholif. 2003. Effect of ruminally protected methionine on the productive performance of lactating goats. Proc. The 1st international Conf. on "Food for Better Health", NRC, 18-20 October, Cairo, Egypt.

Ling E.R. 1963. "Text Book of Dairy Chemistry" Vol. II, Practical Chapman and Hall, L.T.D., London 3rd Edition, pp. 140.

Nichols, J.R., D.J. Schingoethe, H.A. Maiga, M.J. Brouk & M.S. Piepenbrink. 1998. Evaluation of corn distillers grains and ruminally protected lysine and methionine for lactating dairy cows. J. Dairy Sci., 81: 482-491.

Noftsger, S. & N.R. St-Pierre. 2003. Supplementation of methionine and selection of highly digestible rumen undegradable protein to improve nitrogen efficiency for milk production. J. Dairy Sci., 86: 958-969.

Noftsger, S., N.R. St-Pierre & J.T. Sylvester 2005. Determination of rumen degradability and ruminal effects of three sources of methionine in lactating cows. J. Dairy Sci., 88: 223-237.

Rulquin, H. & R. Verite. 1993. Amino acids nutrition of dairy cows: productive effects and animal requirements. In : Garnsworthy P.C. and Cole D.Y.A. (Eds). Recent Advances in Animal Nutrition. Nottingham University Press, Nottingham, pp. 55-77.

Rulquin, H., R. Verite, G. Guinard & P.M. Pisulewski. 1995. Dairy cows requirements for amino acids. In: Center for Food and Animal Research (Ed.) Animal Science Research and development. Agriculture and Agric. Food. Ottawa, Canada, pp. 143-155.

Rulquin, H., R. Vérité, J. Guinard-Flament & P.M. Pisulewski. 2001. Amino acids digestible in the small intestine. Factors of variation in ruminants and consequences on milk protein secretion. INRA Productions Animals, 14: 201-210.

SAS. 1998. Statistical Analysis System. SAS User's Guide Statistics. SAS Institute Inc. Editors, Cary, NC.

Schwab, C.G., C.K. Bozak, N.L.Whitehouse & M.M.A. Mesbah. 1992. Amino acid limitation and flow to duodenum at four stages of lactation. 1. Sequence of lysine and methionine limitation . J. Dairy Sci., 75: 3486-3502.

Shehata, O. Kh. 1971. Lectures in animal production (In Arabic). Animal production Dept., Fac. Agric.; Ain Shams University, Shoubra El-Khema, Cairo, Egypt.

Xu. S., J.H. Harrison, W. Chalupa, C. Sniffen, W. Julien, H. Sato, T. Fujieeda, K. Watanabe, T. Ueda & H. Suzuki 1998. The effect of ruminal by pass lysine and methionine on milk yield and composition of lactating cows. J. Dairy Sci., 81: 1062-1077.

Younge, B.A., J.J. Murphy, M. Rath & B.K. Sloan. 2001. Effect of dietary absorbable methionine and lysine concentrations on milk production and composition of dairy cows offered grass-silage based diets. Irish Journal of Agricultural and Food Research. 40: 1-11.

Elemental analysis of salted yoghurt produced from goat milk

Z. Güler & H. Şanal

Mustafa Kemal University, Agriculture Faculty, Food Engineering Department, Tayfur Sokmen Campus, 31034 Antakya, Hatay, Turkey

Summary

Salted yogurt is known as winter yogurt which is made mainly from goat milk. We have studied gross chemical composition and 13 element contents in 4 salted yogurt samples. The mean values for chemical composition of salted yogurt were: total solids 28.81%; fat 10.50%; crude protein 11.26%; salt 2.51%; ash 2.62% and titratable acidity 2.36%. Ca, Cd, Co, Cr, Cu, Fe, As, Mg, Mn, Mo, P, Pb and Zn elements in salted yogurt were determined. The most abundant elements were Ca, Mg and P, respectively.

Keyword: salted yogurt, elemental analysis, chemical properties.

Introduction

The salted yogurt is one of the most popular variety of dairy products manufactured in Hatay which is located between 36° North latitude and 36° East longitude in the Eastern Mediterranean region. Due to no production at industrial scale, which means that there are no reference standards, no use of packing materials and undetermined shelf – life. The salted yogurt is therefore a local product. Manufacturing techniques varies slightly from place to place, but the most common home made salted yogurt production method is as follows: raw milk is subjected to a heat treatment at 80-85 °C for 1-15 min, and left for cooling to 40-45° C. As starter culture, one day old yogurt is added to the milk, which is then incubated for 4-6 hours. Depending on the outside temperature, the yogurt container is wrapped with a piece of cloth or blanket to it warm. After yogurt gel forms, it is kept one night at room temperature. Thereafter it is usually drained to remove the extra ligned using a cloth bag. Draining also helps to reduce the cooking time. Drained yogurt is transferred to a tinned copper container with a flat bottom under which a fire has been set. For fire, dried bushes are used. Yogurt is stirred continuously until it boils. Continuous stirring is believed to be essential. The cooking stage is continued until the yoghurt thickens enough to start splashing, which takes about 45 to 60 minutes for 5-8 kg yoghurt depending on the fire. At this point, salt is added at the level of 3-5%, and stirring is carried out for another 5 minutes or so. When the cooking stage is completed, salted and cooked yoghurt is transferred to another container and left for cooling. Finally, salted yoghurt is filled into jars and olive oil is added on top of it to prevent mould growth. It is then sold in plastic trays at market. It is consumed directly as snack or at breakfast, lunch, and dinner with olives oil, red pepper and mint.

A variety of minerals are essential for a good health, especially during growth and for elder people. Deficiencies and/or excesses may result in metabolic disorders and pathological problems. Food processing can modify trace element content (Concon, 1988; Reilly, 1980) and external factors such as the environment, eating habits and stress can alter their absorption.

Milk and milk products contains most of the required nutrients for a healthy diet, and in some groups, such as children and the elderly, it may constitute the main or even the only food. For this reason, these trace elements must be detected and measured, in order to ensure the adequate concentrations of these essential elements. The composition of the mineral fractions as well as the nutrients or toxic levels in salted yogurt has not yet been investigated. We have measured the following elements: As, Ca, Cd, Co, Cr, Cu, Fe, Mg, Mn, Mo, P, Pb and Zn .

Materials and methods

Salted yoghurt samples

Four samples of salted yogurt made from goat milk that were obtained from different manufactures in Hatay were analysed.

Determination of gross composition

The samples were analysed for total solids, fat and titratable acidity, crude protein and salt. The following procedures described in Turkish standards (Anonymous, 1991, 1999, 2001)

Determination of trace elements in salted yoghurt

The analytical technique used for the quantification of the trace elements, used the inductively coupled plasma optical emission spectrometry (ICP-OES). Its attributes (multi-element capability, high detection power, wide dynmic range and reduced dependence on matrix composition) make it ideal for this kind of measurement (Caroli, 1988).

Apparatus

The analyses were carried out by radial system Varian model VISTA-MPX Simultaneous ICP-OES (Inductively Coupled Plasma- Optical Emission Spectroscopy) (Varian –Australian) controlled with a computer.

Reagents

All solutions were prepared with ultrapure water. Nitric acid and hydrogen chloride were obtained from Merck (D-64293, D- 6100 Darmstadt 1, Germany). For each element, the working solution series of 5 ppm, 10 ppm and 20 ppm were prepared by using available standards (SCP science, Canada).

To eliminate background problems due to detergents and previous samples, all glasswares were washed with tap water after each use, soaked ind 6M-nitric acid solution and rinsed several times in ultrapure water.

Procedure

Method A

All samples were homogenized and a portion (about 0.5 g) weighed and taken into a volumetric flask of 100 ml, then 10 ml nitric acid was added and shaken. After that it was dilueted to a total volume of 100 ml with ultrapure water.

Method B

All samples were homogenized and a portion (about 0.5 g) weighed and taken into a volumetric flask of 500 ml then 25 ml nitric acid:HCl (3:1) was added and shaken. After that it was dilueted with ultrapure water to a total volume of 500 ml.

Two samples were prepared one to a final volume of 100 ml and the other to a final volume of 500 ml. To evaluate method selectivity, standard addition graphs were prepared for blanks and samples and used to calculate the blank-to-sample slope ratio for each element in the samples. All determinations were performed in triplicate, and the same procedure was used to run the blanks. Everyday a fresh blank solution was prepared.

Results and discussion

The gross composition of salted yoghurt was given in table 1. As it can be seen in table 1, yoghurt contains mainly fat and crude protein. Total solid content was increased during cooking as a result of evaporation of water, compared to set yoghurt. According to Turkish standard, set yoghurt should be contained maximum 15% total solid (Anonymous, 1999). Gross composition of salted yogurt, except for salt and ash, was similar to this reported by Biçer *et al.* (1996).

Results of trace element analysis were shown in table 2. As can be seen in table 2, the contents of thirteen trace elements of nutritional or toxicological importance were found in salted yogurt samples. Na and K elements were not determined. Ca and P were the major ones. Ca comes from the blood, it is difficult to change the milk Ca concentration by changing dietary Ca. The Zn, Mg, Fe, Cu, Mn and Mo minerals are required by enzymes as cofactors. They were also determined in the present study. The amount of each element determined in method B, was higher than the ones obtained by method A. This could be due to extraction conditions. Nitric acid and hydrogen chloride for extraction may be better than just nitric acid. Predominant elements were Ca, P and Mg. Mg, Fe , Mn, Cr and Zn values were higher than those reported by Coni *et al.* (1999) and Garcia *et al.* (1999) for raw milk, curd and different milk slurries respectively. Toxic element contents of salted yogurt such as

Table 1. The gross composition (%) of salted yoghurt (n=4).

Composition	Mean	St. dev.
Total solids	28.81	0.09
Fat	10.50	0.35
Crude protein	11.26	0.60
Salt	2.51	0.18
Ash	2.62	0.16
Titratable acidity	2.36	0.05

Table 2. The trace element content (mg/kg) of salted yoghurt (n=4).

Elements	Method A[1] nm	Method A[1] Mean	Method A[1] St. dev.	Method B[2] nm	Method B[2] Mean	Method B[2] St. dev.
Ca	317.933	1991	84.40	396.840	2 452.00	25.83
Cd	226.502	3.39	1.49	214.439	21.60	0.21
Co	228.615	3.32	2.27	228.615	18.76	0.94
Cr	286.510	5.61	3.22	286.510	33.96	7.88
Cu	327.395	4.61	2.29	327.395	31.12	1.26
Fe	238.204	15.67	11.00	238.204	52.27	0.006
As	216.552	7.12	3.56	216.552	37.41	9.87
Mg	280.270	281.65	14.05	280.270	430.00	0.40
Mn	257.610	3.96	2.01	259.371	21.24	0.15
Mo	204.598	4.51	2.03	204.598	22.99	4.40
P	213.618	236.12	7.92	213.618	257.14	2.02
Pb	220.353	18.04	5.43	220.353	95.17	21.81
Zn	202.548	20.88	8.22	202.548	90.58	6.69

[1]Extraction with 10 ml HNO_3.
[2]Extraction with 25 ml HNO_3 : HCl (3:1).

Cd, Pb and As were high, when compared with the values obtained in raw milk and curd. Elements bound to caseins and fat (Ca, Fe, Mg, Mn, P and Zn) are usually concentrated during the cooking process. Elements such as Cu and Cr may result from a contamination milk container or the cooking container. Barrionuevo *et al.* (2002) found that goat milk contained at higher level of Ca followed by P, Mg, Fe, Cu and Zn, when compared with these in cows milk. In the present study, Ca was the predominant element, followed by Mg, P, Zn, Fe and Cu. Regarding Fe and Cu contents, salted yogurt could be an important source since these elements have nutritional significance. It might prevent iron deficiency and thus the incidence of nutritional ferropenic anaemia. Copper is also essential to mobilize the iron in the synthesis of haemoglobin.

From these results, it can be concluded that all the major minerals have a higher content in salted yoghurt. Therefore, it may be recommend as a good nutritional source in terms of some minerals, but their bioavailability must be measured. Concerning trace element analytical methods, it seems that open systems can be less precise because of the possibility of loss material. The closed systems such as microwave heating should be preferred for digestion of elements. Previous studies reported that precision was greater using microwave procedures using as open vessels than procedures using closed vessel systems (De La Fuente *et al.*, 1997).

Thereoretically this reduces the loss analytical material. In addition, accurate measurement of elements content in food may be related to their fat and casein contents, extraction technique, digestion of food, reagents used and ICP conditions.

References

Anonymous, 1991. Caseins and caseinates. Determination of protein contents. TS 8866. Turkish Standards Institution, Ankara.

Anonymous, 1999. Yoghurt. TS 1330. Turkish Standards Institution, Ankara.

Anonymous, 2001. Cheese and processed cheese products-Determination of chloride content: Potentiometric titration method. TS 4708. Turkish Standards Institution, Ankara.
Barrionuevo, M., M.J.F. Alferes, I. Lopez- Aliaga, M.R. Sanz- Sampelayo & M.S. Campos, 2002. Beneficial effect of goat milk on nutritive utilization of ýron and copper in Malabsorption Syndrome. J. Dairy Sci. 85: 657-664.
Biçer, O., M.B. Güler, M. Keskin & Ş. Kaya,1996. Goat breeding in Hatayda and importance of goat milk in production of some local products. Hatay Folklor Arastirmalar Dernegi, 9: 1-10.
Caroli, S, 1988. The role of ICP spectrometry in human healt and environmental protection. Spectrochimica Acta. 43: 371-380.
Concon, J.M. (Editor). 1988. Food toxicology: Concaminants and Additives. Marcel Dekker, New York, pp. 230.
Coni, E., B. Bocca & S. Caroli, 1999. Minor and trace element content of two typical Italian sheep dairy products. J. Dairy Res. 66: 589-598.
De La Fuente, M.A., C. Belen & J. Manuella, 1997. Determination of major Minerals in dairy Products Digestion in Closed Vessels Using Microwave Heating. J. Dairy Sci. 80: 806-811.
Garcia, E.M., M.L. Lorenzo, C. Cabrera, M. C. Carmen & J. Sanchez, 1999. Trace element determination in different milk slurries. J. Dairy Res. 66: 569-578.
Reilly, C. (Editor), 1980. Metal contamination of food. .Applied Science Publishers, London , pp. 226.

Small ruminants projects in LEADER-PRODER European Rural Development Programs in the South of Spain: A survey analysis

S. Sayadi & J. Calatrava

I.F.A.P.A, Dpt. Agricultural Ecomomics and Rural Studies, Apdo. 2027,
18080 Granada, Spain

Summary

Mountain livestock systems, and particularly those based on breeding small ruminants, constituted traditionally a mainstay element of the local economy in Southeastern Spain's mountainous areas. With the rural exodus which began in the early sixties, these systems were in critical danger of disappearing, due to, among other causes, the scarcity of shepherds and goatherds, the breakdown of an equilibrium between agriculture and animal farming, the small herd sizes, management and handling problems, etc.

In spite of this, there have recently been several attempts by the local population to develop the mountain livestock sector by recovering autochthonous breeds, solving sanitary and livestock handling problems, and the commercial revalorisation of dairy products and their derivatives. EU aid to the small ruminant sectors, through, among others, the LEADER and PRODER European Rural Development Programmes, has no doubt helped to promote these efforts.

This paper, after a brief description of the small ruminants systems in the Penibetic region, comments on the main results of a survey performed to the Local Action Groups (LAG) representatives, and analyses the extent of the inclusion of Southeastern Spain's mountain area livestock projects in the EU rural development programmes (LEADER and PRODER).

Finally some conclusions and main strategic actions to take full advantage of the potential of small ruminant projects for the design of sustainable rural development processes in less favoured rural areas, are mentioned.

Keywords: mountain livestock system, small ruminants, Leader, Proder, rural development.

Introduction

The rural areas of Spain began to manifest deep transformations through the Leader I European programme in 1991 characterised, among other changes, by the following:

- An economic growth based on social transferences of capital and an increase in extra-agrarian activities and incomes.
- An increase in social cohesion and dynamism at the local level.
- A revalorisation of endogenous resources.
- A cultural transformation, in the sense of adopting certain urban cultural traits while at the same time reinforcing other rural cultural values as marketable goods.
- A slowing down and sometimes an end of the exodus.

- A noticeable reduction in gender asymmetry, an especially marked feature in the traditional rural culture.
- An increase, often quite outstanding, of rural tourism and associated activities.

The impact of this transformation process on agrarian systems in general, and on livestock in particular, although it remains to be studied in depth, one may dare say, has not always been a positive one, particularly in those most depressed Mediterranean mountain areas, where, in the best of cases, the agrarian-livestock component of the LEADER and PRODER programmes has been practically incidental.

Within the framework of EU Programmes, such as the Integrated Development Programmes, LEADER, and others, the development policy in mountainous areas has often been oriented towards the secondary and tertiary sector, while the primary sector has been largely left behind, particularly in relation to the small ruminants livestock. The philosophy behind such programmes is to encourage internal development or self-development, as it is said. These programmes refer mainly to the support of tertiary sector activities, and in particular rural tourism, but also to the secondary sector with emphasis on the start-up and running of small handcraft shops and other businesses.

This hypothesis is proposed as the starting point of this paper, which objective is to analyse to what the extent the UE rural development policies have generated or supported projects aimed at improving, revamping, or transforming and developing products derived from breeding small ruminants in Southeastern Spain (hereinafter Penibetic region). The Penibetic region is one of the most mountainous and less developed areas in the European Union.

Methodology

A survey was carried out in December 2001 among those in charge of 17 Local Action Groups (LAG) located in the Penibetic area. They were asked questions regarding the nature of the projects backed by the LAG since 1995 (Leader II and Proder), in addition to their personal opinions regarding the relative importance of livestock in the rural development of their respective geographic areas. Based on this information, the participation level in projects related to small ruminants included in the strategic development plans was analysed.

For a better exposition of the data and for statistical clarity, the agrarian zones used as the basis were those defined in 1983 by the *Dirección General de Investigación y Extensión Agraria* of the *Junta de Andalucía*, although the agrarian zone thus defined are, generally, larger than the

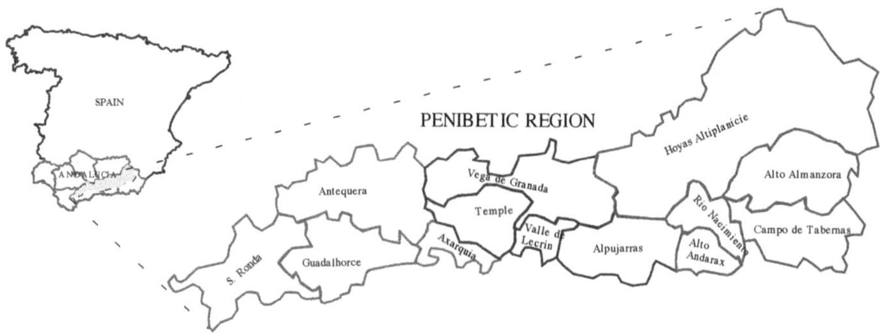

Figure 1. Area of the study.

territories in which the LAG groups operate today. So, in the region of our study there are 13 agrarian zones and 17 LAG groups distributed between the provinces of Granada (GR), Almeria (AL) and Málaga (MA) (South of Spain) (Figure1). The Penibetic region encompasses almost 2 million hectares, most of them highly mountainous.

Results

Small ruminants livestock in the region

In the region of the study there are almost 1.3 million sheep and goats. The number of small ruminants has grown of 9.3% in the 80's and by 50.55% in the 90's (1990-2000). One fourth of the herds are less than 50 head, amounting to less than 4% of the total livestock, whereas less than one fifth of the breeders, with more than 300 heads each, possess over half of the total livestock. The average sized herd in the region, of 187 heads, is quite variable among the zones, oscillating between 99 and 260. Mixed herds are a rarity in the region, due to the species' diverse behaviour in the open field. Mixed herds tend to be the smallest ones.

Livestock projects in LEADER and PRODER Rural Development Programmes: a survey analysis

Of the 17 existing LAG groups in the region, 8 are LEADER II and 9 are PRODER; as per location, 3 are in Almería, 7 in Granada, and 7 in Málaga.

According of the survey, the total number of projects financed have been 2 856[1] with an average of 178 projects per LAG. The LEADER II project average is much higher (214.5) than that of the PRODER (126.6). Of these projects, barely 1% are activities related to small ruminants, which contrasts with the large number (1.3 million heads) of goat and sheep in the region, and with a cultural and gastronomic tradition (meat, cheese, etc.) largely based on the breeding of small ruminants in the region. Furthermore, 23.5% of those responsible for the LAG believe that promoting activities associated with small ruminants at present would not only help develop the area, but are in fact key factors in its development, and 47% consider that, although they are not key factors, they are important. Considering also the future, 35% think that activities connected with small ruminants are, or could potentially become, key activities in the zone's development if measures are taken to direct the livestock sector toward modernisation.

For strategic purposes, the elements which should receive attention with the greatest priority, as identified by the regional LAG representatives of the small ruminant population, are reported in table 1 (three priority elements per answer).

These priority actions, although each zone is influenced, logically, by its own situation, present, in general, a significant coincidence, especially in the three first points, which can be summarised as follows:

- Closer, stronger association among breeders.
- More emphasis on training and publicity.
- Greater returns through elaboration of dairy products.

[1] In the administrative language in reference to rural development, a project is an activity of any type and size (training courses, fair organisation, cheese factories, improving farmlands, etc.).

Table 1. Elements which should receive attention with the greatest priority, as identified by the regional LAG representatives of the small ruminant population.

	% of the total LAG groups
▪ Create livestock co-ops for milk collection, cheese making, trading, etc.	64.1
▪ Implement training programmes aiming to change breeders' attitudes, arousing social interest in the activity, attracting young breeders, informing on the aid options, on dairy products, on markets, etc.	52.9
▪ Generate greater returns through the elaboration of the dairy products by the breeders themselves.	47.1
▪ Improve marketing and distribution both of milk and cheeses.	29.4
▪ Organise the marketing of the meat by taking advantage of local and regional markets, and tourism.	29.4
▪ Preserve autochthonous breeds best adapted to the natural resources of the area.	23.5
▪ Include in the rural tourism and agro-tourism programmes, already in operation, livestock breeding related activities: visiting animal farms and cheese factories, tourist participation in farm activities and in the elaboration of handcrafted cheese, tasting of dairy products typical of the region, etc.	17.7
▪ Modernise the present herds: mechanical milking, introduction of new, highly specialised productive breeds, etc.	17.7
▪ Upgrade the assistance and direct aid to breeders.	5.9

Let us compare this "desideratum" as manifested by the LAG representatives as activities for development, and the scarce projects actually financed. The nature of the projects funded in connection with small ruminants is summarised in table 2.

Regarding the cheese factories, their proliferation, is a consequence, in our opinion, of the following factors:

- The existence of a growing demand for goat cheese in Spain, both at the level of major markets and in local areas.
- The rural tourism boom and increasing demand by visitors for typical products of the area, which cheese being the most sought after.
- Availability of quality raw material locally.
- The local pre-existing tradition of cheese making for self consumption in the area, even if it never grew to an industrial scale.

Table 2. The nature of the projects funded in connection with small ruminants.

Nature of the small ruminant's projects funded by the LAG	% total
▪ Cheese factories	38.8
▪ Technological modernisation of animal farms	33.4
▪ Livestock market fairs and seminars	16.6
▪ Milk collection and treatment plant	5.6
▪ Meat treatment and commercialisation facilities (slaughterhouses, etc.)	5.6

- The existing subsidies to undertake this type of investments.
- The promotional efforts of the LAG.

In relation to the last two factors, they are not very decisive, especially the latter one, since, in most cases, they were not induced investments, but investments that the owner had already decided to make, and in which the role of the LAG was rather of manager of the subsidy than of encourager and promoter.

None of these projects involves a livestock breeder association; they are all practically individual projects or involving small family limited societies. Of the three strategic lines considered to be priorities by the LAG representatives: Close Producer Association, Training, and Local Elaboration of Dairy Products, only the latter is clearly reflected in the activities financed by the LEADER II and PRODER programmes. This is so not because the LAG representatives are actually going against what the priorities of the area demand, but because of the enormous difficulties in managing to carry out activities that would involve the creation of associations among the small ruminant breeders, as well as the very project funding dynamics of the LEADER II and PRODER programmes. In order to maximise efficiency in assistance management, projects and activities which already have "rural entrepreneurs" willing to implement them are given preference over those projects which are basically non-existent, and whose inception would require an extensive, long-term and in-depth awareness programme of "rural extension", in its most classical sense, with very real possibilities that it may all come to naught, as would occur when collective decisions were required.

In this sense, it is surprising that none of the activities financed for the elaboration of cheese has resulted in an association of smaller and middle-sized breeders in order to obtain better returns from the milk they produce, in the elaboration and marketing process.

Conclusions

From the survey of the LAG representatives the following conclusions can be drawn:

- In spite of the importance of goat and sheep livestock holdings in the Penibetic mountains, the number of projects related to small ruminants that are financed is extremely small.
- However, and paradoxically, a consensus exists among the Local Action Groups (LAG), Leader and Proder, on the key role that the livestock activities of these small ruminants could play in the local sustainable development of the mountain areas of South-eastern Spain.
- Creating associations, training and encouraging livestock breeders, and elaborating dairy products from milk locally, are among the measures with the highest priority if the small ruminant sector is to be revamped. All the local GAL coincide in this fact. However, it is mainly in the latter measure (that of the local elaboration of dairy products) that the Leader and Proder have clearly taken any steps in financing projects.
- There is a lack of in-depth studies on the livestock systems in the various LAG areas.
- The projects on small ruminants respond almost exclusively to individual initiatives, no social economy related projects appear, involving livestock associations.
- The initiative of the projects does not derive, in general, from the GAL, but from the farmers themselves.
- The projects are aimed mainly to cover the local (provincial) markets, without views to a wider, more daring share of the market as regards expansion (national and international).
- What has just been said is due not so much to the lack of marketing strategy as to the fact that we are dealing with small industries and the existence of an abundant local demand.

In view of these conclusions, several recommendations could be put forward:

- Efforts should be made to prioritise and generate projects involving the creation of breeder associations either for processing and/or trading processes.
- Besides their role as subsidy managers, the LAG's role in promoting small ruminant related projects should be emphasised. It is important not only to back the existing local initiatives, but to promote and induce the initiatives, of a priority character, which do not come about spontaneously, particularly in a sector which is in a such a crisis as that of the small ruminants.
- It would be interesting to carry out and analysis and detailed studies of the small ruminants livestock systems in order to be able to better emphasise its effective participation in the sustainable development processes in south-eastern Spain.
- Training activities should be increased both considering technical aspects of breeding small ruminants and processing milk into cheese and other products.
- Also institutional features have to be included into training activities, emphasising in associated and co-operative farm management, processing and trading.

Introduction of legume plants in irrigated pastures, with no tillage, to avoid nitrogen fertilization

R. Amaro[1], A. Oliveira[1], A. Arantes[2], G. Feio[1] & A.J.D. Ferreira[1]

[1]Centro de Estudos em Recursos Naturais, Ambiente e Sociedade, Escola Superior Agrária de Coimbra, Bencanta, 3040-316 Coimbra, Portugal
[2]Departamento de Ciências Zootécnicas, Escola Superior Agrária de Coimbra, Bencanta, 3040-316 Coimbra, Portugal

Summary

One of the main objectives of Life Environment Project (EMAS@SCHOOL – Environmental Management and Audit Scheme implementation at a complex school - LIFE 03 ENV/P/000501), currently being implemented at the "Escola Superior Agrária de Coimbra" is the reduction of soil tillage and nitrogen fertilization. This is expected to reduce nitrate lixiviation risk, decreasing aquifer contamination and avoiding organic matter oxidation in the soil. This approach is thought to contribute to a higher sustainability at farm level.

Within a 4 years old irrigated pasture, dominated by grass plants, two different treatment strips, were installed, with 3 replicates. Treatments were as following: Witness (W), without tillage and with no legume introduction; Seed throwing (S), without tillage and Brillion seedling with *Trifolium repens* and *Trifolium fragiferum*. During the first year (October 2003 to September 2004), cover fertilization was applied at the rate of 82 units of nitrogen per ha/yr in W and 84 units of de P_2O_5 ha/yr in S. Rotational grazing was performed by 6 to 8 adult Lusitanian mares with their foals. Due to technical limitations in the spring and the rainy summer, the amount of irrigation water used was only 4 000 m^3/ha. Pasture productivity was evaluated in terms of quantity and quality, assessed by the floristic composition and chemical composition.

The productivity results do not show significant differences ($P \leq 0.05$) between W and S for the entire year (W=10 037 kg DM/ha and S=9 529 kg DM/ha) or in each of the seasons. On what concerns the qualitative parameters, there is a significant improvement of the available food ($P \leq 0.05$) in terms of CP (W=15.50 *vs.* S=16.43), UFL (W=1.00 *vs.* S=1.05), UFC (W=0.68 *vs.* S=0.70) and DMD (W=75.52 *vs.* S=79.12). Analyzing the different seasons, the presence of legume plants appeared to have an influence only during the summer, when all the parameters (CP, CF, DMD, UFL and UFC) improved considerably as a consequence of the higher legume proportion in S (27.26%) compared with W (0.16%). We speculate that the nitrogen fertilization performed in March prevented the observation of qualitative differences during the spring, when legumes also presented significant differences between the two treatments (W=10.43% *vs.* S=36.23%).

Due to the evolution observed during the first year, we expect the DM production to become increasingly higher for S, while witnessing an improvement on food quality. This implies that sustainability is compatible with the quality improvement and quantity increase of livestock production in the Mediterranean region.

Keywords: pasture, legume, nitrogen, no tillage.

Introduction

Pastures are an important part of farming systems Europe wide. They allow the production of animal feeds with an excellent quality/price ratio. Nevertheless the associated environmental problems increasingly determine their adoption as a suitable farming system. Soil fertility, carbon retention, protection against erosion, nitrogen lixiviation reduction, fire breaks and their contribution for landscape diversity are benefits provided by pastures and other extensive farming systems that should be taken into consideration in the definition of a sustainable strategy for the rural world.

The environmental concern in agriculture was introduced by the agro-environmental schemes (Reg. CEE n.° 2078/92) under the support of the Common Agriculture Policy (CAP), to foster environmentally friendlier practices, aiming resources conservation and reduction of pollutant production in farm activities (Rizov, 2004). In the EU directives, the protection of groundwater against pollution caused by certain dangerous substances (80/68/CEE) and concerning the protection of waters against pollution caused by nitrates from agricultural sources (91/676/CEE), are examples of this policy. Nutrients lost from non point sources, and especially the nitrogen fertilization, is the main cause of nitrate pollution in European freshwaters (EEA, 2003).

The Mediterranean climate is considered excellent for pasture legume plants, capable of supporting an intensive grazing (Crespo, 2004). Legume plants establish a symbiotic relation with atmospheric nitrogen fixing bacterias (*Rhizobium, Bradyrhizobium, Azorhizobium* and *Cinorhizobium)*, which store nitrogen in root nodules. Part of this nitrogen is made available for the host plant and the other part can be used by the associated grass plants. This turns nitrogen fertilization unnecessary (Simpson and Culvernor, 1985; Garcia and Muslera, 1991; Ferreira *et al.*, 2002). On the other hand, the reduction of soil mobilization increases the amount of carbon that the soil can capture and store, increasing its organic matter content and fertility on the long run. While reducing the amount of CO_2 released by oxidation to the atmosphere, soils with high organic matter content maintain their productivity, are less prone to erosion and have a positive effect on freshwater chemical pollution, since they can retain a higher amount of water, promoting therefore a higher rate of decay and immobilization of dangerous substances (ECAF, 1999).

This work, produced under the scope of Life Environment LIFE03 ENV/P/000501 "Environmental Management and Audit Scheme Implementation at a complex school" project, currently under way at the Escola Superior Agrária de Coimbra (ESAC), is part of a strategy of sustainability for farm's activities, through the reduction of nitrogen fertilization and soil mobilization, aiming the reduction of freshwater contamination by nitrates and soil's organic matter oxidation.

Material and methods

Within a 4 years old irrigated pasture, dominated by grass plants (as a result of successive applications of nitrogen fertilization and Triclopir for weed control), two different treatment strips, 11 meters (m) wide and 60 m long, were installed, with 3 replicates. Treatments were as following: Witness (W), without tillage and with no legume introduction; Seed throwing (S), without tillage and Brillion seedling with 2.5 kg/ha of *Trifolium repens* and 1.5 kg/ha of *Trifolium fragiferum*. During the first year (October 2003 to September 2004), cover fertilization was applied at the rate of 82 units of nitrogen per ha/yr (41 in March and 41 in September) in W and 84 units of de P_2O_5 ha/yr in S (54 in May and 30 in September). Rotational grazing was performed by 6 to 8 adult Lusitanian mares with their foals during 7 days, in average, followed by 3 to 5 weeks without grazing. Due to technical limitations in the spring and the rainy summer, the amount of irrigation water used was only 4 000 m^3/ha.

During the experimental year reported, nine grazing periods were held with sample collection immediately before mare's introduction in the pastures. Square method (0.25 m^2) was used to sample nine randomly chosen spots for each treatment and in each sampling period. Pasture productivity was evaluated in terms of quantity (kg DM/ha) and quality, assessed by the floristic composition and chemical composition [CP, CF and DMD, as a % of DM, UFL/kg of DM (according to Jarrige, 1980) and UFC/kg of DM (according to Vermorel and Martin-Rosset, 1997)].

Data statistic analysis was performed using the GLM model and the SAS statistic programme.

Results

The production values, expressed in kg DM/ha, for the entire monitored period and the four seasons, present no significant differences ($P \leq 0.05$) for the two treatments (Table 1). Table 2 shows a significant improvement on the annual average qualitative parameters, namely in what concerns CP, DMD, UFL and UFC.

The comparison of the quality indicators for the various seasons reveals significant differences ($P \leq 0.05$) in all parameters for the last season monitored (Summer) (Table 3).

The comparison between legume and grass composition (Figure 1 and Figure 2), shows that legume plants acquire an important weight in the S treatment that becomes increasingly different of W ($P \leq 0.05$) in the subsequent seasons.

Discussion

The annual values obtained, around 10 000 kg DM/ha, may be considered low when compared with those found by Crespo (1966), Salgueiro (1984) and Garcia and Muslera (1991). The reasons for the lower production can be ascribed to: i) legume slow installation (Piñeiro, 2002); ii) insufficient irrigation (Crespo *et al.*, 1980, Garcia and Muslera 1991), iii) a far too short interval between grazing periods (Crespo and Romano, 1972) and iv) the lack of an annual resting period in the pasture use (Crespo and Romano, 1972; Crespo *et al.*, 1980). Nevertheless some literature refers similar results for similar conditions in the Portuguese centre region (Alarcão and Gama, 1992; Amaro *et al.*, 2004)

The low values can be ascribed to the initial pasture conditions and to the fact that the experience was conducted in real exploitation conditions, which implied an intensive grazing. These reasons together with the high nutritive value of the pasture imply that the productive values are acceptable for the first year of the experiment. In fact, the pasture quality improvement is notorious in S (Table 2), due to the increase of legumes and the decrease of grass plants (Figure 1) and to the short grazing intervals. These results agree with the literature, in terms of the level of CP (Jarrige, 1988; Salgueiro and Diaz, 1990; Garcia and Muslera, 1991; Piñeiro, 2002), DMD and energetic value (Jarrige, 1988; Moreira *et al.*, 2000; Amaro *et al.*, 2004).

Despite the fact that legume proportion starts to increase from the winter onwards (S with higher legume proportion than W), the differences only become significant in the summer ($P \leq 0.05$) for all the qualitative parameters. We speculate that the nitrogen fertilization performed in March prevented the observation of qualitative differences during the spring, when legume plants also presented significant differences between the two treatments (W=10.43% *vs.* S=36.23%).

Table 1. Annual and seasonal dry matter production (kg DM/ha ± standard error) in the two treatments (S and W).

Treatments	Anual	Autumn	Winter	Spring	Summer
S	9529.1±511.5^a	1368.6±193.9^a	1510.5±169.4^a	3043.8±215.6^a	3606.2±238.4^a
W	10037.1±511.5^a	1528.1±193.9^a	1915.9±169.4^a	3325.0±215.6^a	3268.1±238.4^b

Different index in the same column indicate significant differences (*P*?0.05).

Table 2. Mean annual qualitative values (CP, CF and DMD, in % of DM, UFL/kg of DM and UFC/kg of DM ± standard error) in the two treatments (S and W).

Treatments	CF	CP	DMD	UFL	UFC
S	20.52±0.63^a	16.43±0.21^a	79.12±0.87^a	1.05±0.02^a	0.70±0.005^a
W	21.82±0.63^a	15.50±0.21^b	75.52±0.87^b	1.00±0.02^a	0.68±0.005^b

Different index in the same column indicate significant differences (P?0.05).

Table 3. Mean seasonal qualitative values (CP, CF and DMD, in % of DM, UFL/kg of DM and UFC/kg of DM ± standard error) in the two treatments (S and W).

Treatments	Autumn	Winter	Spring	Summer
		CF (% of DM)		
S	21.46±0.97^a	18.39±0.67^a	20.55±0.79^a	21.68±0.60^b
W	22.84±0.97^a	19.26±0.67^a	21.02±0.79^a	24.18±0.60^a
		CP (% of DM)		
S	15.55±0.66^a	16.25±0.36^a	16.11±0.47^a	17.83±0.58^a
W	14.44±0.66^a	16.23±0.36^a	16.79±0.47^a	14.53±0.58^b
		DMD (% of DM)		
S	85.51±1.43^a	79.24±0.83^a	76.30±1.27^a	75.42±0.81^a
W	82.67±1.43^a	77.05±0.83^a	75.66±1.27^a	66.71±0.81^b
		UFL/kg of DM		
S	1.08±0.02^a	1.11±0.02^a	1.03±0.02^a	0.97±0.02^a
W	1.02±0.02^a	1.07±0.02^a	1.03±0.02^a	0.86±0.02^b
		UFC/kg of DM		
S	0.69±0.006^a	0.73±0.003^a	0.70±0.011^a	0.69±0.007^a
W	0.67±0.006^b	0.72±0.003^a	0.70±0.011^a	0.64±0.007^b

Different index in the same column indicate significant differences (*P*?0.05).

Conclusion

The introduction of legume plants in irrigated pastures, with no tillage, to avoid nitrogen fertilization allowed to keep the level of production and increased pasture nutritive value. To evaluate the reduction of nitrate lixiviation, we are currently monitoring solute lixiviation from pastures. Due to the evolution observed during the first year, we expect the DM production to become increasingly higher for S, while witnessing an improvement on food quality. This implies that sustainability is compatible with the quality improvement and quantity increase of livestock production in the Mediterranean region.

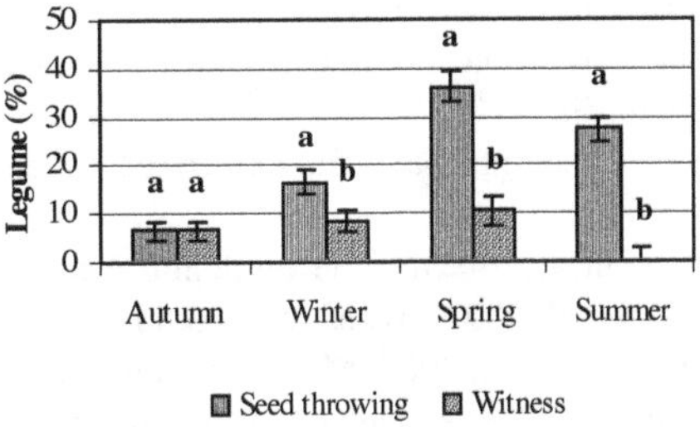

Figure 1. Mean seasonal legume and grass proportions ± standard error in the two treatments (S and W). Different index indicates significant differences (P≤0.05).

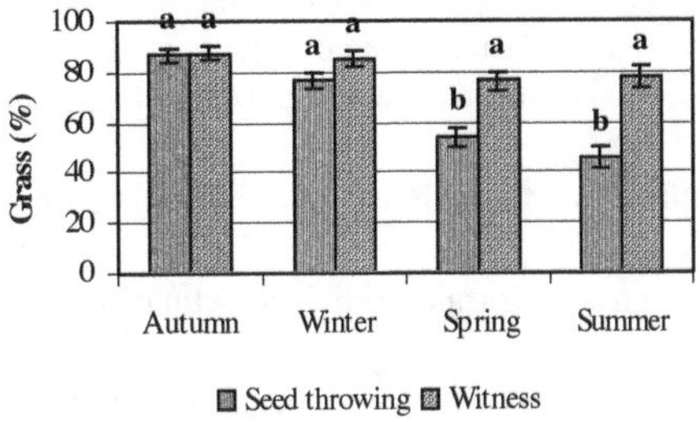

Figure 2. Mean seasonal legume and grass proportions ± standard error in the two treatments (S and W). Different index indicates significant differences (P≤0.05).

Acknowledgements

This work was supported by the European Union's LIFE Environment programme, through the LIFE03 ENV/P/000501 "Environmental Management an Audit Scheme Implementation at a Complex School – EMAS@SCHOOL" project.

References

Alarcão, C. & Gama, J., 1992. Avaliação do potencial produtivo e interesse agronómico de consociações forrageiras e pratenses de regadio. Série Estudos, 3: 7-38.

Amaro, R., Feio, G., Lopes, N. & Oliveira, A., 2004. Contributo para o estudo do comportamento de sete consociações de pastagens temporárias de regadio. Resumos da XXV Reunião da SPPF: 53-53.

Crespo, D.C., 1966. Alguns ensaios de produção de forragens em regadio. VII Reunião da SEEP: 51-57.

Crespo, D.C., 2004. O papel das pastagens e forragens no uso da terra portuguesa. Bases para o seu desenvolvimento sustentável. XXV Reunião da SPPF: 17-19.

Crespo, D.G. & Romano, A.M., 1972. Influência da frequência de corte na produção e qualidade da erva de prados temporários de regadio. Melhoramento, 24: 95-124.

Crespo, D.C., Antunes, J.H. & Dias, J.M., 1980. Influência dos encabeçamentos na produção de carne de bovino em prados de regadio. SPPF, 1: 96-109.

ECAF, 1999. Agricultura de Conservação na Europa: Aspectos ambientais, económicos e Políticos na U.E.. Bruxelas, Bélgica. www.aposolo.pt.

EEA, 2003. Europe's Environment: the third assessment. European Environmental Agency. http://reports.eea.eu. int/environmental_assessment_report_2003_10/en/kiev_eea_low.pdf.

Ferreira, J. C., Srecht, A., Ribeiro, J. P., Soeiro, A. & Cotrim, G., 2002. Manual de agricultura biológica: Fertilização e Protecção das plantas para uma agricultura sustentável, Agrobio (Editors), Lisboa, Portugal, pp. 435.

Garcia, C.R. & Muslera E.M. , 1991. Praderas y forrajes. Ediciones Mundi-Prensa, Madrid, Spain, pp. 702.

Jarrige, R., 1988. Alimentation des ruminants. INRA. Paris, France, pp. 621.

Moreira, O.C., C.C. Belo & Ribeiro, J.R., 2000. Evolução do valor nutritivo de uma pastagem. 1. Constituintes orgânicos da pastagem. Reunião Ibérica de Pastagens e Forragens, 3: 511-516.

Piñeiro, J., 2002. Efecto de la siembra en superficie y del manejo en la presencia del trébol blanco en praderas. Revista agropecuaria, 71: 472-474.

Rizov, M., 2004. Rural development and welfare implications of CAP reforms. Journal of Policy Modelling, 26: 209-222.

Salgueiro, J. & Dias, M., 1990. Producción de carne con pastos y forrajes. Ediciones Mundi-Prensa, Madrid, Spain, pp. 389.

Salgueiro, T., 1984. A problemática da produção animal e suas relações com a alimentação. Ministério da Agricultura, Florestas e Alimentação, Série Técnica, 13.

Simpson, R. & R. Culvenor, 1985. Photosynthesis, carbon partitioning and herbage yield. In: Temperate pastures - their production, use and management. Wheeler, J.L., C.J. Pearson & G.E. Robards (Editors), CSIRO, Australia, pp. 103-118.

Vermorel, M. & Martin-Rosset, W., 1997. Concepts, scientific bases, structure and validation of the French horse net energy system (UFC). Livestock Production Science, 47(3): 261-275.

Effect of the stocking rate on economic indicators of dehesa's cattle farms

P. Gaspar, A. Rodríguez de Ledesma, M. Martín, M. Escribano, F.J. Mesías & F. Pulido

Escuela de Ingenierías Agrarias. Universidad de Extremadura,
Ctra. de Cáceres, s/n, 06071 Badajoz, Spain

Summary

A group of extensive cattle farms were studied in semiarid rangelands in the southwest of Spain (dehesas) to evaluate the effect of stocking rate on some of their economic indicators. The analysis started with a random survey with a confidence level of 95%. The sampling error was less than 5%, usual in this type of research. The data were obtained from a representative sample in the year 2004. The economic analysis carried out with 69 farmers showed that cost, production and profitability indicators increased with the stocking rate.

Keywords: grazing, dehesa, stocking rate , economic analysis.

Introduction

After the Common Agricultural Policy (CAP) reform in 1992, the management of the extensive systems has suffered important transformations, and their effects, such as an increase in the cattle pressure within the farms, have been observed. Although it may seem contradictory, the CAP policy towards extensification has caused the opposite effect towards an intensification of cattle systems, depending on the characteristics of each specific system.

The Dehesa rangelands, located in the southwestern region of the Iberian Peninsula, representing extensive cattle farming model in Europe suffered from this policy. These Dehesas are spread out over 7 million hectares in Spain, and over 2.5 million hectares in Portugal where they are known as montados (Escribano & Pulido, 1998). The grazing resources and quercus trees (quercus ilex sp. and quercus suber sp., in particular) in these lands are mainly used to feed cattle from local breeds (54% of total cattle population in Spain), sheep (41%), goats (44%) and Iberian pigs (98%).

The dehesa in Extremadura is well preserved, being the core of this kind of cattle system in the Iberian Peninsula. The figures of the dehesas in Extremadura alone add up to 19% of cattle, 18% of sheep, 10% of goat and 54% of Iberian pig of the Spanish census.

Therefore, this study was carried out to evaluate whether increasing cattle pressure/stocking rate in these rangelands affects profitability of farms.

Material and methods

The methodology used in this paper is based on a microeconomic adjustment of the European System of Integrated Economic Accounts, applied to the Economic Accounts of Agriculture and Forestry (EUROSTAT, 1995). We tried to fill the gaps related to production and capital accounting. Thus, we created several synthetic economic indicators to explain the management changes and financial results in the year 2004. These methodological modifications allow to evaluate the economic resources more rigorously, so as to better understand the internal performance of the farms and to study how they have adjusted to the EU regulation mechanisms.

The economic analysis of the dehesa cattle farms was performed by means of surveying dehesa farmers following a random sampling design, including all those farmers with beef and/or sheep cattle listed in the Farm Register of Extremadura. A total of 69 farmers were surveyed, rearing a total of 4,162 cows and 57,108 sheep.

The effects of the stocking rate on those indicators were determined by an analysis of variance procedure. The model used was: $y_{ik} = m + T_{ik} + E_{ik}$, where m is the mean value, T is the effect of the treatment (stocking rate) and E is the random error. The statistical package used was SPSS 10.01 (SPSS inc, 1999).

Three stocking rates were studied: low (< 0.3 LU (livestock units)/ha UFS), medium (0.3-0.5 LU/ha UFS), and high (> 0.5 LU/ha UFS), where UFS is the available farming surface of the farm (excluding buildings, paths, ...).

The sets of variables generated in this study were based on criteria taken from the technical-economic literature or from earlier studies (Pulido and Escribano 1994; Escribano *et al.* 1996, 1997). We define a few indicators as follows:

- *total raw material* is the cost of the production input (feeding, petrol, ...);
- *external services* is the cost of work services performed by outside private enterprises;
- *sales of end product* is the gross income obtained from farming product sales; and
- *intermediate product* is the internal production of the farm that is used again within the farming system.

The homogeneity of the groups was tested by Cluster Analysis. The differences among groups of standarization variables were determined according to the mean of the sub-populations generated by the factor of variation. Group differences were measured by means of Bonferroni's multiple comparison test (Luque, 2000; SPSS Inc., 1999).

Results and discussion

Significant differences were found among the stocking rates in relation with the cost of total raw product (Table 1). Farms with higher stocking rates have also higher cost values. There were differences between high stocking rate farms and both low and medium groups. This means that when stocking rate per ha increases, the needs of the livestock increase too.

In relation to costs, sales of end product and gross product, significant differences were observed among those farms with higher stocking rate values and farms with lower stocking rate values.

Important differences were also found in the asigned subsidies. These go up as the stocking rate levels increase. On the other hand, both the net operating surplus and the net value added show significant differences between farms with high and low stocking rates.

With regard to the economic structure ratio indicators of the farms, the higher re-used raw material versus the total raw material ratio values corresponds to farms with low stocking rates (Table 2). This is due to their relatively lower total raw material values. Higher values of this ratio mean a better use of the environment resources an, as result, a decrease of the feeding expenses.

Farms with high stocking rates also tend to have higher values in the Sales of end products versus Gross products ratio. This is due to higher income from cattle sales. On the other hand, the profitability rate improves as the stocking rate increases.

The final average value of the profitability rate is clearly low, being below 3% in most cases.

Table 1. Cost, production and rent indicators according to their stocking rate (€/ha).

Indicators	Socking rate High	Medium	Low	Sign.	Overall
Total raw materials (RMt)	258 041[b]	165 919[a]	103 259[a]	***	179 799
External Services (EXS)	26 856	16 170	18 891	n.s	20 676
Labour cost (LC)	62 736	56 457	49 923	n.s	56 747
Current Cost (CC)	406 704[b]	299 960[a]	206 826[a]	***	310 093
Total Cost (TC)	453 926[b]	356 371[ab]	247 827[a]	***	358 841
Sales of end product (SEP)	336 528[b]	243 907[ab]	147 192[a]	**	248 090
Intermediate Product (IP)	99 812[b]	69 191[a]	43 817[a]	***	72 487
Gross product (GP)	535 574[b]	390 872[ab]	267 805[a]	***	405 531
Net operating margin (NetM)	81 648	34 501	19 978	n.s	46 691
Operating subsidies (Sb)	129 885c	98 582[b]	66 533[a]	***	100 181
Net operating surplus (NetS)	211 533[b]	133 083[ab]	86 512[a]	*	146 871
Net value added c.f. (NVAcf)	274 269[b]	189 540[ab]	136 435[a]	**	203 618

[a, b, c...] different index in the same row show significant differences between groups for $P>0,05$;
n.s. no significant; * $P<0,05$; ** $P<0,01$; *** $P<0,001$
CC = RMt+EXS+LC+ OC (other costs)
TC = CC+AK (annual amortization of the fixed capital)
GP = IP + FP (final product) = IP + GIFK (gross investment of the fixed capital) + SEP + FPE (final production existences) + OFP (other final productions)
NetM = GP - TC
SNE = Sb - Tx (indirect taxes or tied to productions)
NetS = NetM + SNE = (GP - TC) + (Sb - Tx)
NVAcf = NetS + LC

Table 2. Profitability indicators and principal ratios between cost, productions and rents of the farms in relation with their stocking rate (%).

Ratio indicators	Socking rate High	Medium	Low	Overall
Re-used raw materials/Total raw materials	41.87	43.48	45.31	43.45
External services/Total cost	5.38	4.92	7.98	5.97
Labour/Total cost	12.43	15.95	20.35	16.00
Sales of end product/Gross product	61.42	57.74	56.02	58.52
Profitability rate (Pr)	3.39	2.19	1.39	2.38

Pr = NetS/K (net operating surplus/inmobilized capital).

References

Boyazoglu, J.& J.C. Flamant, 1991. The actual state and future of animal production in mediterranean rangelands. 4th Rangeland Congress. Montpellier, 1017-1025.

Escribano, M. & F. Pulido, 1998. La dehesa. Estructura económica y recursos naturales. SGT. Junta de Extremadura. Mérida, Spain.

Escribano, M., A. Rodríguez de Ledesma, F.J. Mesías & F. Pulido, F., 1997. Índices técnicos de gestión ganadera en espacios agrarios adehesados. ITEA Producción Animal. 93: 99-118.
Escribano, M., F. Pulido, A. Rodríguez de Ledesma & F.J. Mesías, F.J., 1996. Determination of energetic resources that cover cattle food needs in dehesa systems. Arch. Zootech. 45: 379-393.
EUROSTAT, 1995. Sistema Europeo de Cuentas Económicas Integradas (SEC). 2ª edición. Comisión de las Comunidades Europeas. Bruxels.
Luque, T. (coord.). 2000. Técnicas de análisis de datos en investigación de mercados. Ediciones Pirámide, Madrid, Spain. pp. 557.
Pulido, F., 2002. La producción animal en la dehesa extremeña. Nuevas tendencias y estrategias de mejora. Libro Blanco de la Agricultura y el Desarrollo Rural del Ministerio de Agricultura, Pesca y Alimentación. Ponencia de la Jornada Autonómica de Extremadura (Badajoz 12-11-2002).
SPSS inc. 1999. SPSS Base 10.01 Application Guide. SPSS Inc. Chicago. USA

Besnoitia besnoiti impact on fertility of cattle exploited in Mediterranean pastures (Alentejo)

H. Cortes[1], J. Chagas e Silva[2], M.C. Baptista[3], R.M. Pereira[3], A. Leitão[4], A.E.M. Horta[3], M.I. Vasques[3], J.P. Barbas[3] & C.C. Marques[3]

[1]*Laboratório de Parasitologia. N. Mitra. Universidade de Évora, Ap. 947002-554, Portugal*
[2]*Divisão de Selecção e Reprodução Animal-DGV, Venda Nova, 2704-507 Amadora, Portugal*
[3]*Dpt. Reprodução Animal, Estação Zootécnica Nacional, 2005-048 Vale Santarém, Portugal*
[4]*Inst. Invest. Cient. Tropical (IICT; CIISA), Fac. Med. Vet., 1300-447 Lisboa, Portugal*

Summary

Besnoitia besnoiti is a bovine parasite endemic in many tropical and subtropical areas whose prevalence in the Mediterranean countries such as Portugal seems to be increasing. Most infections are mild or subclinical, characterized by the formation of numerous cutaneous and sub-cutaneous microcysts, lowering the quality of skins for the leather industry. Male sterility or impaired fertility is a common sequela in breeding bulls, and is one of the most negative aspects of the disease in animals that survive infection. Our objective was to investigate if asymptomatic Besnoitiosis leads to bovine infertility, by comparing seminal parameters pre and post-thawing, in vitro fertilization (IVF) and embryo rates between asymptomatic infected (n=3) and uninfected (n=5) bulls, exploited in an extensive production system in Alentejo-Portugal. Skin biopsies were submitted to histopathological analyses to identify *B. besnoiti* cysts in sires. Semen was collected by electroejaculation and sperm quality parameters before cryopreservation and after thawing were analyzed using ANOVA. The quality of semen collected from asymptomatic infected and uninfected bulls presented no differences before cryopreservation. From all the post-thawed sperm quality parameters (motility and hypoosmotic swelling test; post-swim-up motility, activity, concentration and agglutination; fertilization and embryo rates) evaluated, only post-thawed (51.0±36.3 vs. 42.3±10.6%, $P \leq 0.05$) and post-swim-up (36.3±18.8 vs 25.1±12.0 %, $P \leq 0.009$) motility were significantly different between asymptomatic infected and uninfected bulls, respectively.

Semen from asymptomatic *Besnoitia besnoiti* infected bulls may maintain fertilization ability. However the presence of these animals in herds represents a risk of spreading the disease leading to further economic losses.

Keywords: Besnoitia besnoiti, bovine besnoitiosis, infertility, Portugal.

Introduction

Besnoitia besnoiti (Marotel 1912) is a coccidian parasite endemic in many tropical and subtropical areas where it causes both acute and chronic bovine besnoitiosis, leading to significant economic constraint to commercial cattle production (Bigalke, 1968, Kumi-Diaka *et al.*, 1981). In Europe, it has been reported only in the Mediterranean countries, France (Besnoit & Robin, 1912), Spain

(Irigoien *et al.*, 2000), and Portugal. In the latter country, prior to 1991, the disease was seldom recognized or reported. At present its prevalence may be increasing (Leitão, 1949, Cortes *et al.*, 2003, 2005).

Cattle regardless of breed, sex and age are affected by this parasitosis, although clinical disease occurs rarely in calves less than 6 months of age. The occurrence is usually sporadic, and only a small proportion of infected animals develops clinical disease (Pols, 1960, Bigalke, 1968). In the acute stage of disease, parasitic tachyzoites invade blood vessels of the skin, subcutaneous tissues, fascia and testes causing widespread vasculitis and thrombosis. The result is a severe generalized reaction which is accompanied by oedema of the skin and acute orchitis. During the chronic stage, large numbers of tissue microcysts containing bradyzoites are formed. The most striking features of this stage are thickening, wrinkling and hair loss of the skin, accompanied by anorexia and severe weight loss. The case mortality rate is approximately 10% (Bigalke, 1968, Basson *et al.*, 1970). Male sterility or impaired fertility is a common sequela in breeding bulls, and is one of the most negative aspects of the disease in animals that survive the acute and chronic stages of infection (Basson *et al*, 1970, Kumi-Diaka *et al.*, 1981, Sekoni *et al.*, 1992, Cortes *et al.*, 2005).

The main objective of this experimental work was to compare fertility parameters between uninfected and infected bulls without clinical signs of disease and to predict some implications of their presence on farm reproduction.

Material and Methods

Animals

Three different farms with cases of bovine besnoitiosis were chosen for this study. All the farms were located in Évora region (Portugal) and exploit beef cattle extensively. Eight animals, seven Limousine and one Charolais bull, aged 2-9 years, were selected. Skin samples collected from the bulls' neck were fixed in 10% formalin solution and stained with Hematoxilin /Eosin to identify the presence of *Besnoitia* cysts in a light microscope (100X). The aim was to find affected males without any clinical symptoms, but with the presence of *Besnoitia* cysts revealed by histopathology examination. This was achieved in three bulls.

Semen collection and cryopreservation

Semen was collected by electroejaculation to asymptomatic infected (n=3) and uninfected bulls (n=5) and immediately stored in a 32°C water bath. Before cryopreservation, semen morphology was evaluated by determination of gross (1 to 5) and individual progressive (0–100%) motility, percentage of abnormal and live (supravital stain eosin–nigrosin) spermatozoa (spz). After determination of sperm concentration, dilutions were performed using TriladylÒ (ref. 13500/0250, Minitub GmbH) with egg yolk and distilled water, previously warmed at 32°C, to obtain progressively motile 20x10^6 spz per seminal dose. Extended semen was immediately loaded into 0.25 ml straws. All the straws were vertically stored in an appropriate device and maintained at 5°C for 4 hours and then frozen by vertical freezing method (Chagas e Silva, personal communication). The rackets with straws were kept inside the RCB_4 Container (Air Liquide, France) in the nitrogen vapours (in a position that the device only touches the liquid phase) for 20 minutes with the lid closed, before being placed into appropriate cryogenic containers and immersed into liquid nitrogen for storage.

Post-thawed semen evaluation

The effect of *B. Besnoiti* on post-thawed semen quality parameters was investigated according to the following experimental design: In experiment 1 (6 replicates), semen from uninfected bulls (n=5) was tested. In experiment 2 (5 replicates), semen from two uninfected bulls, with the worst and the best fertilizing ability selected in experiment 1, and from three asymptomatic infected bulls was used. For the hypoosmotic swelling test (Host), only two straws from each infected and uninfected animals were evaluated.

Post-thawed motility, Host and post-swim-up tests

Individually progressive semen motility after thawing from infected or uninfected bulls was evaluated. To perform the Host test two straws from each bull were thawed. Semen aliquots were incubated with an hypoosmotic solution (100 $mOsmL^{-1}$) for 25 minutes and fixed with 2% glutaraldeyde in BL-1 (Ferreira *et al.*, 2001).

To perform post-swim-up tests thawed semen was incubated for 1 hour in TALP medium without Ca^{++} supplemented with 48.6 mg mL^{-1} caffeine (Sigma, ref. C-0750) in an humidified atmosphere of 5% CO2 at 39°C. Spermatozoa from the upper layer were taken to evaluate their individual motility (0 – 100%), activity (1 to 5), concentration and agglutination (number of agglutinated heads in 100 observed spz).

In vitro fertilization and embryo development

In vitro maturation, fertilization and co-culture procedures have been described previously (Pereira *et al.*, 2005). Briefly, follicles with 2-6 mm diameter were aspirated from abattoir bovine ovaries. Selected cumulus oocyte complexes were matured in TCM199 (GibCo, ref. 22340-020) with 10% superovulated oestrus cow serum (SOCS), 10 μg mL^{-1} FSH (Follotropin, Vetrepham Inc.) and antibiotics (Sigma, ref. P0781) during 22-24 hours. For fertilization, frozen-thawed semen was submitted to swim-up as above. Spermatozoa (10^6 spz mL^{-1}) and 10 oocytes were placed in 40 μL droplets of TALP medium supplemented with 5.4 USP mL^{-1} heparin (Sigma, ref. H-3393), 10 mM penicillamine (Sigma, ref. P-4875), 20 mM hypotaurine (Sigma, ref. H-1384) and 0.25 mM epinephrine (Sigma, ref. E-1635). Following co-incubation for 22 hours, the presumptive zygotes were placed in 100 μL droplets of a granulosa cell monolayer cultured with TCM199, 10% SOCS and antibiotics under paraffin oil. Atmospherical conditions for IVM-IVF and embryo co-culture were 39°C and 5% CO_2 in air with humidified atmosphere. Cleavage was assessed 48 hours post insemination and embryos were morphologically evaluated on day 7 and 8.

Statistical analysis

All data are presented as means ± standard deviations (SD). The mean values were compared by using the ANOVA and LSD and $P<0.05$ was considered significant.

Results

Besnoitia besnoiti diagnosis

Three bulls, 2 Limousine and 1 Charolais, presented *B. Besnoiti* cysts with live bradoyzoites in the skin (Figure 1).

Semen evaluation

Before cryopreservation, there were no differences (P>0.05) in morphological evaluation (Table 1) of semen from asymptomatic infected with *Besnoitia besnoiti* and uninfected bulls.

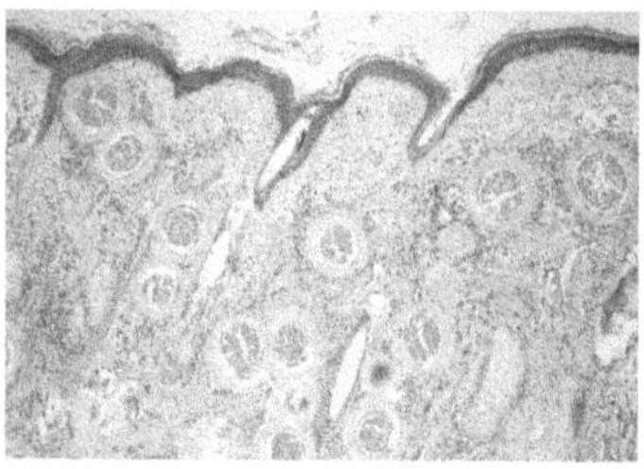

Figure 1. Hematoxilin-eosin stain of skin presenting Besnoitia cysts.

After cryopreservation, there were no significant differences between the 2 uninfected bulls running simultaneously in both experiments. Consequently, post-thawing results from experiment 1 and 2 were treated together (Table 2). From all the sperm quality parameters evaluated only post-thawed (P=0.05) and post-swim-up (P=0,009) motility were significantly different between asymptomatic infected and uninfected bulls, respectively. Cleavage and day 7-8 embryo production rates obtained with semen from uninfected and asymptomatic infected bulls presented no differences (P>0.05).

Discussion

Besnoitia besnoiti infections cause economic losses to cattle owners in endemic areas due to mortality, loss of condition, low market value and temporary or permanent sterility (Bigalke, 1968, Kumi-Diaka *et al.*, 1981). Prevalence of bovine besnoitiosis in Portugal, especially the asymptomatic form, is not yet well determined. In the present study live bradyzoites within *Besnoitia* cyst were identified although all the bulls tested were asymptomatic. Previous studies (Cortes *et al.*, 2001, 2005) in farms located in Évora region diagnosed 42% of *B. besnoiti* infected Limousine bulls without clinical signs of the disease.

Asymptomatic *B. besnoiti* infected bulls presence in herds implies a potential transmission of besnoitiosis although as demonstrated sire fertilitlity is not affected. Infertility or sterility seems to be a late consequence of *B. besnoiti* infection (Sekoni *et al.*, 1992). Semen quality pre and post cryopreservation was not diminished in infected asymptomatic bulls. Neither single parameters of semen quality (motility, sperm number and morphology) having moderate value as predictors of absolute fertility levels (Parkinson, 2004), nor supravital stain and Host tests which can be considered important variables for normal sperm function (Tartaglione & Ritta, 2004), were affected. IVF and

Table 1. Evaluation (mean±SD) of semen collected by electroejaculation to asymptomatic infected with Besnoitia besnoiti *(INF, n=3) and uninfected bulls (control, n=5) before cryopreservation.*

Bulls	Volume (mL)	Gross motility	Individual motility (%)	Concentration (10^5spz mL^{-1})	Dead spz (%)	Morphological abnormalities (%)
Control	5.2±2.4	3.0±0.7	82.0±11.0	10.6±3.7	23.6±13.7	23.4±3.2
INF	6.7±2.8	3.3±0.6	76.7±11.7	8.4±3.6	16.0±1.7	25.0±1.0

Table 2. Evaluation (mean±SD) of post-thawed semen collected to asymptomatic infected with Besnoitia besnoiti *(INF) and uninfected (control) bulls.*

	Post-thawed				Post-swim-up tests				
Bulls	n	Motility (%)	n	Host (%)	n	Motility (%)	Activity	Agglutination (%)	Concentration (10^6spzmL^{-1})
control	45	43.3±10.6[a]	10	54.1±9.6	45	25.1±12.0[a]	2.4±0.7	28.4±12.0	83.6±82.9
INF	15	51.0±18.6[b]	6	45.4±19.1	15	36.3±18.9[b]	2.6±1.1	27.2±10.3	100.5±63.8

Within columns, means with different letters are significantly different ($P<0.05$, ANOVA).

embryo production rates also presented no differences between asymptomatic infected bulls and control. IVF and embryo production rates are useful indicators of bull fertility. Relevant reports correlated those results with the non-return rates using cryopreserved semen (Marquant-LeGuienne *et al.*, 1990) and the conception rates in natural service (Brahmkshtri *et al.*, 1999).

These results suggest that in sub-clinically infected bulls there are no implications on the fertility. However, in advanced stages of infection infertility is a well characterised consequence and, therefore, in endemic farms, the simple elimination of infertile bulls is not recommended since when this manifestation is established, the disease is already in an advanced stage and dissemination to other animals may already have occurred. Under these circumstances early detection of infection and adequate control measures have to be implemented in order to attenuate the economical impact of besnoitiosis. In addition, we strongly disagree with the transport of infected animals, even if subclinically, to other areas or herds without any contact with this parasite. In fact, due to the mechanical transmission of *B. besnoiti* by blood-sucking insects or iatrogenicaly from infected to susceptible cattle (Bigalke, 1968), and the high number of tabanids and *Stomoxis calcitrans* during summer season, moving infected animals to Besnoitia free herds will contribute to the spread of the disease.

Acknowledgements

We wish to thank Andrew Hemphill for fruitful discussions and for revising the manuscript.

References

Basson, P.A., R.M. McCully & R.D. Bigalke, 1970. Observations on the pathogenesis of bovine and antelope strains of *Besnoitia besnoiti* (Marotel, 1912) infection in cattle and rabbits. Onderstepoort J. Vet. Res. 37: 105-126.

Besnoit, C. & V. Robin, 1912. Sarcosporidioses cutanée chez une vache. Rec. Vet. 37: 649.

Bigalke, R.D., 1968. New concepts on the epidemiological features of bovine besnoitiosis as determined by laboratory and field investigations. Onderstepoort J. Vet. Res. 35: 3-138.

Brahmkshtri, B.P., M.J. Edwin, M.C. Jhon, A.M. Nainar, A.M. Krishnan, 1999. Relative efficacy of conventional sperm parameters and sperm penetration bioassay to assess bull fertility in vitro. Anim. Reprod. Sci. 54: 159-169.

Cortes, H., A. Leitão, R. Vidal, M.J. Vila-Viçosa, M.L. Ferreira, V. Caeiro & C.A. Hjerpe, 2005. Bovine Besnoitiosis in Portugal. Vet. Rec. (In press)

Cortes, H., M.L. Ferreira, J.F. Silva, R. Vidal, P. Serra & V. Caeiro, 2003. Contribuição para o estudo da besnoitiose bovina em Portugal. Rev. Port. Cien. Vet., 545: 43-46.

Ferreira, G., J.P. Sousa, J.P. Barbas & A.E.M. Horta, 2001. Buck semen Hos test of Serrana breed. In: Proc. 3rd Iberian Congr. Anim. Reprod, SPRA (Ed.), Porto, Portugal, 559-564.

Irigoien, M., E. Del Cacho, M. Gallego, F. Lopez-Bernad, J. Quilez & C. Sanchez-Acedo, 2000. Immunohistochemical study of the cyst of Besnoitia besnoiti. Vet. Parasit. 91: 1-6.

Kumi-Diaka, J., S. Wilson, A. Sannusi, C.E. Njoku & D.I.K. Osoru, 1981. Bovine besnoitiosis and its effect on the male reproductive system. Theriogenology 16: 523-530.

Leitão, J.L.S., 1949. Globidiose bovina por Globidium besnoiti (Marotel 1912). Rev. Med. Vet. 330: 152-158.

Marquant-LeGuienne, B., P. Humblot, M. Thibier & C. Thibault, 1990. Evaluation of bull semen fertility by homologous in vitro fertilization tests. Rep. Nutr. Dev. 30: 259-266.

Parkinson, T.J., 2004. Evaluation of fertility and infertility in natural service bulls. Vet. J. 168: 215-229.

Pereira R.M., C.C. Marques, M.C. Baptista, M.I. Vasques & A.E.M. Horta, 2005. Influence of supplementation of arachidonic acid and cyclooxygenase/lipoxygenase inhibition on the development of early bovine embryos. Rev. Braz. Zoot. (In press).

Pols, J.W., 1960. Studies on Bovine besnoitiosis with special reference to the aetiology. Onderstepoort J. Vet. Res. 28: 265-356.

Sekoni, V.O., A. Sanusi, M.O. Abata, E.O. Oyedipe, P.I. Rekwot & L.O. Eduvie, 1992. Loss of libido and terminal sterility in a Frisean bull naturally infected with Besnoitia besnoiti in Northern Nigeria: A case report. Theriogenology 37: 533-549.

Tartaglione, C.M. & M.N. Ritta, 2004. Prognostic value of spermatological parameters as predictors of invitro fertility of frozen-thawed bull semen. Theriogenology 62: 1245-1252.

Reproduction in the ovine Saloia breed: seasonal and individual factors affecting fresh and frozen semen performance, in vivo and in vitro fertility

C.C. Marques[1], J.P. Barbas[1], M.C. Baptista[1], C. Cannas Serra[2], M.I. Vasques[1], R.M. Pereira[1], S. Cavaco-Gonçalves[1] & A.E.M. Horta[1]

[1]*Department of Animal Reproduction, Estação Zootécnica Nacional, 2005-048 Vale Santarém, Portugal*
[2]*Direcção Regional de Agricultura do Ribatejo e Oeste, 2000 Santarém, Portugal*

Summary

Our objectives were to characterise seminal traits from fresh and frozen semen and in vivo and in vitro (IVF) fertility rates in Saloia breed. Five rams from this breed were used for semen collection with artificial vagina in spring, summer and autumn during three years. Eighty two ejaculates of good quality were collected, refrigerated (4°C) and frozen in nitrogen vapours and its seminal traits evaluated. Cervical artificial insemination (AI, n=440) 55 hours after sponge removal and in vitro fertilisation of in vitro matured oocytes (n=1636) with semen frozen in autumn and winter were performed. In fresh semen significant differences (P<0.05) among seasons were detected for individual motility, live sperm, normal and tail abnormalities. Individual variations were detected for volume, concentration, live and normal sperm. For frozen/thawed semen traits, no differences among seasons were observed. No differences in fertility (15%) and fecundity (20.2%) were observed between semen frozen in autumn or winter. One of the rams presented higher fertility and fecundity rates when season of semen was not considered. No correlations were found for semen traits and AI results. Oocyte maturation rate was 79.47±9.6%, a higher maturation rate being observed during the shortening day length semester (83.5% vs. 74.61%; P<0.05). Following in vitro swim-up procedure, only sperm motility was different between two males in autumn (56.3% vs. 36.3%; P<0.02). No differences in cleavage rates between males or seasons were found (overall mean: 40.9±16.5%). In one of the males, embryo rates at Day 7 were higher when fertilisation with autumn rather than winter frozen semen was performed (38.1% vs. 20.3%; p=0.01). Only post-thawed sperm head abnormalities and embryo cleavage rates were highly and negatively correlated (r = -0.99; P<0.05). This study shows that seminal traits either in fresh or frozen semen were not predictable of in vivo or in vitro fertility rates with thawed semen in the Saloia breed.

Keywords: ovine Saloia breed, frozen semen, seminal traits, AI, in vitro fertilization.

Introduction

The ovine Saloia breed is exploited for milk transformed to cheese in the Lisboa and Portalegre areas of Portugal. The main breeding season occurs during spring and a secondary one in the autumn (Sobral *et al.*, 1987). Between January and April there is a marked seasonal ovarian inactivity in this breed (Pombo *et al.*, 1993). During April, there are reports on good fertility rates (50%) using hormonal synchronisation treatments and artificial insemination (AI) (Barbas *et al.*, 2002b). Using natural service during April and May, fertility is usually 72%-81% (Barbas *et al.*, 1992), but when the breeding period is extended until the end of June, a 94% fertility rate is reached (Goddard *et al.*, 1993). A slightly higher fertility rate is obtained in summer/autumn than in spring with artificial insemination (Barbas *et al.*, 2004). Prolificacy in this breed varies between 1.05 and 1.2 (Sobral *et al.*, 1987) and no differences seem to exist between seasons (Chagas e Silva & Cidadão, 1997). In the male a general tendency for a worse performance in seminal traits during the increasing day length season were found (Cruz e Silva, 1989). The first results using artificial insemination with frozen semen showed a 17.3% fertility rate compared with a 40% using refrigerated semen (Barbas *et al.*, 2003). A marked difference between males concerning their ability to fertilise with frozen semen after AI was also detected (Barbas *et al.*, 2004). No reports of IVF in Saloia breed are available, but with the local Serra da Estrela and Merino breeds good quality frozen semen was necessary to achieve acceptable results (Baptista *et al.*, 2002).

The objectives of this work were the characterisation of seminal traits from fresh and frozen semen during spring, autumn and winter and the determination of fertility rates after AI and after IVF using semen frozen in autumn and winter in the Saloia breed.

Material and methods

Five rams of Saloia breed, located at Santarém were used for semen collection with artificial vagina in spring, autumn and winter during three years. Seminal traits of fresh and frozen/thawed semen were evaluated according to Evans and Maxwell (1990). Eighty two ejaculates of good quality based on individual motility (IM>65%) were collected. Fresh semen was diluted and packed in 0.25 mL Cassou straws (300x10^6 spz/straw) and refrigeration lasted 4 h. Refrigerated semen (+4°C) was frozen in a cryo-chamber before LN_2 immersion. Semen extender composition for freezing was described by Evans and Maxwell (1990). Frozen thawed semen was used for artificial insemination (AI) in 440 ewes of the same breed. Oestrus synchronization was done with fluorogestone acetate (40 mg) vaginal sponges (Intervet®) during 12 days and 500 IU of eCG i.m. at sponge removal. Cervical AI was performed 55 h after sponge removal using 2 straws per ewe, with sperm motility >40% after thawing. Fertility and fecundity were evaluated in the inseminated ewes.

Semen from three Saloia males frozen in autumn (4 replicates) and winter (15 replicates) were also tested for in vitro fertilisation. Ovaries from slaughtered prepubertal ewes belonging to different breeds were collected. Oocyte aspiration from 2 to 6 mm diameter ovarian follicles was performed. Selected cumulus-oocyte complexes were matured in TCM-199plus 5 mg mL^{-1} FSH, 5 mg mL^{-1} LH, 10% superovulated oestrus sheep serum (SOSS) and antibiotics in a 5% CO_2 incubator at 39°C for 22 hours. Capacitation of frozen-thawed spermatozoa was processed by swim up after incubation in modified Bracket's medium plus 20% (v/v) SOSS at 39°C for 1 h in the same conditions (Crozet *et al.*, 1987). Post-thawed sperm motility (%) and post-swim up sperm characteristics, such as sperm motility (%), agglutination rate (percent of agglutinated or capacitated spermatozoa) and sperm concentration were evaluated. Spermatozoa (10^6 spz mL^{-1}) were added to IVF medium droplets (modified Brackett´s medium plus 7.5 mM calcium lactate) containing 10 oocytes each and incubated altogether for 22 h at 39°C in 5% CO_2. After co-incubation, zygotes were cultured in HEPES buffered synthetic oviduct fluid (SOF) droplets supplemented with 10% SOSS in an atmosphere of 5% CO_2, 90% N_2 and 5% O_2 at 39°C. Cleavage at 48 h post fertilisation (cleaved embryos/inseminated oocytes) and embryo (embryos/ cleaved oocytes) rates at day 7 (D7) of culture were determined.

Average found for individuals or seasonal fresh and frozen semen traits were compared by ANOVA or multifactor ANOVA for interactions. Fertility results (in vivo and in vitro) were also compared by ANOVA.

Results

Average and 95% confidence limits (CL) of fresh and thawed seminal parameters computed independently of seasonal and individual factors (ejaculates from 5 rams) are presented in Table 1. Freezing significantly ($P<0.05$) decreased all semen traits, exception made for the intermediate piece abnormalities (IPAbn). Intermediate piece abnormalities were the only semen parameter without significant correlation between fresh and frozen semen (homologous traits).

For all ejaculates, significant differences ($P<0.05$) among seasons for individual motility, live sperm, normal and tail abnormalities were found in fresh semen with best performances in the autumn (Figure 1). For frozen/thawed semen traits, no differences among seasons were observed. The difference of values for live sperm between fresh and thawed semen, was higher in the autumn than in winter or spring (19.6% vs. 11.6% or 10.1%, respectively; $P<0.03$). Differences among rams ($P<0.008$) were found only in fresh semen, namely for volume, concentration, individual motility, live and normal sperm (spz). Interactions of season with semen type (fresh or frozen) were found only in individual motility ($P<0.05$) where differences between autumn and winter were only evident in fresh semen.

In fresh semen, the interaction of season (autumn and winter) with rams (n=4) was significant for volume and concentration (P<0.03). Higher volume and lower concentration in winter depended each on one ram. No interactions were found for thawed semen.

Results on fertility after artificial insemination performed during spring (n=440), with semen frozen in autumn or winter from 5 males are presented. No differences between seasons were found, a 15% fertility rate (5% conf. limits: 11.7% - 18.4%) and 20.2% fecundity rate (5% conf. limits: 15.4% - 25.1%) were observed. When season of semen freezing was not considered, one of the rams presented higher fertility and fecundity rates (Table 2).

Data selected from rams with semen frozen in autumn and winter, allowed to study 3 males (902, 903 and 907) running in 2 seasons. In these three males, no specific effect of season or male for fertility and fecundity were found ($P>0.05$). No interaction of male x season of sperm freezing was found for fertility or fecundity ($P>0.05$). However, differences among these males were observed only in winter sperm freezing season, either for fertility (23.8% vs. 7.8% and 9.8%; $P<0.03$) and fecundity (33.3% vs. 12.5% and 13.7%; $P<0.03$). Male 907 was found the ram with better resistance for freezing but no correlations were found for semen traits and artificial insemination results.

Table 1. Overall results for fresh and frozen/thawed semen traits.

	Fresh semen			Thawed semen			Difference (thawed – fresh)
	N	Mean	95% C L	N	Mean	95% C L	
Volume (mL)	81	0.94	0.87-1.02				
Conc. (x 10^9 mL^{-1})	81	4.89	4.53-5.25				
Ind Motility (%)[2]	81	66.8	65.8-67.9	82	48.8	47.4-50.2	-17.2
Live spz (%)[1,2]	80	71.7	69.8-73.6	82	54.6	52.5-56.8	-15.3
Normal spz (%)[1,2]	80	84.2	83.2-85.3	82	73.6	71.8-75.4	-8.6
Head Abn. (%)[1,2]	80	7.8	7.1-8.6	82	14.4	13.3-15.6	+6.6
IP Abn (%)[1]	80	3.2	2.6-3.8	82	3.9	3.2-4.6	+0.7
Tail Abn (%)[1,2]	80	4.7	4.0-5.4	82	7.5	6.4-8.7	+2.8

[1]Data lost from 2 fresh ejaculates but included in thawed semen.
[2]Traits differing significantly between fresh and frozen semen (ANOVA; $P<0.05$).

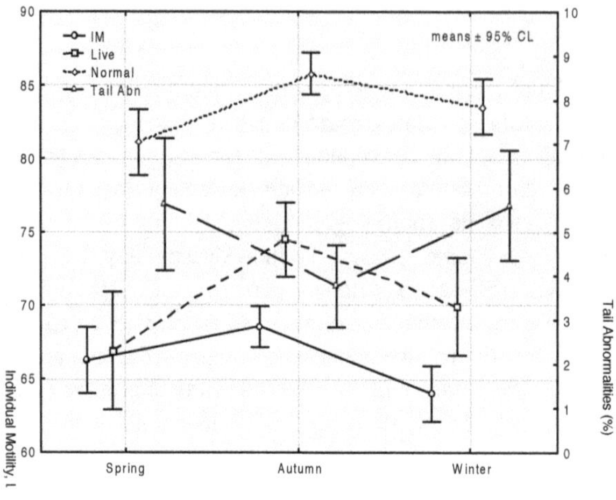

Figure 1. Fresh semen parameters with significant seasonal variations.

Table 2. Fertility and fecundity rates of ewes inseminated in spring with thawed semen collected from different males (cervical AI's).

Male id	AI (n)	Fertility (%)		Fecundity (%)	
		Mean	Std Err	Mean	Std Err
811	30	6.7[b]	4.6	10.0[b]	7.4
902	121	12.4[b]	3.0	18.2[b]	4.7
903	94	11.7[b]	3.3	13.8[b]	4.2
906	46	10.9[ab]	4.6	10.9[b]	4.6
907	149	22.2[a]	3.4	30.9[a]	5.1
Total	440	15.0	4.6	20.2	2.5

[a,b] Within columns, means with different letters are significantly different (P<0.05, ANOVA).

From a total of 1633 ovaries, 5003 good quality oocytes were collected and used for in vitro maturation (3.17±1.40 oocytes/ovary). No differences between seasons (increasing and decreasing daylight semesters) were found for the number of good quality oocytes collected. The overall oocyte maturation rate was 79.47 ± 9.6%, a higher maturation rate being observed during the decreasing daylight semester (83.5% vs. 74.61%; P<0.05). Motility of thawed sperm before swim up was higher in autumn than winter (seasons of freezing) in one male (56.3% vs. 41.3%, respectively; P=0.01). Post swim-up sperm motility was different between two males in autumn (56.3% vs. 36.3%; P<0.02). Mean sperm agglutination rate was 47.5±16.3% and post swim-up concentration was 94.0 ± 84.8 spz x 10^6 mL^{-1}. No differences in cleavage rates between males or seasons were found (overall mean: 40.9±16.5%), but D7 embryo rates were higher in the group fertilised with autumn than winter frozen semen for one of the males (38.1% vs. 20.3%; P=0.01). The mean values of all seminal and fertility traits found for each of three rams, from both in vivo and in vitro studies, were computed for correlations. Only post-thawed sperm head abnormalities and embryo cleavage rates were highly and negatively correlated (r = -0.99; P<0.05).

Discussion

For fresh semen traits, a significant variation between animals was found for volume, concentration, live and normal sperm. Between seasons, best performances on individual motility, live and normal sperm, and tail abnormalities, were found in the autumn. This is in accordance to previous results using electro ejaculation in this breed (Cruz e Silva, 1989). No differences among animals or seasons were found for seminal traits in frozen thawed semen. Saloia breed seems to be less sensitive to depressive seasonal effects on semen frozen during winter than Serra da Estrela and Merino breeds (Barbas *et al.*, 2002a). The significant difference between fresh and frozen semen found in the autumn for live sperm, was due to higher values found in fresh semen and does not traduce a real depressive effect on frozen semen. Except for intermediate piece abnormalities, all seminal traits were worst in frozen semen than in fresh semen. The semen freezing season did not influence fertility and fecundity. Although no correlations were found between seminal traits and fertility, there was a ram among five showing better results after AI indicating to be a good "male freezer". A previous work also identified a significant difference between two males in this breed used in AI concerning fertility only in the winter freezing season (Barbas *et al.*, 2004).

Oocytes collected from several sheep breeds showed better ability to mature in vitro during the shortening day length season. This was not observed with bovine oocytes (Marques *et al.*, 2002). Semen frozen in the autumn allowed higher D7 embryo rates than in winter. Although there were differences between two males for post swim up sperm motility, this was not evident for cleavage or D7 embryos. Only post-thawed sperm head abnormalities and embryo cleavage rates were highly and negatively correlated. No other correlations between in vitro and in vivo performances were found, which is in accordance to results recently published for northern European rams (Papadopoulos *et al.*, 2005).

This study shows that seminal traits, either in fresh or frozen semen, were not predictable on fertility rates achieved either in vivo or in vitro with thawed semen in the Saloia breed.

References

Baptista, M.C., C.C. Marques, M.I. Vasques, R.M. Pereira, J.P. Barbas & A.E.M Horta, 2002. Implementation of ovine in vitro fertilization technique in Portugal. In: Proceedings of the Veterinary Sciences Congress 2002, SPCV (Editor), Lisboa, Portugal, 465-466.

Barbas, J, R. Mascarenhas, A.E.M. Horta, H. Tavares, C. Cannas Serra, L. Cardigos, R. Moura & J. Cannas Simões, 2002b. Influência do diluidor, época e exploração, nos resultados de IA em ovelhas de raça Saloia. In: Proceedings of the Veterinary Sciences Congress 2002, SPCV (Editor), Lisboa, Portugal, 469.

Barbas, J.P., A.E.M. Horta, M.C. Baptista & C. Marques, 2003. Variação na capacidade fertilizante do sémen refrigerado e congelado em ovelhas de raça Saloia. In: Ponencias e Comunicaciones. IV Congresso Ibérico de Reproduccion Animal, Federación Ibérica de Reproducción Animal, AERA (Editor), Arucas, Las Palmas, España, 63.

Barbas, J.P., L. Cardigos & R. Mascarenhas, 1992. Aumento da produtividade ovina através da imunização anti-androstenediona. In. Libro de Comunicaciones. VI Jornadas Internacionales de Reproduccion Animal e Inseminação Artificial. AERA (Editor), Salamanca, Spain, 285-293.

Barbas, J.P., M.C. Baptista & A.E.M. Horta, 2002a. Comparison of two freezing methods for Merino Regional and Serra da Estrela ram semen. In: Proceedings of the Veterinary Sciences Congress 2002, SPCV (Editor), Lisboa, Portugal, 473-474.

Barbas, J.P., M.C. Baptista, C. Cannas Serra, C. Marques, S. Cavaco Gonçalves, M. Vasques & A. Horta, 2004. First results in Portugal of artificial insemination by cervical route using ram frozen semen on reproductive performance. In: Proceedings of the 15 th International Congress on Animal Reproduction, ICAR (Editor), Porto Seguro, Brazil, Abstracts Vol. 2, 378.

Chagas e Silva, J.N. & M.R. Cidadão, 1997. Superovulação em ovelhas de raça Saloia. In: Proceedings of the 1st Iberian Congress on Animal Reproduction, SPRA (Editor), Lisboa, Portugal, Vol. 2, 448-458.

Crozet, N., D. Huneau, V. De Smedt, M-C. Theron, D. Szollozi, S. Torres. & C. Sevellec, 1987. In vitro fertilisation with normal development in sheep. Gam. Res. 16: 159-170.

Cruz e Silva, J.C., 1989. Variações sazonais das características dos ejaculados obtidos por electroejaculação de ovinos Saloios. In: Livro de Resumos do IV Simposium Internacional de Reprodução Animal, SPRA (Editor), Lisboa, Portugal, 137-157.

Evans, G. & W. Maxwell, 1990. Inseminacion Artificial de Ovejas e Cabras. Acribia S.A. (Editor), Zaragoza, España, pp. 192.

Goddard, S., J. Barbas & R. Mascarenhas, 1993. The response of Merino Branco, Serra da Estrela and Saloia sheep to treatment with subcutaneous melatonin implants. In: Publicações do 5° Simpósio Internacional de Reprodução Animal, SPRA (Editor), Luso, Portugal, Vol. 2, 270.

Marques, C.C., R.M. Pereira, M.C. Baptista, M.I. Vasques & A.E.M. Horta, 2002. Seasonal variation on in vitro bovine embryo production between 1996 and 2002. In: Proceedings of the Veterinary Sciences Congress 2002, SPCV (Editor), Lisboa, Portugal, 455-456.

Papadopoulos, S., J.P. Hanrhan, A. Donovan, P. Duffy, M.P. Boland & P. Lonergan, 2005. In vitro fertilization as a predictor of fertility from cervical insemination of sheep. Theriogenology 63: 150-159.

Pombo, M., A. Oliveira, J. Chagas e Silva & J. Robalo Silva, 1993. Puberdade em borregas da raça Saloia: efeito da época de nascimento. In: Publicações do 5° Simpósio Internacional de Reprodução Animal, SPRA (Editor), Luso, Portugal, Vol. 2, 241-250.

Sobral, M., C. Antero, J. Borrego, & A. Domingos, 1987. Recursos Genéticos. Raças Autóctones. Espécies Ovina e Caprina. Direcção Geral da Pecuária (Editor). Lisboa, pp. 207.

Reproduction in the Serrana goat breed: seasonal and individual factors affecting fresh and frozen semen performance, in vivo and in vitro fertility

J.P. Barbas[1], C.C. Marques[1], M.C. Baptista[1], M.I. Vasques[1], R.M. Pereira[1], S. Cavaco-Gonçalves[1], R.M. Mascarenhas[1], N. Poulin[2], Y. Cognie[2] & A.E.M. Horta[1]

[1]Department of Animal Reproduction, Estação Zootécnica Nacional, 2005-048 Vale de Santarém, Portugal
[2]Unité de Physiologie de la Reprod. et des Comportements, INRA, 37380 Nouzilly, France

Summary

Seven bucks of the Serrana goat breed were used for semen collection with artificial vagina throughout the year. Two hundred and eighteen ejaculates of good quality were collected, washed to remove seminal plasma, refrigerated (4°C) and frozen in nitrogen vapours. Seminal traits were evaluated in fresh and frozen semen. Cervical insemination with frozen semen (n=35) or refrigerated semen (n=251) was done 43-45 hours after sponge removal and fertility and fecundity were evaluated. Frozen semen of two bucks x two seasons was tested on in vitro fertilization. In fresh semen individual differences were detected for volume, concentration, live, normal and sperm abnormalities and seasonal variation was found for volume, normal and mid piece abnormalities. Freezing decreased significantly all sperm traits except midpiece abnormalities. In frozen/thawed semen, individual variations were detected in almost all seminal traits with exception of mid piece sperm abnormalities and seasonal variations were observed for sperm abnormalities, and for Host values at 5 and 25 minutes. In goats inseminated with refrigerated semen, fertility, fecundity and prolificacy were respectively 60.2%, 106.0% and 176.1% with differences among males. Positive correlations between individual motility (r=0,82), live (r=0,99) and normal sperm (r=0,795) each with fertility were found but only significant for live sperm (P<0,05). Reproductive parameters in goats inseminated with frozen semen were respectively 27%, 56% and 208%. At 24 hours post in vitro insemination (pi), independently of season of freezing, one of the bucks showed higher cleavage rates (39.9% vs. 25,3%, respectively, P< 0.05) and produced significantly higher embryo rates at D6, D7 and D8 (P< 0.05). At 48 hours post-fertilisation, these differences among animals were no longer significant. Cleavage at 24 h (48.0% vs. 22.9%) and D6 (58.1% vs. 14.7%) rates of autumn frozen semen were also higher in the former best buck (P<0.05). Higher cleavage rate was observed in autumn than in winter frozen semen in one buck. At D7 and D8, seasonal effect on embryo rates was not significant.

Keywords: Serrana goat breed, frozen semen, seminal traits, AI, in vitro fertilization.

Introduction

The Serrana goat breed is found in the north and centre of Portugal under semi extensive conditions and is exploited for milk and cheese production. Ovarian inactivity occurs between January and the end of May, with a deeper anoestrus in February and March (Mascarenhas *et al.*, 1992). There are two breeding periods, one in May-June and the other in August-October. In goats delivering in March/April, seasonal anoestrous is confounded with lactational anoestrous till May/June. Puberty in males and females is influenced by seasonal conditions (Horta *et al.*, 1987, Baptista *et al.*, 1991). Fertility rates obtained with natural service are 89.3% (Azevedo *et al.*, 1993) and with artificial insemination using refrigerated semen varies from 47 to 63% with most values around 58-60% (Azevedo *et al.*, 1993, Mascarenhas & Barbas, 2000). No reports on artificial insemination (AI) with frozen semen are available in Serrana breed. There are individual variations in most seminal traits in fresh semen and some of them are influenced by season with higher values for live and normal sperm during the decreasing photoperiod (Sousa *et al.*, 2001). In vitro fertilization has been performed in bulls and rams in order to investigate its potential as predictor of the in vivo fertility with frozen-thawed semen (Papadopoulos *et al.*, 2005). No reports are available of in vitro fertilisation tests carried out for this purpose with buck semen from Serrana breed. The objectives of this work were to characterize seminal traits from fresh and frozen semen around the year, estimate fertility rates using AI with refrigerated and frozen semen and investigate the ability of semen frozen in autumn and winter to fertilize oocytes matured in vitro, in the Serrana breed.

Material and methods

Seven bucks from Serrana breed, were used for semen collection with artificial vagina throughout the year during two years, 2003 and 2004. Seminal traits of fresh and frozen/thawed semen were evaluated according to Evans & Maxwell (1990). Two hundred and eighteen ejaculates of good quality, individual motility (IM>65%) were collected. After evaluation, krebs-ringer-phosphate-glucose solution was used to wash semen according to Leboeuf (2001). Semen extender composition for freezing was described by Evans & Maxwell (1990). Fresh semen was diluted and packed in 0.25 mL Cassou straws (200 x 10^6 spz/straw) and refrigeration lasted 4 h. Refrigerated semen (+4°C) was frozen in a cryo-chamber before LN_2 immersion. Host test (Hypoosmotic swelling test) was performed in thawed semen (5, 25 and 40 minutes) according to Artiga (1994). If thawed semen was in good conditions, namely ≥40% IM and <20% sperm abnormalities, it was used for artificial insemination (AI). In this work fresh ejaculates were prepared to make AI with refrigerated semen (15°C) according to Chemineau *et al.* (1993). Oestrus synchronization was done with fluorogestone acetate (45 mg) vaginal sponges during 11 days according to Leboeuf (2001). AI with refrigerated semen (15°C) or with frozen semen was done respectively in 251 or 45 Serrana goats. Cervical AI with frozen semen (2 bucks) was performed 43-45 h after sponge removal using 2 straws or with refrigerated semen (8 bucks) using one straw per goat. Fertility, fecundity and prolificacy were evaluated in the inseminated goats.

Abbatoir goat ovaries in Ribatejo region are not abundant. Consequently, in vitro fertilisation of in vitro matured oocytes with autumn (n=363 oocytes) and winter (n=373) frozen semen from two Serrana bucks was performed at INRA, Nouzilly-France. Aspirated oocytes from adult French goat ovaries were maturated in TCM199 plus 100μM cysteamine and 10 ng mL^{-1} epidermal growth factor (EGF) in a 5% CO_2 incubator at 39°C for 22 h (Cognié *et al.*, 2004). Motile thawed spermatozoa were obtained by centrifugation on a Percoll density gradient. Sperm was resuspended (10^7 spz mL^{-1}) with synthetic oviductal fluid (SOF) plus 10% estrus sheep serum (ESS) and 4 μl mL^{-1} gentamicine and incubated for 30 min at 39°C in 5% CO_2. Capacitated spermatozoa (10^6 spz mL^{-1}) were co-incubated with matured oocytes in SOF plus 10% ESS for 18 h at the same conditions. After fertilisation, zygotes were transferred to SOF droplets with 10% bovine serum albumin in an atmosphere of 5% CO2, 90%N_2 and 5%O_2 at 39°C. At 48 h post insemination (pi), SOF was supplemented with 10% ESS. Cleavage at 24 and 48 h pi (cleaved embryos /inseminated oocytes), embryos at day 6 (D 6), D7 and D8 of culture (embryos /cleaved embryos) and hatched embryo rates (hatched embryos/ D7 embryos) were defined.

Results

Average and 95% confidence limits (CL) of fresh and thawed seminal parameters computed independently for seasonal and individual factors (ejaculates from 7 bucks) are presented in Table 1. Freezing decreased significantly all semen traits with exception of tail piece abnormalities. All the corresponding traits between fresh and frozen semen were significantly correlated, particularly individual motility (IM), tail and mid piece abnormalities.

Seasonal variations in fresh semen were detected for volume, normal sperm and mid piece abnormalities with better performances in the autumn. Individual variations ($P<0.03$) were observed for volume, concentration, live, normal, and all sperm abnormalities. In thawed semen all seminal traits with exception of mid piece sperm abnormalities were different among animals ($P<0.006$). In frozen/thawed semen, seasonal variations were observed for all sperm abnormalities, Host 5 and Host 25 values. In the autumn there were higher sperm head abnormalities and during spring and winter more mid piece and tail sperm abnormalities. Higher values of Host 5 and Host 25 were observed in winter and spring. The difference of means between fresh and thawed semen for normal sperm were significantly higher in the autumn than in summer, winter or spring (18.9% vs. 14.5% or 14.3 or 11.5%, respectively; $P<0.001$). Interactions of season with semen type (fresh or frozen) were found for normal sperm ($P<0.01$), head ($P<0.002$) and mid piece sperm abnormalities ($P<0.05$) with different seasonal variations between fresh and frozen semen particularly for normal sperm and head sperm abnormalities.

In fresh semen the interaction of seasons (autumn, winter, spring and summer) with bucks (n=3) was significant for IM, live sperm and tail sperm abnormalities ($P<0.03$). In frozen semen the interaction of seasons with bucks was significant for IM, live and normal sperm, mid piece abnormalities and Host 25-40 values. Fertility, fecundity and prolificacy after artificial insemination performed during spring (n=251) with refrigerated semen (15 °C) from eight Serrana bucks were significantly different among bucks. Globally these reproductive parameters were 60.2%, 106.0 and 176.1%, respectively. Higher fertility and fecundity rates were determined in three bucks. In three bucks used in AI we did not see individual variations (fresh semen) concerning IM, live and normal sperm. Positive correlations between IM (r=0.82), live (r=0.999) and normal sperm (r=0.795) with fertility were determined,

Table 1. Overall results for fresh and frozen/thawed semen traits.

	Fresh semen			Thawed semen			Difference (thawed – fresh)
	N[1]	Mean	95% C L[2]	N	Mean	95% C L	
Volume (mL)	218	1.09	1.04-1.15				
Conc. (x 10^9 mL^{-1})	214	4.04	3.85-4.23				
Ind Motility (%)[2]	218	65.1	64.4-65.8	218	38.4	37.3-39.4	-23.1
Live spz (%)[3,4]	216	72.2	70.8-73.6	218	44.5	43-46	-27.7
Normal spz (%)[3,4]	216	81.1	80.1-82.1	218	64.9	63.8-66	-16.2
Head Abn. (%)[3,4]	216	9.3	8.6-9.9	218	26.8	25.7-27.9	+20.8
IP Abn (%)[3,4]	216	5.2	4.6-5.7	218	2.4	2.0-2.8	-2.8
Tail Abn (%)[1]	216	4.4	3.9	218	5.8	5.1-6.4	+1.9

[1]Number of samples.
[2]Confidence limits.
[3]Data lost from 2 fresh ejaculates but included in thawed semen.
[4]Traits differing significantly between fresh and frozen semen (ANOVA; $P<0.05$).

with significant correlations between live sperm and fertility. Forty five goats were inseminated with frozen thawed semen from two bucks. Fertility, fecundity and prolificacy were, respectively, 27%, 56% and 208%, with no significant differences among bucks.

At 24 hours post in vitro insemination (pi), independently of the freezing season, buck 711 showed higher cleavage rates than buck 131 (39.89% ± 11.68% vs. 25.26% ± 7.12%, respectively, $P < 0.05$). The former buck achieved higher cleavage rates at the same stage when his semen was frozen in autumn than in winter (48.0% ± 10.98% vs. 31.78% ± 4.70%, respectively, $P < 0.05$). In autumn, frozen semen from buck 711 was also superior to buck 131 (cleavage rates: 48.0% ± 10.98% vs. 22.91% ± 8.01%, respectively, $P < 0.05$). At 48 hours post-fertilisation, these differences among animals and seasons were no longer significant. Buck 711 also produced significantly higher embryo rates at D6, D7 and D8 than buck 131 ($P < 0.05$), regardless the season of semen preservation. No differences between the two bucks were found for hatched embryo rates. In vitro fertilisation with autumn semen from buck 711 produced significantly higher number of embryos than autumn semen from buck 131 (58.08% ± 11.42% vs. 14.70% ± 3.81%, respectively, $P < 0.05$). At D7 and D8, seasonal effect on embryo rates was not significant.

Discussion

In fresh semen all seminal traits were different among animals with exception of IM which also varied in other studies (Sousa *et al.*, 2001). Significant seasonal variations were found for volume, normal sperm and mid piece abnormalities with higher performances in autumn followed by summer in agreement with Sousa *et al.* (2001). Positive correlations between live and normal sperm were found in this and other studies (Ferreira, 2000). There are no reports in this breed about seminal traits evaluation in frozen/thawed semen. Individual variations were observed in frozen semen for all seminal traits. In all sperm abnormalities, Host 5 and Host 25 values showed seasonal variations. Tail and mid piece abnormalities in frozen semen were higher in spring and winter while head abnormalities were higher in autumn. Higher values of Host 5 and 25 were detected in winter and spring but with fresh semen no seasonal variations were found (Ferreira, 2000). The difference between fresh and frozen semen found in the autumn for normal sperm was due to higher values found in fresh semen and does not reflect a depressive effect on frozen semen. Except for tail abnormalities, all seminal traits were poorer in frozen semen than in fresh semen which is in accordance with other authors (Donavan *et al.*, 2004). Interactions of season with semen type (fresh or frozen) were found for normal sperm, head and mid piece abnormalities. In winter fresh semen there was an increase in mid piece abnormalities and a decrease in normal sperm according with others (Azerêdo *et al.*, 2001, Gillan *et al.*, 2004). Results on fertility of inseminated goats with refrigerated semen are in accordance with previous reports (Azevedo *et al.*, 1993, Mascarenhas & Barbas, 2000) showing individual variation between bucks. In this work fertility results (27.0%) obtained in Serrana does with frozen semen are encouraging and did not show individual variations between bucks. Results of artificial insemination with refrigerated semen were higher than with frozen/thawed semen which are in accordance with others (Gillan *et al.*, 2004).

Differences found in cleavage rates between the two bucks at 24 hours post in vitro insemination, although not present at 48 hours, persisted further in culture through the number of embryos produced at D6, D7 and D8. These differences are consistent with those shown by Morris *et al.* (2003) in rams, suggesting that other intrinsic sperm factors besides motility are not excluded by the Percoll gradient or swim up techniques. With autumn frozen semen, differences between males were only evident in cleavage rates at 24 h pi and embryo rates at D6 of culture, suggesting that individual rather than seasonal factors influence in vitro embryo production.

References

Artiga, G., 1994. Teste de endosmose en ovino. In: VII Jornadas Internacionales de Reproduccion Animal. AERA (Editor), Ponencias e Comunicaciones, Sesion de andrologia, Murcia, 77-81.

Azerêdo, G., C. Esper & K. Resende, 2001. Evaluation of plasma membrane integrity of frozen-thawed goat spermatozoa with or without seminal plasma. Small Ruminant Research, 41: 257-263.

Azevedo, P., M.C. Baptista, A. Simões Nunes & R. Mascarenhas, 1993. In: 5° Simpósio Internacional de Reprodução Animal, SPRA (Editor), Luso, Portugal, Vol. 2, 231-237.

Baptista, M.C.., M. Vasques & R. Mascarenhas, 1991. Determinação da puberdade em cabritos de raça Serrana nascidos em duas épocas diferentes. In: 1 as Jornadas Internacionais de Ovinos e Caprinos, Colectânia SPOC (Editor), Évora, 103-115.

Chemineau, P., Y. Cognié, Y. Guérin, P. Orgeur & J.C.-Vallet, 1993. In: Manuel de Formation pour l´Insemination Artificielle chez les Ovins et les Caprins. Etude FAO (Editor). Production et Santé Animale, n° 83, pp. 222.

Cognié, Y, G. Baril, N. Poulin & P. Mermillod, 2003. Current status of embryo technologies in sheep and goat. Theriogenology, 59: 171-188.

Donovan, A., J. Hanrahan, E. Kummen, P. Duffy & M. Boland, 2004. Fertility in the ewe folowing cervical insemination with fresh or frozen-thawed semen at a natural or synchronized oestrus. Animal Reprod. Sci, 84: 359-368.

Evans, G. & W. Maxwell, 1990. Inseminacion Artificial de Ovejas e Cabras. Editorial Acribia, S.A. Zaragoza (España), pp. 192.

Ferreira, G., 2000. Características seminais em bodes de raça Serrana. Relatório de estágio do curso de Medicina Veterinária. Universidade do Porto, Instituto de Ciências Biomédicas Abel Salazar, 1-54.

Gillan, L., W. Maxwell, & G. Evans, 2004. Preservation and evaluation of semen for artificial insemination. Reproduction, Fertility and Development, 16: 447-454.

Horta, A.E.M., L. Ribeiro, S. Paula & I. Vasques, 1987. Study of the onset of puberty on Serrana Goats by plasma progesterona profiles-First approach. XXXVIII Annual Meeting of the EAAP (Editor), Lisbon, Vol. 2: 918.

Leboeuf, B., 2001. Insemination artificielle caprine: etat de l´art. In : III Congresso Ibérico de Reprodução Animal (Editor), Livro de Resumos, Porto, Portugal, 387-392.

Mascarenhas, R. & J. Barbas, 2000. Melhoramento da eficiência reprodutiva em caprinos de raças autóctones. Relatório Intercalar, Projecto Pamaf 3042 (INIA). Instituto Nacional de Investigação Agrária, EZN, Pamaf-IED, pp. 8.

Mascarenhas, R., M. Baptista, M. Vasques & R. Maurício, 1992. Variação da testosterona plasmática no bode em diferentes épocas do ano. O Médico Veterinário, 30: 15-19.

Morris, L.H.A., A.E. Randall, W.A. King, W.H. Johnson & B.C. Buckrell, 2003. The contribution of the male to ovine embryogenesis in an in vitro embryo production system. Animal Reprod. Sci., 75: 9-26.

Papadopoulos, S., J.P. Hanrhan, A. Donovan, P. Duffy, M.P. Boland & P. Lonergan, 2005. In vitro fertilization as a predictor of fertility from cervical insemination of sheep. Theriogenology, 63: 150-159.

Sousa, J., J. Barbas, G. Ferreira & A.E.M. Horta, 2001. Variação anual das características seminais em bodes de raça Serrana. Revista Portuguesa de Zootecnia, Associação Portuguesa de Engenheiros Zootécnicos (Editor), ISSN: 0872-7098; pp. 297-311.

The effect of ram exposure previous to progestagen oestrus synchronization on corpus luteum function and fertility in crossbred ewes

M.I. Vasques, S. Cavaco-Gonçalves, C.C. Marques, J.P. Barbas, M.C. Baptista, T.P. Cunha, & A.E.M. Horta

Departamento de Reprodução Animal, Estação Zootécnica Nacional-INIAP, 2005-048 Vale de Santarém, Portugal

Summary

Merino and crossbred ewes were used to evaluate whether previously ram exposure in late winter could improve the response to further progestagen oestrus synchronization treatment in terms of luteal function and fertilization rate.

Fifty-six adult Merino and crossbred ewes were used randomly and allocated in two groups: control (C; n= 27) and male effect (ME; n= 29). All ewes had been isolated from rams for at least 2 months. On the 20th of February and for 5 days (D0-D5), 15 males were introduced into ME group to induce male effect. Oestrous synchronization of animals from both groups began on D20 by the introduction of vaginal sponges containing 40 mg of FGA for 12 days and the administration of 500 IU of eCG on the day of sponge withdrawal (D32). Cervical AI with refrigerated semen (400 x 10^6 spz) was performed 55 h (D34) after sponge withdrawal. Progesterone levels were measured by RIA on blood samples collected twice a week for two weeks before D0, and on days 0, 3, 5, 12, 20, 27, 32, 34, 42 and 52 for ovarian activity evaluation. Blood samples were also collected each 4 hours, during 24 hours, starting 44 hours after sponge withdrawal to identify LH preovulatory surge in 5 animals of each group.

On ME group the number of cyclic ewes on D12 and D20 was significantly (P< 0.05) higher than on D0. On D12, the number of cyclic ewes on ME group was significantly (P< 0.05) higher than in C group. There were no differences between both groups for the number of ewes lambing and not lambing, as well as for the number of ewes lambing as a result of AI or natural service.

Introduction of rams enhanced the number of ewes cycling but this advantage relatively to the control group did not last until the synchronisation treatment. In spite of this stimulatory effect of rams on ovarian activity, it was not enough to improve the lambing rate achieved at the end of the essay probably because the number of acyclic animals at synchronisation treatment was higher than cyclic ewes. It was concluded that when ewes of this region are in deep anoestrus they do not respond favourably to the ram effect and no improvements on fertility are expected to occur.

Keywords: sheep, ram effect, fertility.

Introduction

Animals living in temperate latitudes have to face seasonal climatic changes in temperature and food availability. This cyclic variation in natural resources leaded to the development of seasonal reproduction of species, in order to give birth at the optimal time of year, usually spring, allowing the new-born to grow under favourable temperature and food availability conditions before the next winter (Thiéry *et al.*, 2002). Domestication has attenuated or suppressed some of the physiological expressions of seasonality, but domesticated small ruminants still have seasonal reproduction (Rosa & Bryant, 2003). The seasonality of reproduction in sheep breeds in temperate latitudes is controlled mainly by photoperiod, with ewes responding to shortening days in autumn by initiating breeding activity, and inhibiting breeding activity during lengthening days in spring. However, the dates of onset and cessation of anoestrous vary widely with breed, location and management (Notter, 2002). Breeds from intermediate latitudes, such as Mediterranean breeds, have a short anoestrous during which a proportion of ewes ovulate spontaneously.

The duration of seasonal anoestrous in the ewe may be modified by introduction of males to the flock, a process usually known as male effect, which is a suitable tool for out-of-season oestrus induction mainly because of its negligible cost. In anoestrous ewes that have been isolated from rams, introduction of males commonly induces a rise in LH pulse frequency within few minutes. This pattern of LH release causes ovulation, usually within 1 to 2 days of ram introduction. Two periods of oestrous activity, at about 19 and 24 days after ram introduction, will be detected, depending on the kind of corpus luteum formed (Martin *et al.*, 1986; Rosa & Bryant, 2002). Induction of sexual activity during anoestrous can also be achieved by hormonal means. The association of intravaginal progestagen sponges with eCG in oestrous cycle synchronization treatments is extensively applied in reproductive management of sheep flocks during both breeding and non-breeding season.

The majority of research investigating the interaction between male effect and controlled breeding programmes has focused on the influence of the ram during or after removal of the artificial progestagen. Exposure of ewes to rams on the last three days of a progestagen synchronization programme induced an increase in LH concentrations in response of pre-mating ram exposure as well as a more rapid onset of oestrous, shorter oestrous period and earlier LH surge and ovulation, compared to ewes isolated from rams prior to breeding, although a decrease in fertility was observed (Evans *et al.*, 2004; Hawken *et al.*, 2005). Exposure of ewes to rams post sponge removal can reduce variation and hasten LH surge, ovulation and the onset of oestrous (Lucidi *et al.*, 2001).

This work aimed to study whether previously ram exposure in late winter could improve the response to further progestagen oestrus synchronization treatment in terms of luteal function and fertilization rate.

Material and methods

During late winter 16 Merino and 40 crossbred ewes, which had been isolated from rams for at least 2 months, were randomly allocated in 2 groups: control (C, n= 27, 8 Merino, 5 Merino X Serra da Estrela, 6 Merino X Romanov and 8 Merino X Ille de France) and male effect group (ME, n=29.8 Merino, 6 Merino X Serra da Estrela, 6 Merino X Romanov and 9 Merino X Ille de France). On the 20th of February and for 5 days (D0-D5), 15 Merino males were introduced into ME group to induce male effect. Oestrous synchronization of animals from both groups began on D20 by the introduction of vaginal sponges containing 40 mg of FGA for 12 days and the administration of 500 IU of eCG on the day of sponge withdrawal (D32). In the two weeks before male introduction, ovarian activity was assessed by plasma progesterone (P4) quantification on blood samples collected twice a week. During the assay, corpus luteum function was assessed by measuring plasma P4 levels on samples collected on days 0, 3, 5, 12, 20, 27 32, 34, 42 and 52. Blood samples were also collected each 4 hours, during 24 hours starting 44 hours after sponge withdrawal to identify LH preovulatory surge in 5 animals of each group. Animals were submitted to cervical artificial insemination 55 hours after sponge withdrawal (D34) using refrigerated semen (400 x 10^6 spz). Rams were introduced to both flocks 4 weeks later allowing all ewes to become pregnant.

Progesterone concentrations were determined by RIA using a commercial kit (Count-a-Count Progesterone, DPC, Los Angeles, USA). Plasma LH was measured by ELISA using a commercial kit (LH Detect®, INRA, France). Mean concentrations of progesterone were compared between treatments by ANOVA. Other values were compared between groups by chi-square test (Statistica 6.0, Statsoft Inc.)

Results

There were no differences between groups C and ME for the number of cyclic and acyclic ewes on D0 and D20 (Table 1). On D12, seven days after finishing male effect, the number of cyclic ewes on ME group was significantly higher than in C group (Table 1). In ME group the number of cyclic ewes on D12 and D20 was significantly higher than on D0 (Table 1).

Plasma progesterone concentrations on D12 were higher (P=0.07) on animals from group ME than those observed on group C (Figure 1). On AI day (D34), all animals but one from group C, presented plasma progesterone concentration below 0.5 ng/ml.

There were no differences between both groups for the number of ewes lambing and not lambing, as well as for the number of lambing ewes as a result of AI or natural service (Table 2).

Table 1. Number of cyclic and acyclic ewes on D0, D12 and D20.

		Groups		
		C (n=27)	ME (n=29)	
D0	Cyclic (%)	7 (25.93)	4 (13.79)	P>0.05
	Acyclic (%)	20 (74.07)	25 (86.21)	
D12	Cyclic (%)	3 (11.11)	11 (37.93)	P>0.05
	Acyclic (%)	24 (88.89)	18 (62.04)	
D20	Cyclic (%)	6 (22.22)	12 (41.38)	P>0.05
	Acyclic (%)	21 (77.78)	17 (58.62)	
		P>0.05	P<0.05 (D0 vs. D12 and D20)	

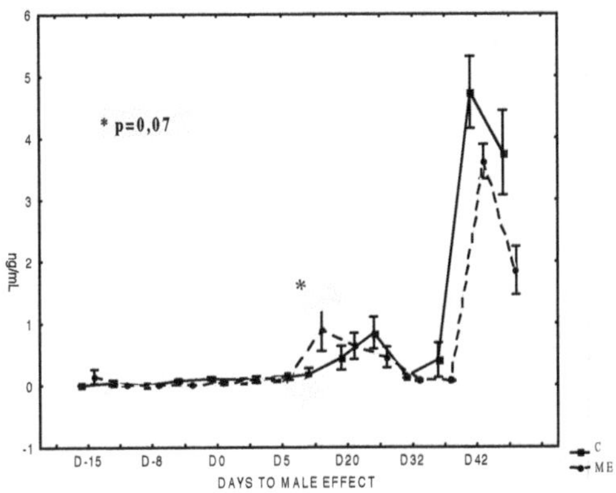

Figure 1. Plasma progesterone concentrations on animals from groups C and ME during assay (mean ± standard error).

Table 2. Number of ewes lambing (from AI and natural service) and not lambing.

Groups	Total no.	Ewes lambing		Ewes not lambing (n)
		AI (n)	Natural service (n)	
C	27	15	3	9
ME	29	11	8	10

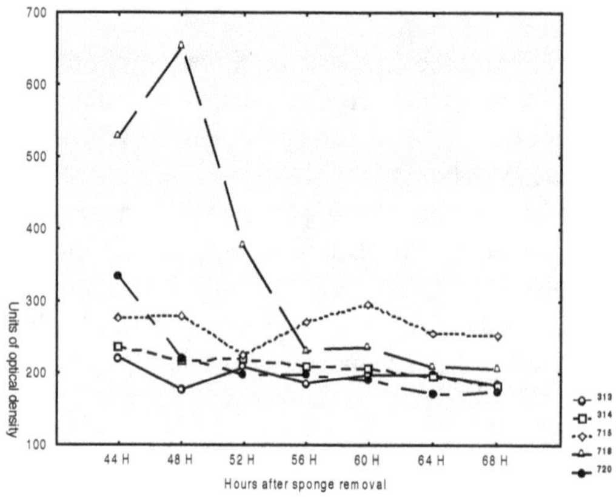

Figure 2. LH plasma levels recorded in 5 ewes belonging to C group.

The LH preovulatory surge was only detected in one ewe of each group. In the animal of group C it was detected 48 hours after sponge withdrawal (Figure 2), while in the animal of group ME it was detected just before 44 hours (Figure 3).

Discussion

Plasma levels of progesterone on D12 on animals from group ME confirm the occurrence of ovulations induced by the presence of males. LH preovulatory surge was detected only in one ewe from each

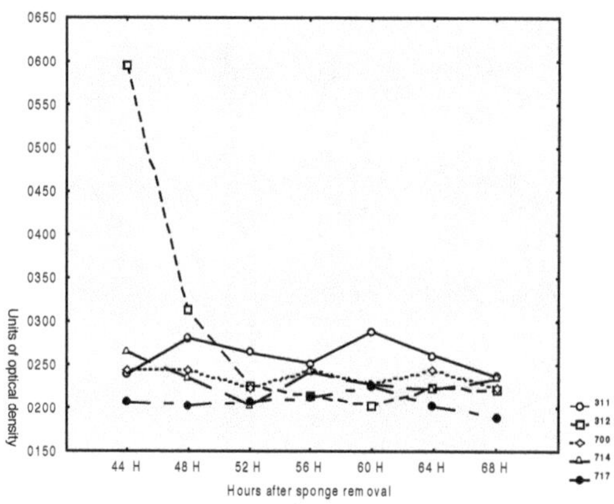

Figure 3. LH plasma levels recorded in 5 ewes belonging to ME group.

group, and even in those cases it was observed only the descendent part of the surges. This means that preovulatory surge of LH occurred before the expected time which, according to several authors (Barbas, 1999; Menegatos *et al.*, 2003) is observed 48.3 ± 2.7 or 56.5 ± 3.6 hours after sponge removal in animals submitted to a similar synchronization treatment.

Introduction of rams to Merino and crossbred ewes enhanced the number of ewes cycling, but this advantage relatively to the control group did not last until the synchronisation treatment. In spite of this stimulatory effect of rams on ovarian activity, it was not enough to improve the lambing rate achieved at the end of the essay, probably because the number of acyclic animals at synchronisation treatment was higher than cyclic ewes.

In fact, in a previous work by Horta *et al.* (2004) using a similar methodology but with a greater percentage of cyclic animals at the beginning of the assay, it was obtained a higher fertility rate in animals from ME group. Plasma progesterone concentrations 7 days after the synchronized ovulations were also significantly higher ($P<0.05$) on animals belonging to ME group, which might result from male effect inducing more competent ovulations.

It is concluded that when ewes of this region are in deep anoestrus they do not respond favourably to the ram effect and no improvements on fertility are expected to occur.

References

Barbas, J.P., 1999. Contribuição para o melhoramento da eficiência reprodutiva das raças ovinas nacionais através da imunização com líquido folicular bovino. PhD Thesis, University of Trás-os-Montes e Alto Douro, Portugal.

Evans, A. C. O., P. Duffy, T.F. Crosby, P.A.R. Hawken, M.P. Boland & A.P. Beard, 2004. Effect of ram exposure at the end of progestagen treatment on oestrus synchronisation and fertility during the breeding season in ewes. Anim. Reprod. Sci., 84: 349-358.

Hawken, P.A.R., A.P. Beard, C.M. O'Meara, P. Duffy, K.M. Quinn, T.F. Crosby, M.P. Boland & A.C.O. Evans, 2005. The effects of ram exposure during progestagen oestrus synchronisation and time of ram introduction post progestagen withdrawal on fertility in ewes. Theriogenology, 63: 860-871.

Horta A.E.M, C. Santos, J.P. Barbas, M.C. Baptista, R. Ricardo, M.I. Vasques & T. Cunha, 2004. Male effect 22 days prior oestrus synchronization enhances corpus luteum function and fertility in Merino ewes. 15th Int. Cong. Anim. Reprod., CBRA (Editor), Porto Seguro-Brazil, Abstracts Vol 2, p. 344.

Lucidi, P., B. Barboni & M. Mattioli, 2001. Ram-induced ovulation to improve artificial insemination efficiency with frozen semen in sheep. Theriogenology, 55: 1797-1805.

Martin, G.B., C.M. Oldham, Y. Cognié & D.T. Pearce, 1986. The physiological responses of anovulatory ewes to the introduction of rams: a review. Livest. Prod. Sci., 113: 219-247.

Menegatos, J., S. Chadio, T. Kalogiannis, T. Kouskoura & S. Kouimtzis, 2003. Endocrine events during the periestrous period and the subsequent estrous cycle in ewes after estrous synchronization. Theriogenology, 59: 1533-1543.

Notter, D.R., 2002. Opportunities to reduce seasonality of breeding in sheep by selection. Sheep and Goat Research Journal, 17: 21-32.

Rosa, H.J.D. & M.J. Bryant, 2002. The "ram effect" as a way of modifying the reproductive activity in the ewe. Small Ruminant Research, 45: 1-16.

Rosa, H.J.D. & M.J. Bryant, 2003. Seasonality of reproduction in sheep. Small Ruminant Research, 48: 155-171.

Thiéry, J.C., P. Chemineau, X. Hernandez, M. Migaud, & B. Malpaux, 2002. Neuroendocrine interactions and seasonality. Domestic Animal Endocrinology, 23: 87-100.

Consumer's choice of small ruminants products in relation to sustainability in Lebanon

R. El Balaa[1,2], *M. Marie*[1] *& S. Abi Saab*[2]

[1]*URAPA, ENSAIA-INPL-Nancy, B.P. 172, 54505, Vandœuvre-lès-Nancy, France*
[2]*Faculty of Agricultural Sciences, Lebanese University, Beirut, Lebanon*

Summary

In order to identify the position of Lebanese consumers towards small ruminants' local production and sustainability issues, a survey have been conducted on 250 individuals over the five districts of Lebanon (Beirut, Northern Lebanon, Mount Lebanon, Southern Lebanon and the Bekaa valley). The purchasing behaviour prefers local products (96.8 %) for their taste (75.2%), health benefits (72.8%), hygienic properties (70%), and quality (63.2%). Almost 88% of the interviewees are ready to bear extra cost for local traditional products. The most widely consumed products are Kesheck (powdery mixture of fermented milk and crushed wheat), Halloumi, Akkaoui and fermented cheeses; lamb meat is widely consumed for grilling and cooking purposes. The preferred sources for buying traditional food products are the producers themselves. The guarantee that customers get what they look for is based on trusting the source (85%) and direct purchasing from the producer (59%). The consumers are interested by labelling information like expiry date, ingredients & additives, animal origin etc.

The awareness of the Lebanese consumers concerning the sustainability status of small ruminants farming systems is covered by a series of questions treating the economical, social and environmental facets.

Keywords: consumer, food, small ruminants, sustainability.

Introduction

With 297 892 sheep and 408 933 goats, small ruminants contribute significantly to the dairy (51 493 tons) and meat (11 400 tons) production in Lebanon (Ministry of Agriculture, 2002).

Sustainable animal production systems are systems that facilitate continuing benefits from land, water and biological resources to meet human populations' current needs while preserving and improving the base of all natural resources for the generations to come (Sirados,1994). The small ruminant production in the Middle Eastern region has been experiencing difficulties and facing many obstacles during the last decade (Nordbolm & Shomo, 1995), which brought the sustainability of these systems to a risk (Steinfeld *et al.*, 1998). Studies show a clear deterioration of the agro-environmental factors in Lebanon small ruminants' production (Srour *et al.*, 2004).

Typical products of small ruminants such as dairy specialities as well as fresh and cooked meats are in most cases the result of the evolution in time of agrarian production systems and consumption traditions (Morand-Fehr *et al.*, 1998). The current situation of these products is difficult if not marginal because they are accused of belonging to an obsolete culture facing the new trends to "modern" values such as product standardization, continuity of the product's taste through the year and a sort of aversion to the more marked typical tastes (Boyazoglu & Morand-Fehr, 2001).

It would be insufficient and even imprecise to try to find a solution for the sustainability problem without taking into consideration the interest of the consumers and their preference for a safe product, sound environmental practices and animal friendly production system (Boehlje *et al.*, 1995). There is a great need of marketing research to identify future trends of small ruminants sector in Lebanon (Hamadeh *et al.*, 1996). This survey will contribute to analysing the Lebanese consumers purchasing behaviour of local typical products and their awareness of the sustainability problems facing local small ruminant production systems.

Materials and methods

The survey covered a sample of 250 individuals from all five districts of Lebanon. Fifty individuals were selected from Beirut, the capital; for each of the other four districts (Northern Lebanon, Mount Lebanon, Southern Lebanon and the Bekaa Valley), 30 individuals were chosen from their respective major cities (Tripoli, Jounieh, Saida and Zahle) representing the urban lifestyle and 20 individuals from three different villages representing the rural lifestyle. In the cities, 5 sectors (according to the administrative distribution) were chosen in a draw, and individuals were equally distributed in these sectors (the first floor of every 10th right hand building of the main road). In villages, the same sampling plan was used while considering it as one administrative sector.

The questionnaire was divided into three main parts: The first covered the purchasing behaviour (frequency of consumption and purchasing source) of 31 small ruminant products (18 milk products and 13 meat products). This section also included 13 different questions covering the place and duration of residence, local and imported products preference, significance and source of purchasing of Baladi (native) products and variation of taste across the generations. The second part covered the conception of sustainability problems; it included 31 questions covering the contact with the rural environment, the knowledge of economic, social and environmental sustainability of small ruminant production systems, preferred labelling information, quality guarantee and willingness to pay extra cost for traditional products from sustainable production systems. The final part covered information about the household and contained 6 questions: age, profession and religion of every household member, educational level of the household decision makers, monthly revenue and percentage of monthly revenue spent on food.

The interviews were performed in a 20-25 minutes period each, and the results were analysed using the GenStat 8th edition program.

Results

Only 23.5% of urban areas residents have previously lived in rural areas for at least one year; however, most of the urban population (61.2%) still have a direct contact with the rural areas: 36.5% stay in rural areas during the summer, 12.9% during the week ends, and 10.6 % on a monthly basis.

About 96.8 % of the interviewees preferred local small ruminants' products, the main reasons being their taste (75.2%), health benefits (72.8%), hygienic properties (70%), and quality (63.2%), while the least considered reasons are price (9.5%), brand name (4.5%) and prestige (3.3%). Most of the interviewees (86%) are willing to pay extra cost for good quality small ruminant local products; 29.3 % of them are ready to pay 20% or more and 12.6% are willing to pay more than 25%.

In general, traditional local products are bought directly from the producers (64.8%), followed by local stores (30.8%) and hyper-markets (23.6%), which are both preferred in urban areas.

The dairy questions covered 18 local products; for most of them, they are mainly consumed once or more per week (43.6%) and purchased directly from the producers (43.2%). They are appreciated in rural as much as in urban areas. They are mostly favoured in the Bekaa region (28.2% of the responses). The preferred products are Kesheck (powdery mixture of fermented milk and crushed wheat) and Halloumi cheese with 72.4% each, followed by Double Cream cheese (71.2%), Akkaoui cheese (66%) and Chanklish (fermented cheese) (46%). The Kesheck's frequency of consumption is at least once a week in 70.7% of the cases; it is generally (82.4%) purchased directly from the producers. Halloumi, Akaoui and Double Cream cheese are consumed mostly once or more per week (50.3%, 50.3 and 54.5%). They are mostly consumed in Beirut and in Southern Lebanon, and purchased from hyper-markets (44.6%, 44% and 42.8%) or local stores (40.2%, 40.4% and 43.3%). Chanklish is consumed occasionally (40%) and purchased directly from the producers.

The meat group of questions covered 13 local products; the mean frequency of consumption of meat products is at least once a week (41.2%) or a month (77.9%). Most of them are purchased from local butchers (93.6%); they are preferred in rural area and in the Bekaa region. The most widely consumed meat products are grilled sheep meat (72%), sausages (66.8%) and cooked sheep meat (48%). Grilled sheep meat is consumed at least once per week (56.1%) and is preferred in the rural area; however, sausages are mostly consumed in urban areas. Sheep or lamb meat is used for cooking once or more per week (64.2%) with no difference between the urban and rural areas.

The guarantee that customers get what they look for is based mostly on trusting the source (85%), direct purchasing from the producer (59%), and geographical sources (44%); the least used guarantees are external appearance (28%), brand name, and cost (18%). The brand name is mostly considered in Beirut (42%), the cost in Southern Lebanon (28%) and Northern Lebanon (26%), whereas the external appearance is preferred in Southern Lebanon.

As shown in table 1, the best rated element is the expiry date with 94.4% of the highest notation (5), followed by the ingredients and additives (71.7%), and the animal source (69.8%). The elements that were the most rated with the lowest notation (1) are the brand name with 38.3%, calorific value with 30.2% and the weight with 25.0%.

Around 75.4% of the individuals agree that the small ruminants' producers are in a good financial condition, while 20% think that the financial situation of the producers is bad (77% of those live in urban areas). The social situation of the producers is considered as bad by over 46.2% of the individuals.

Table 1. Percentage of notation for each labelling element (1: less valued;5: most valued).

	1	2	3	4	5
Brand name	38.3	0.8	6.9	4.0	50.0
Expiry date	3.2	0.4	1.6	0.4	94.4
Production origin	17.3	3.6	12.9	3.2	62.9
Ingredients & additives	8.1	2.0	8.9	9.3	71.7
Animal source	8.9	4.0	12.9	4.4	69.8
Calorific value	30.2	4.0	13.3	4.4	48.0
Production centre	19.0	3.6	15.0	4.0	58.3
Weight	25.0	7.3	8.9	4.0	54.8

Table 2. Consumers' evaluation of small ruminants impact on environmental factors.

	Positive	Negative	No Impact
Soil quality	13.0%	24.3%	62.8%
Erosion	67.9%	7.6%	24.5%
Water pollution	1.2%	44.8%	53.9%
Vegetative cover	16.9%	47.0%	36.0%
Biodiversity	13.7%	41.0%	45.3%
Air pollution	1.2%	21.8%	77.0%
Global Warming	2.2%	15.9%	81.9%

Most of the consumers believe that small ruminants production systems have no impact on soil quality (62.8%, table 2), water pollution (53.9%), biodiversity (45.3%), air pollution (77%) or global warming (81.9%). On the other hand, 67.9% believe that small ruminants have a positive effect against erosion and 47% think that they have a negative influence on the vegetative cover. The results show that the consumers have insufficiently clear information about the environmental impact of small ruminant production systems.

Discussion

The results show that the traditional products are preferred mainly because of their good taste, which is also shared by consumers in France, for example, where traditionally-made products have an image of being tasty (Siret and Issanchou, 2000). In fact, Grunert (2002) noticed that consumers use colour of meat to infer tenderness, or consistency of yoghurt to infer taste.

The survey shows that 86% of the individuals are willing to pay more money for products from sustainable production systems. In a similar survey on sustainability-labelled wood (Hansmann *et al.*, 2004) the same trend was shown; however, the author concluded that these are only declarations of intent, which are not necessarily redeemed in real purchasing behaviour. In fact, Funck (1997) showed that the most prominent barrier preventing consumers from ecologically positive consumption behaviour are high prices.

Lebanese consumers give a great importance to certain label information such as the origin of production and the expiry date, which is in accordance with the results of a recent study that was carried in Europe (Bernues *et al.*, 2003). Trusting the source of purchasing and direct purchasing from producers are the two most important guarantees that consumers look for in this study. This highlights the importance of positive experience allowing the purchasing source to gain the consumers confidence. It would be interesting to turn this trusting relationship between the consumer and the purchasing source into a relationship between the consumer and the brand which could allow the consumers to draw on their previous experience with the product (Grunert, 2002).

In a survey conducted in Canada, Teng *et al.* (2004) noted that the main reason why consumers did not purchase cheese from farmer markets was the distance from home, whereas in our case, 64.8% of the sample individuals preferred direct purchasing from the producer. This is mainly due to the fact that Lebanon is a small country (10 452 km^2) with a very diversified topography and with mountains very close to the sea shore.

Conclusion

The Lebanese consumers' notion of sustainability is questionable and sometimes surprising, for example, the positive effect of small ruminants against erosion; this highlights the need for further societal communication and education on this subject.

The consumers showed an attachment to the traditional products and a deep knowledge of these products, which should lead to a standardisation process bringing more information, especially through labelling schemes based on elements contributing to the sustainability of the food chain.

References

Bernués, A., Olaizola, A. & Corcoran, K., 2003. Labeling information demanded by European consumers and relationships with purchasing motives, quality and safety of meat. Meat Science. 65: 1095-1106.

Boehlje, M., Akridge, J., & Downey, D., 1995. Restructuring agribusiness for the 21st century. Agribusiness. 11: 6.

Boyazoglu, J. & Morand-Fehr, P., 2001. Mediterranean dairy sheep and goat products and their quality: A critical review. Small Ruminant Research 40: 1-11.

Funck, D., 1997. Ökologische Eigenmarken im Handel. In: Bruhn, M. (Editor), Handelsmarken. Entwicklungstendenzen und Zukunftsperspektiven der Handelsmarkenpolitik, (2nd Edition). Schäffer-Poeschl Verlag, Stuttgart, pp. 199-214.

Grunert, K.G., 2002. Current issues in the understanding of consumer food choice. Trends in Food Science & Technology, 13: 275-285.

Hamadeh, S.K., Shomo, F., Nordbolm, T., Goodchild, A. & Gintzburger, G., 1996. Small ruminant production in Lebanon's Beka'a Valley. Small Ruminant Research. 21: 173-180.

Hansmann, R., Koellner, T. & Scholz, R., 2004. Influence of consumers' socioecological and economic orientations on preferences for wood products with sustainability labels. Forest Policy and Economics. (In press).

Ministry of Agriculture in Lebanon, 2002. Available on http://www.agriculture.gov.lb/production99/anani02htm. Date of consultation [15.06.2005]

Morand-Fehr, P., Rubino, R., Boyazoglu, J., & Le Jaoven, J.P., 1998. Reflexion sur l'historie, la situation actuelle et l'evolution des produits animaux typiques. In: Basis of the quality of typical Mediterranean animal products. EAAP Publication no. 90, Wageningen Press, Netherlands, pp. 17-29.

Nordbolm, T.L. & Shomo, F., 1995. Food and feed prospects to 2020 in the West Asia/North Africa region. ICARDA 2 Social Science Papers, Aleppo, Syria, pp. 55.

Sirados,G., 1994. Sustainable agriculture and agricultural development. Medit. 5, 13.

Siret, F. & Issanchou, S., 2000. Traditional process: influence on sensory properties and on consumers' expectation and liking. Application to 'pâté de campagne'. Food Quality and Preferences. 11: 217-228.

Srour, G., Marie, M. & Abi Saab, S., 2004. Agro-environmental sustainability of small ruminant production in Lebanon. Proceedings of the 55th EAAP Annual Meeting, 5-9 September 2004, Bled, Slovenia.

Steinfeld, H., Haan, C.D. & Blackburn, H., 1998. Livestock Environment Interactions, Issues and Options, WRENmedia, UK pp. 56.

Teng, D., Wilcock, A., & Aung, M., 2004. Cheese quality at farmers markets: observation of vendor practices and survey of consumer perception. Food Control. 15: 579-587.

The economic incidence of adopting a new feeding system in Aragonese sheep farms

A. M. Olaizola Tolosana, T. Chertouh & E. M. Persiva

Departamento de Agricultura y Economía Agraria, Universidad de Zaragoza, Miguel Servet 177, 50013 Zaragoza, Spain

Summary

Innovation in products, techniques and processes has been relatively limited in Spanish farming sheep systems, especially if compared with other species. Innovation in traditional feeding systems consists of replacing conventional feeds with alternative ones and possibly, the modification of distribution and animal access systems. The labour-feeding interrelation has been considered one of the basic problems faced by meat sheep production, not just in Spain, essentially due to the high costs and the social problems associated with them. Hence, the rationalization of feeding has become a key element in the profitability/survival of farms.

The aim of this paper was to evaluate the economic-financial viability of adopting a new, self-service, complete diet feeding system on sheep farms.

The information used was obtained from a sample of 101 meat sheep farms integrated in the Technical-Economic Management Network of a co-operative company. A typology of farms that have not adopted the new feeding system was established using structural indicators (herd size, labour availability), feeding costs (particularly grazing costs) and reproduction results, by means of multivariate analysis. A financial analysis has been carried out on the adoption of the new feeding system, in farm groups and under three different hypotheses. The greatest economic profitability is, in general, obtained in all of the groups of farms if the implementation of the new feeding system reduces labour needs and increases the technical results obtained by the farms. Hypothesis A (reduction in labour needs) is that which generates least profits and profitability in the groups of farms established and, in some groups, the investment is not recommended in these conditions. For this reason if the implementation of the new feeding system only implies a reduction in labour needs, i.e. if it does not imply an improvement in the technical results obtained by farm, it is not recommended as an investment for most of the groups of farms considered.

Keywords: financial analysis, innovation, multivariate analysis, sheep farming systems.

Introduction

In the Spanish sheep production sector in general, innovation in products, techniques and processes (genetics, reproduction, feeding, differentiated quality, etc.) has been relatively limited, especially if compared with other species. This fact, together with the technical and structural conditions of the farms could explain the reduced increase in labour productivity that is a feature of this sector (Manrique *et al.*, 2003).

Innovation in traditional feeding systems consists of replacing conventional feeds with alternative ones and possibly, the modification of distribution and animal access systems. Complete diet feeding systems for ruminants have generally focused on simplifying feeding procedures bearing in mind aspects that affect efficient feeding (Owen 1980). In addition, it has been pointed out that the application of a self-service, complete diet feeding system on sheep farms, allows labour costs to be reduced as well as lessening the social implications of work (Sierra, 1990, 1996). Numerous authors highlight the labour-feeding interrelation as one of the basic problems faced by meat sheep production, not just in Spain, essentially due to the high costs (grazing, distribution of feeds, etc.) and the social problems associated with them. Hence, the rationalization of feeding has become a key element in the profitability/survival of farms due to its impact on the farmer's quality of life and its incidence both in the farm's economic results and the homogeneity/quality of its products.

Over the past few years the use of complete diet feeding systems has grown to a certain degree amongst meat sheep farms in the region of Aragon, although these systems are mostly used for certain periods throughout the year or at certain stages of production. As has occurred with other technical innovations in the sector (control of reproduction, insemination, genetics, etc.) the diffusion of this innovation is still far from generalised. The speed and amplitude with which an innovation is diffused has highly diverse origins, from the level of information that farmers receive, its interest in economic terms or its objective advantages, not forgetting the specific circumstances of the farms and the subjectivity of farmers.

Within this framework, the aim of this paper was to evaluate the economic-financial viability of adopting a new, self-service, complete diet feeding system (SSCDF) on sheep farms. The study has not therefore considered costs and intangible benefits for the farms, the environment or other aspects.

Material and methods

The information used was obtained from a sample of 101 meat sheep farms located in the Autonomous Community of Aragon integrated into the Technical-Economic Management Network of a co-operative company. A typology of the 79 farms that have not adopted the new feeding system was established using structural indicators (herd size, labour availability), feeding costs (particularly grazing costs) and reproduction results (mean values of the 1999-2001 campaigns). By means of a Principal Component Analysis (PCA) and a Cluster Analysis, four groups of farms have been established plus one farm that constitutes a single group which has not been commented on. Considering the four groups obtained in the typology, a financial study has been carried out on the adoption of the new feeding system, considering three different hypotheses. The hypothesis A supposes the reduction in labour needs, hypothesis B the reduction in labour needs and improvement in technical results and hypothesis C the increase in flock size and improvement in technical results. In addition, in order to establish some of the parameters of the financial evaluation, we have used information obtained by means of a direct survey carried out on a sample of 23 sheep farms (forming part of the management network) that have been using the new feeding system for some years. The criteria employed in the financial analysis have been those of Net Present Value (NPV) and the Internal Rate of Return (IRR). The NPV of an investment is defined as the difference between the sum of the present value of its cash flows and the initial investment. The discount rate considered is 4.25% and the length of the project's economic life is12 years.

The expression of this criterion is:

$$NPV = -A + \frac{C_1}{(1+k)} + \frac{C_2}{(1+k)^2} + ... + \frac{C_n}{(1+k)^n}$$

Where:

A: The initial investment

C_n: Cash flows generated by the investment over its lifetime (difference between benefits and costs each year)

k: discount rate

The IRR is defined as the discount rate that makes the investment have a zero NPV, i.e. it is the value of *r* such that:

$$0 = -A + \frac{C_1}{(1+r)} + \frac{C_2}{(1+r)^2} + ... + \frac{C_n}{(1+r)^n}$$

Results and discussion

The groups of farms that do not use the new SSCDF system, obtained by means of a multivariate analysis, are:

- *Group 1. Farms with small flocks and high grazing costs on own land.* This group is formed by 23 farms that have the lowest average availabilities of labour and flocks, without the presence of salaried labour (Table 1). At the same time, these farms have the lowest costs of forage rentals per sheep whilst the cost per sheep for their own pastures is the highest of all of the groups.
- *Group 2. Large flock size and high fecundity.* These are farms with a large average flock size and high fecundity rate in relation to the other groups. They have high grazing costs per sheep on own pastures, own feeding and rented forage for grazing.
- *Group 3. The group with the highest costs per sheep in terms of own feeding.* This group comprises those farms with the highest average costs for own feeding per head, but this is not the case of grazing on own pastures. These farms thus have crops on their Utilised Agricultural Area (UAA), which are basically cereal crops destined for feeding flocks, but not grazing areas.
- *Group 4: Large flock size and low fecundity.* This group is formed by eleven farms with a large average flock size but lower rate of fecundity than the other groups established. Total labour availability (AWU) is high, as is the presence of salaried labour. The costs of forage rental per head are very high, whilst the cost of own feeding per sheep and the cost of grazing on own land are lower than in the other groups.

A financial analysis has been carried out on these four types of farms to evaluate, in economic terms, the acquisition of the new SSCDF in each of the groups. The adopting of the new feeding system in general involves an initial investment by the farms that consists of the cost of acquiring the system and the adaptation of a farm building to install it. Over the past few years grants have been available to purchase the new system so that the obtaining of said grants has been considered. In some of the hypotheses the increase in flock size with implementation of the new system has been considered, in which case the acquisition of these animals is included as an initial investment (Table 1).

Table 1. Average characteristics of farm groups that do not use the new SSCDF system.

	Group 1	Group 2	Group 3	Group 4
Number of farms	23	17	27	11
Sheep	309	768	541	1 186
Grazing costs on own land/ewe (€)	3.98	2.00	1.74	1.41
Own feeding costs/ewe (€)	10.24	9.21	11.56	6.98
Anual Working Unit (AWU)	0.85	1.65	1.35	2.22
Fecundity (Lambs born/ewe and year)	1.45	1.57	1.51	1.44
Rented grazing costs/ewe (€)	2.82	5.46	3.60	6.41
Unpaid Labour input (AWU)	0.85	1.5	1.17	1.66
Paid Labour input (AWU)	0	0.15	0.18	0.56
Cost of salaried labour (€)	0	2 152	1 939	6 942
Total feeding costs/ewe (€)	23.60	24.57	26.46	20.99
Sanitary costs/ewe (€)	2.37	3.06	2.76	2.71
Reproduction costs/ewe (€)	0.18	0.23	0.19	0.21
Lambs sold/ ewe and year	1.10	1.19	1.10	1.10

- *Hypothesis A: Reduction in labour needs.* It has been considered that the implementation of the new feeding system allows savings in labour needs on the farm, whilst maintaining the same flock size (Table 2). This saving means that there is less use of salaried labour and consequently a saving of the cost of this labour in some groups of farms as well as reducing the need for family labour. With regard to family labour, an opportunity cost equivalent to the minimum inter-professional salary (6 447€/year: Ministry of Employment and Social Affairs, 2004) has been considered. To establish the reduction in labour needs brought about by the new system, it has been estimated that an AWU can handle 600 heads (Olivan *et al.*, 2002). It has also been considered that the acquisition of this new system represents a reduction in the cost of forage rentals and an increase in the costs of flock feeding. To estimate the reduction in forage rental costs, the difference between the average cost of the group of farms that use the new SSCDF system (3.23 €/sheep) and the average cost of each farm group established in the typology has been calculated and the same procedure has been used to estimate the increase that arises in the cost of feeding per head.
- *Hypothesis B: Reduction in labour needs and improvement in technical results.* In this hypothesis, as in the previous one, not only have we considered that adopting the new feeding system reduces labour needs, but, also that there is an improvement in the farms' technical results, i.e. an increase in the number of lambs sold; also considering the increase in sanitary and reproduction costs that this implies (Olivan *et al.*, 2002) and an increase in the cost of feeding lambs due to the increase in lambs sold.
- *Hypothesis C: Increase in flock size and improvement in technical results.* This hypothesis considers that the SSCDF system allows an increase in flock size in the groups established. The increase in flock size for each group has been estimated according to the average flock size and total labour availability (AWU) and taking into account the same criterion as in the previous hypotheses.

The economic profitability indicators obtained in hypothesis A, indicate that in the case of groups 1 and 4 the acquisition of the new feeding system would not be feasible in the conditions established. In hypothesis B, the acquisition of the SSCDF system generates high profits in all of the farm groups

Table 2. Principal assumptions considered in three hypotheses for each farm groups.

	Group 1	Group 2	Group 3	Group 4
Hypothesis A				
Reduction Paid Labour (AWU)	0	0.15	0.18	0.56
Reduction Family Labour (AWU)	0.35	0.25	0.25	0
Opportunity cost family labour (€)	2 256	1 612	1 612	0
Increase in feeding costs/ewe (€)	6.61	5.64	3.76	9.22
Hypothesis B				
Increase in lambs sold/ewe	0.5	0.3	0.4	0.2
Increase in sanitary costs/ewe (€)	0.52	0.49	0.52	0.35
Increase in reproduction costs/ewe (€)	0.36	0.24	0.30	0.12
Reduction AWU as hypothesis A				
Hypothesis C				
Increase flock size	201	222	269	146
Increase in lambs sold/ewe	0.4	0.25	0.3	0.2
Increase in sanitary costs/ewe (€)	1.46	1.18	1.44	0.65
Increase reproduction costs/ewe (€)	0.43	0.29	0.36	0.14

established. The highest IRR is obtained in the groups with the highest increase in lambs sold and the lowest for the groups with the lowest increase in lambs sold. The results obtained in hypothesis C indicate that the acquisition of the new SSCDF system in the established conditions would be profitable for all of the groups of farms, except for group 4. If the results obtained in the different hypotheses are compared, it can be seen that the acquisition of the new feeding system in the conditions of hypothesis B, i.e., a reduction in labour needs with an improvement in technical results, is profitable for all of the types of farm groups established. The greatest economic profitability is, in general, realized in all of the groups of farms if the implementation of the SSCDF system reduces labour needs and increases the technical results obtained by the farms. In the case of group 1, hypothesis C generates greater profitability than hypothesis B, which could indicate that the acquisition of the new feeding system requires a minimum farm size, as this is the group with the smallest flock size of all of the groups established. Hypothesis A is that which generates least profits and profitability in the groups of farms established and, in other cases, the investment is not recommended in these conditions

Table 3. The economic profitability indicators obtained in the hypotheses.

	Group 1	Group 2	Group 3	Group 4
Hypothesis A				
Net Present Value (€)	-2 191	391	9 739	-21 334
IRR (%)	-	5	25	-
Hypothesis B				
Net Present Value (€)	64 040	104 510	103 617	81 574
IRR (%)	197	131	180	67
Hypothesis C				
Net Present Value (€)	76 543	86 571	97 219	-2 826
IRR (%)	54	46	50	2

(Table 3). For this reason if the implementation of the new feeding system only implies a reduction in labour needs, i.e. if it does not imply an improvement in the technical results obtained by farm, it is not recommended as an investment for most of the groups of farms considered.

References

Manrique, E., A. Olaizola & T. Chertouh, 2003. Costes de pastoreo y de alimentación forrajera en explotaciones ovinas en proceso de adopción de innovaciones. In: Pastos y ganadería extensiva, García Giado *et al.* (Editors). Actas de XLIV Reunión Científica de la Sociedad Española para el Estudio de los Pastos (SEEP): 691-696.

Oliván, A., J. Yañiz, F. Muñoz & J. Donato, 2002. Nuevo sistema de alimentación a libre disposición para ganado ovino: COME–RUM. Institución Fernando El Católico, Zaragoza, Spain, pp. 84.

Owen, J.B., 1980. Sistemas de alimentación integral para vacuno y ovino. Mundi-Prensa. Madrid, Madrid, Spain,pp. 174.

Sierra, I., 1990. Problemática de la alimentación-mano de obra en la explotación ovina; nuevo sistema de alimentación con "concentrado" *ad libitum*. Actas de las XV Jornadas de la Sociedad Española de Ovino y Caprino. Córdoba, Spain, 43-53.

Sierra, I., 1996. Nueva alternativa en alimentación ovina. II Ración completa, granulada y *ad libitum*. Archivos de Zootecnia, 169: 51-63.

Relationships between the variability of the global quality of cow's milk in Morocco and milking conditions

M.T. Sraïri[1], O. Nebia[2], A. Hamama[3], S. Messad[2] & B. Faye[2]

[1]Hassan II Agronomy and Veterinary Medicine Institute, Department of Animal Production, P.O. Box 6202, Madinate Al Irfane, 10101, Rabat, Morocco
[2]International Agronomy Research Cooperation Center for Development (CIRAD), Animal Production Program, Campus international de Baillarguet, TA30/A, 34398 Montpellier Cedex 5, France
[3]Hassan II Agronomy and Veterinary Medicine Institute, Department of Hygiene and Industry of Animal Feedstuffs, P.O. Box 6202, Madinate Al Irfane, 10101, Rabat, Morocco

Summary

The evaluation of the quality of bulk milk obtained from 76 dairy farms in Morocco has shown a great variability of its physical (pH and density) and chemical (fat content, proteins and urea) parameters. In numerous cases, the value of these parameters has revealed poor rearing practices and even frauds (watering, skimming). The evaluation has also indicated that milk hygienic criteria used in this study (aerobic plate count, total and fæcal coliforms) were even more variable. A follow-up study of farms and their housing conditions has allowed differentiating four types of milking conditions: "mechanical and clean", "manual and clean", "manual and relatively clean", and "manual and dirty". These different types of practices were found to have an impact on the hygienic quality of milk, particularly on the fæcal coliforms count, which increased significantly from « mechanical and clean » to « manual and dirty » milking conditions. These results suggest a closer monitoring of the milking conditions as means of classifying milks and the interest of using milk pH, as an easy and cheap indicator of the milk hygienic quality because of its high correlation with the hygienic criteria.

Keywords: milk, quality, milking conditions, hygiene, Morocco.

Introduction

In Morocco, dairy production has been intensively encouraged because of an important demand due to demographic growth and changing nutritional habits. This is why a "National Dairy Plan" has been launched in the early seventies. Thirty years later, an increase in the annual milk production has been noticed, even at the scale of smallholders, due to all the state incentives provided to stockmen: dairy heifers imports, forages irrigation, artificial insemination ... (Sraïri, 2004). Meanwhile, very few studies have tried to characterise milk global quality (physical, chemical and hygienic) in farms and its relationship with milking conditions, as milk is mainly paid in relation to the quantities delivered and to basic quality criteria (milk fat and milk density). Recent studies in Morocco show that milk quality is very variable due to huge fluctuations in milk hygienic parameters (Sraïri *et al.*, 2005). Milk hygienic quality is getting more and more interest in other Southern countries of the Mediterranean (Tunisia for instance), as it represents the limiting factor for a sustainable dairy production (Djemali and Kayouli, 2003).

This article deals, at farm level, with the relationship between milk global quality and cattle milking and rearing conditions in Morocco.

Material and methods

A survey dealing with milk quality and milking conditions and rearing practices was conducted in the suburban belt of Rabat city. Seventy six farms reflecting the diversity of dairy cattle systems were enquired. In each farm, the level of milk production per cow, the type of milking (manual vs. mechanical), milking conditions (teat dipping, use of antibacterial solution ...) and rearing practices (type of forages and of concentrates) were looked at.

A sample of bulk milk from all milked cows was taken from each farm. It was then divided in two distinct fractions:

- 500 ml for physical and chemical analysis (pH, temperature, density, milk fat, milk proteins, milk urea);
- 90 ml taken in a sterile bottle and used for hygienic analyses, which relied on aerobic plate count (IDF, 1987) and total and fæcal coliforms (IDF, 1974).

Milk analyses were made at the laboratory. Reference methods were used to determine milk fat (Gerber method) and milk proteins (Kjeldahl method) (AFNOR, 1977). Milk urea was determined by a colorimetric method using dimethyl amino benzaldehyd (DMAB).

Statistical analyses were achieved to describe milk quality and its variability. In a second step, statistical multivariate tools such as Multiple Factorial Analyses (MFA) and Cluster Analyses (CA) were used to design a typology of milk quality and a typology of milking and rearing conditions. The relationships between those two typologies were determined by the use of a co-inertia method, as presented by Faye and Grelet (1990). R statistical package was used (R-Project, 2004).

Results and discussion

Primary results show a wide variability in milk productivity per cow, coming from a minimal value of 2.7 kg to a maximal value above 25 kg. This is in agreement with general trends of dairy production in Morocco, which show a huge diversity in production system i.e. milk and meat, milk without concentrates, milk with important amounts of concentrates. etc. (Sraïri, 2004). The genetic type of herds was very similar as 92% of cows were of Friesian or Holstein breeds and only 8% were crosses between Holstein or Friesian with local strains.

Milking was manual in 84% of cases and it is made twice daily in 59% of farms. In fact, some farms only milk cows once a day, even if they have pure Holstein cows, as they consider that the afternoon production should only for the calf.

Table 1 shows the values of milk physical, chemical and hygienic parameters. It is possible to notice a wide variability in milk quality parameters which is even more important for hygienic parameters. Those hygienic parameters were all far from international standards (Plusquellec, 1991). This confirms the poor hygienic conditions at farm level. However physical and chemical parameters were closer to milk standards. But milk fat and milk urea were by far the most variable due to the fact that those parameters are the most affected by feed variations: concentrates/forage ratio for milk fat (Labarre, 1994) and degradable/total nitrogen ratio for milk urea (Whitaker *et al.*, 1995).

By MFA treatment, milk quality variability was assessed (Figure 1). The first axis is related to hygienic quality (total and fæcal coliforms), whereas the second one is correlated to physical and chemical (pH, milk fat, milk proteins). At the exception of pH, all the physical and chemical variables

Table 1. Physical, chemical and hygienic parameters of the 76 milk samples (Sraïri, 2004).

Parameters	Minimum	Means±SD	Maximum
pH	6.28	6.60±0.13	6.87
Temperature (°C)	11.5	21.05±4.52	33.0
Density at 20°C	1.0219	1.0278±0.002	1.0336
Milk fat (g/kg)	21.5	37.53±8.46	63.5
Milk proteins (g/kg)	22.1	31.73±4.49	46.5
Milk urea (mg/100 ml)	165	354.5±106.5	690
Aerobic plate count (10^7 cfu/ml)	5.6×10^{-3}	5.9±30.2	260
Total coliforms (10^6 cfu/ml)	10^{-5}	1.1±3.1	25.6
Fæcal coliforms (10^4 cfu/ml)	0	5.2±31.0	2.52×10^3

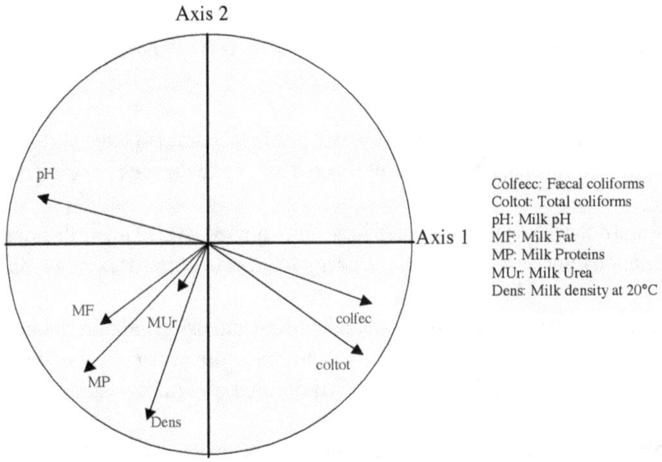

Figure 1. MFA graphical results: milk variables projection on the factorial plan.

were independent from hygienic ones, which indicates the interest to collect all those data to characterise milk quality more precisely. On another hand, milk quality is mainly affected by hygienic parameters, as they represented 65% of the weight of the MFA variability.

While studying milking and rearing conditions, discrimination between dairy intensification variables (use of concentrates, level of productivity per cow, type of milking ...) and hygienic variables was made. Besides, three types of variables related to hygienic conditions had to be taken into account:

1. hygienic conditions before milking (hand washing, type of containers, ...);
2. hygienic conditions during milking (udder washing, use of disinfectant, ...);
3. hygienic conditions after milking (milk filtration and refrigeration).

MFA and CA allowed distinguishing the following four profiles of milking and rearing conditions:

- PROFILE 1, with a manual and clean milking conditions associated with an intermediary level of intensification (12.7 kg of milk per cow and 12,6 Mcal of Net Energy from concentrates per cow);

Table 2. Milk global quality according to types of milking and rearing conditions.

	Type 1	Type 2	Type 3	Type 4
N	22	6	16	32
Milk pH	6.67±0.12[a]	6.63±0.13[b]	6.56±0.12[c]	6.57±0.13[c]
Milk density at 20°C	1.028±0.002[b]	1.029±0.002[a]	1.028±0.001[b]	1.027±0.002[c]
Milk fat (g/kg)	39.36±5.78	40.08±5.37	35.81±9.43	36.66±9.45
Milk protein(g/kg)	33.19±3.30	31.78±1.54	31.01±3.65	31.08±5.53
Milk urea (mg/100 ml)	326±90	352±35	415±133	344±97
Aerobic Plate Count (x 10^6 CFU/ml)	1.70±1.19	1.10±0.74	2.45±1.08	2.52±1.24
Total Coliforms (x 10^4 UFC/ml)	5.13±1.32	7.08±1.00	15.49±3.32	11.22±1.48
Fæcal Coliforms (CFU/ml)	420±1 055[b]	5±12[c]	460±1 120[b]	122 152±473 372[a]

[a,b,c]: Means in the same line with different superscript letters are significantly different (P<0.005).

- PROFILE 2, with a mechanical and clean milking conditions with the highest level of intensification (15.2 kg of milk per cow per day and 16,3 Mcal of Net Enrgy from concentrates per cow per day);
- PROFILE 3, with a manual and rather clean milking conditions with a limited level of intensification (11.5 kg of milk per cow per day and 7.6 Mcal of Net Energy from concentrates per cow per day);
- PROFILE 4, with a manual and dirty milking conditions and with an extensive milk production (7 kg of milk per cow per day only 1.5 Mcal of Net Energy from concentrates per cow per day).

The use of a co-inertia method showed a relationship between the typology of milking and rearing conditions and milk quality (Table 2).

When considering milk pH, fæcal coliforms and milk density, it appears that those differences were significant. This is in agreement with the data published by Bonfoh *et al.* (2006) which showed, in developing countries, that the type of milking and the use of inappropriate container (plastic vs. metal) usually alter milk hygienic quality with greater amounts of fæcal coliforms. Our results also confirm the fact that the dirtier the milking conditions the more important the fæcal coliforms counts, as suggested by Tourette *et al.* (2002).

Conclusion

This study has revealed the extreme variability in milk global quality in Morocco. It has also shown its close dependence to hygienic parameters, which effect is more important than the effect of physical or chemical factors. As milk price is still based on its chemical and physical parameters, it is necessary to take into account its hygienic quality. On another hand, this study has shown the wide variability in milking and rearing conditions and their influence on milk quality. It is therefore obvious that an intensive promotion of good milking hygienic practices is required in order to improve milk quality at farm level.

References

Association française de Normalisation (AFNOR), 1977. Dosage de l'azote en vue de la détermination de la teneur en protéines brutes dans le lait. Editions AFNOR, Paris, France, pp. 7.

Bonfoh, B., C. Roth, A.N. Traoré, A. Fané, C.F. Simbé, I.O. Alfaroukh, J. Nicolet, Z. Farah & J. Zinsstag, 2006. Effect of washing and disinfecting containers on the microbiological quality of fresh milk sold in Bamako (Mali). Food Control. 17: 153-161.

Djemali, M., & C. Kayouli, 2003. L'élevage laitier en Tunisie. In: Les filières lait en Méditerranée: enjeux pour un future durable, M. Djemali & Guellouz, M. (Editors), Wageningen Pers, EAAP publication no. 99, The Netherlands, 98-105.

Faye, B., & Y. Grelet, 1990. Breeding type and health profile. Two statistical strategies. Presented at the 1st Conference on: "Agro industry & statistical methods". Angers, 14-15 June, pp. 111-125.

International Dairy Federation (IDF), 1974. Milk bacteria count. International Dairy Federation Standards 73/74. Brussels, p. 7.

International Dairy Federation (IDF), 1987. Milk coliforms count. International Dairy Federation Standards 14/87. Brussels, p. 9.

Labarre J.F., 1994. Nutrition et variation du taux de matières grasses du lait de vache. Rec. Méd. Vét. 170: 381-389.

Plusquellec, A., 1991. Techniques d'analyse et de contrôle dans les industries agro-alimentaires: le contrôle microbiologique. In Laits et produits laitiers, Bourgeois C.M. & J.Y. Leveau (Editors), Editions Lavoisier, Paris, 334-353.

R-Project, 2004. Version 2.0.0. Statistical analyses package. Online Access 20 September 2004. www.r-project.org.

Sraïri, M.T., 2004. Typology of dairy cattle systems in Morocco to analyse their performances. Ph. D. Thesis. Gembloux Agricultural Science University. Belgium, pp. 200.

Sraïri, M.T., I. Hasni Alaoui, A. Hamama & B. Faye, 2005. Relationship between rearing practices and cows' milk global quality in suburban dairy farms in Morocco. Rev. Méd. Vét. 156: 155-162.

Tourette, I., S. Messad & B. Faye, 2002. Impact des pratiques de traite des éleveurs sur la qualité sanitaire du lait de chamelle en Mauritanie. Rev. Elev. Méd. Vét. Pays Trop. 33: 229-233.

Whitaker D.A., J.M. Kelly & H.F. Eayres, 1995. Assessing dairy cows diets through milk urea tests. Vet. Rec. 136: 179-180.

Utilization of maize silage by growing finishing Bísaro pigs (50-100 kg LW)

J. Santos e Silva[1], *J. Pires da Costa*[3], *J. Ramalho Ribeiro*[3] *& J.M. Abreu*[2]

[1]*DRAEDM – DPA, Quinta do Pinhó, S. Torcato, 4800, 875 S. Torcato, Portugal*
[2]*Pólo Ciências Agrária da Univ. do Porto, R. Padre Armando Quintas 4485, 661 Vairão, Portugal*
[3]*INIAP, Estação Zootécnica Nacional, Fonte Boa, Vale de Santarém, Portugal*

Summary

The aim of this work was to study the effects of maize silage intake on the growing finishing period of Bísaro pigs (50-100 kg) during winter/spring and spring/summer. A factorial design (2 treatments x 2 seasons) was used for twenty-two (22) Bísaro pigs weighing 44 kg live weight in an experiment carried out in an open air system in northern Portugal. Two diets were used in this experiment: diet C (100% concentrate) and diet CSM (75% concentrate + 25% maize silage on DM base). Maize silage was given daily and total intake was registered. Pigs were slaughtered at ± 100 kg LW and linear carcass measurements were taken after slaughter. At the same time, samples of *Longissimus dorsi* were collected and used for muscle and meat quality assessment.

When compared to pigs fed 100% concentrate (421 g/day), pigs fed maize silage showed slow growth rate (336 g/day). However, a seasonal effect was observed in pigs fed diet C in summer (C = 565g *vs*. CMS = 367 g/day). These pigs presented fast growth rate, while pigs fed both diets did not show differences in growth rate during the winter. The effect of treatment on carcass composition showed no significant difference, while the killing out percentage and fat deposition were higher in summer than in winter. These results suggest that the metabolizable energy (ME) in winter was mainly used for thermoregulation due to a bad winter with heavy rain and very low temperatures (-2 to 5° C). Results show that, although seasonal variations were observed, diets had no influence on the studied traits of meat (% water, nitrogen, fat, collagen and MFV).

Keywords: maize silage, growth, carcass, meat, Bísaro pig, season.

Introduction

The Bísaro pig production in the northern part of Portugal either in the traditional confined system (lojas) or in the outdoor, are still using a feeding regime based in vegetables, cereal grains (maize, rye, tritticalis) potatoes and fruits grown in their local regions. This feeding regime is normally complemented in the finishing period of fattening with commercial concentrate. At this time this typical familiar system of pig production can be seen in the small farms in the north of Portugal.

In general as in the old times, the output of this system, the pig meat is consumed by the farmer's family as fresh meat, or in form the typical cured ham and different varieties of sausages. It's represents a larger amount of the total auto consumption of pig meat in Portugal which has no statistical indication. The increments of production of these high quality products (hams, sausages "*alheira*") deriving from pure and registered Bísaros pigs is of great evidence today, as it as been shown by the National

Bísaros Pig Association (A.N.C.S.U.B) as well as by the number of small artisan industries already licensed.

The social and economic contribution to the development of the local population due to our work is well known and recognised by those regions where the people's income is growing, helping is standard of life. Beside this economic impact this study also made a good contribution to fix the people in the rural areas, having a better quality of life.

The aim of the present study, integrated in a wider project (Agro 339), linked to other in the same area of pig production, is a research on fibre diets based on maize silage and its utilisation on the growing finishing period by Bísaros pig and its effect on carcass quality.

Material and methods

In this experiment, 22 Bísaros pigs (barrows) were used. They were born in an outdoor system and after weaning pigs remained in the same place and system, and the first growth control occurred at about 50 kg LW. Growing finishing phase took place in an outdoor paddock (65 m^2/animal) and the feeding regimen consisted of concentrate and maize silage in 2 different proportions: (C) = 100% concentrate (2 700 g animal/day); (CSM) = 2/3 concentrate + 1/3 Maize Silage, on DM base. Concentrate and Silage chemical composition is presented on table 1.

Feed intake and performances

During the finishing period, feed consumption (silage and concentrate) was registered on a daily basis. Every 14 days, animals were weighted and the subcutaneous fat deposition (P_2mm) in vivo was simultaneously measured. Throughout this phase, ADG, subcutaneous fat deposition and feed intake were evaluated. Pigs were sent to a private slaughterhouse in Ponte de Lima when they reached ±100 kg LW.

Carcass and meat quality

After slaughter, carcasses weights were taken and from 1/2 left carcass of each pig the following parameters were determined: Muscle area and backfat thickness (between 13th and 14th vertebras in D, P_1, P_2 e P_3 spots. All carcasses were stored for 24 hours in a refrigerated chamber, at approximately 4°C. According to the Portuguese traditional carving method, also used in Ponte de

Table 1. Chemical composition (%DM) of the maize silage and concentrate.

	Maize silage	Concentrate
DM %	36	94
Crude fibre (CF) %	17	4
NDF g/kg DM	435	
Digestible energy (DE) MJ /kg MS*	10.4	13.02
Crude Protein (CP) %	8.6	15.0
Calcium (Ca)	0.2	0.8
Lysine %		0.7
Phosphorus (P)	0.14	0.6

Lima (Minho Fumeiro), the half carcass was dissected in to six joins (shoulder, ham, boned loin, belly and head) and were weighted. At the same time, samples of *Longissimus dorsi* muscle were taken, for analysis of meat quality: % water, ash, nitrogen, fat, collagen (total and soluble) and the myofibrils fragmentation value (MFV).

Statistical analysis

The results of the growing test and of the carcass parameters were submitted to a general linear model analysis considering the fixed effect treatment and season, and the interaction treatment : season.

Results and discussion

Feed intake, growth rate and carcass traits

The individual daily feed intake during the finishing period in terms of dry matter (DM) and digestible energy (DE) is on table 2.

As observed on table 2, differences of energy intake between treatments were estimated as following: C = 4 922 MJ; CSM = 4 591 (-7% energy). The DE intake on diet CSM, in which silage was included, is 7% lower than with the concentrate diet. This maize silage feed to the pigs has a good DM composition. This DM is largely built up with maize grain.

Control group were benefited in the total energy intake (+7%): The total dry matter intake only comes from concentrate (a total of 4 922 MJ DE/animal). The over all the maize silage intake was not balanced, in terms of energy and protein requirements. This may be responsible by the differences observed on the growth rate registered between the two treatments (Table 3). In a similar experience (Santos Silva *et al.*, 2004) using a diet with 75% of concentrate and fresh grass *ad libitum*, with the same breed of pigs, showed that carcass obtained were thinner (1 to 2%) and the ADG was -27% lower, comparing with the pigs on the concentrate diet.

Performance, carcass composition of pigs fed the two diets C and CMS.

The pig fed maize silage showed slow growth rate (336 g/day), when compared to pigs fed 100% concentrate (431 g/day), and live weight (kg) at slaughter was directly proportional to ADG and to the concentrate intake (C = 111.3; CSM=101.0; $P<0.01$). However, a seasonal effect was observed in pigs fed diet C in summer (565 g vs. 367 g/day). These pigs presented fast growth rate, while pigs fed both diets did not show differences in growth rate during the winter (Figure 1).

Table 2. Feed and energy intake during finishing period.

	Treatments	
	C	CSM
Concentrate (kg DM/animal)	378	252
Energy intake DE (MJ)	4 922	3 281
Maize silage (kg DM/animal)	0	126
Total DM ingested (Kg /animal)	378	378
Total energy intake DE (MJ)	4 922	4 591 (-7%)

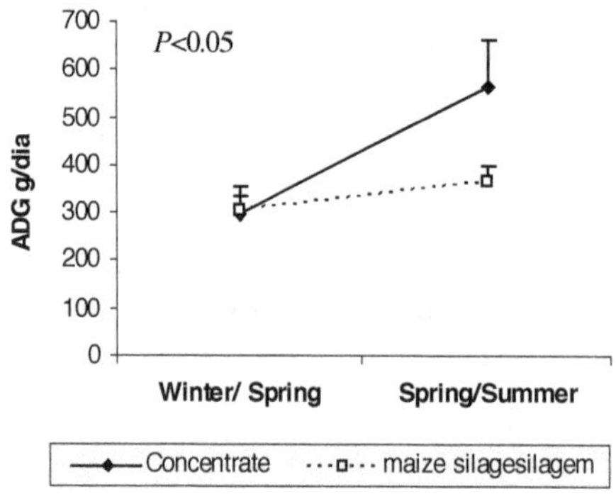

Figure 1. Interaction season : treatment for ADG.

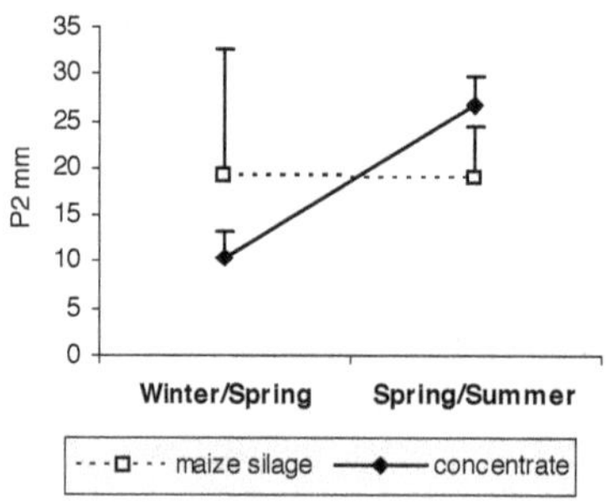

Figure 2. Interaction season : treatment for subcutaneous fat (P_2mm).

Table 3. Performances and carcass traits of the pigs feed diet C and CMS.

	C	CMS	
	LSM±SE	LSM±SE	*P*-value
Age $_{initial}$ (days)	189.0±3.0	182.0±3.0	NS
Weight $_{initial}$ (kg)	57.7±3.0	52.1±3.0	NS
Age $_{final}$ (days)	329.0±3.2	322.0±3.1	NS
Weight $_{final}$ (kg)	111.3±2.1	101.1±2.2	**
ADG g/day	431.0±0.02	336.0±0.02	**

In what concern the carcass composition, no significant differences were observed among the different traits, although the killing out percentage presents a little difference (0.8%) between treatments. However when we look at the data related seasonal effect, significant differences were registered in the carcass traits (Table 4).

Regarding the qualitative carcass aspects, the subcutaneous fat deposition and muscle percentage shows no differences between treatments, while a seasonal and interaction effect were registered

Table 4. Seasonal effect on carcass traits.

	Winter/Spring	Spring/Summer	*P*-value
Slaughter age days	370.0±3.0	280.0±3.0	
Slaughter weight kg	109.3±3.2	103.1±2.5	NS
Percentage lean	48.5±1.6	48.7±1.2	NS
Killing out percentage	72.4±0.6	74.4±0.6	*
Percentage ham	29.2±0.3	27.3±0.2	***
Percentage shoulder	22.3±0.2	20.3±0.2	***
Percentage loin	24.1±0.4	28.1±0.3	***
Percentage belly	15.8±0.4	15.9±1.3	NS
Percentage head	8.6±0.2	8.3±0.2	NS
Subcutaneous Fat P_2 (mm)	15.0±2.0	23.0±2.0	*

(Figure 2): i.) Subcutaneous fat deposition on pigs fed on diet C was lower in winter/spring (C = 10 vs. CSM = 19 mm) and higher in spring/summer (C = 27 vs. CSM = 19 mm) comparing with the pigs fed on maize silage. ii.) Percentage of muscle was higher for treatment C, only in winter/spring (C = 51.5% vs. CSM = 45.5%);

A striking result was obtained in this experiment carried out during the winter/spring time and the pigs sent to slaughter in June. A similar daily gain was observed in the two diets C and CSM in winter (Figure 1 and 2). The carcasses of the pigs fed on maize diet were fatter ($P<0.05$; subcutaneous fat deposition was 8 mm higher). We can not see any metabolic pathway that can explain these results, unless the pigs fed on silage spend less energy on thermoregulation and behaviour (Gerrits *et al.*, 2003). In the table 4 data observed in fat and carcass dissection of Bísaro pigs slaughter in two season of the year are present.

Meat quality

The effect of diets and season on meat quality is represented in table 5. The results showed that, although seasonal variations were observed, diets had no significant influence on the studied traits of meat (table 5; % water, nitrogen, fat, collagen and MFV). However when we look at seasonal effect on meat quality, we can see that some parameters affected mainly fat, collagen and the myofibrils fragmentation value (MFV). These parameters showed a high concentration value in summer (Fat = 12.5%, Total collagen = 5.2%, Soluble collagen = 9.8%, and MFV = 31.2 kg) compared with slaughtered pigs in spring (Fat = 6.4%, Total collagen = 3.6%, Soluble collagen = 28.3%, and FMV = 20.2 kg). Pigs sent to slaughter in summer time had more collagen concentration (total and soluble) than in winter, this may be due to its faster growth rate observed and their young age at slaughter (280 vs. 370 days).

Conclusions

In the present study it was found that the results obtained were not clear enough and some more experiments along the year should be done using more pigs per treatments. Besides this limitations on material and climatic conditions the observed data suggest:

1. It is possible to include maize silage up to 1/3 of the DM intake in the diet for growing finishing Bísaro pigs without affecting the efficiency of pig production in the open air system.

Table 5. Effect of diets on meat quality (Longissimus dorsi).

	Diets		Season			
	C	CSM	Winter/ Spring	Spring/ Summer	Treat	Seas.
N	10	12	8	14		
DM %	26.1±0.4	26.0±0.3	26.6±0.4	26.3±0.3	NS	NS
Pigments %	6.3±0.6	5.2±0.3	5.6±0.6	5.7±0.3	NS	NS
Ash %	4.5±0.1	4.5±0.1	4.7±0.08	4.3±0.07	NS	**
Nitrogen %	13.7±0.2	14.0±0.2	14.1±0.2	13.8±0.2	NS	NS
Fat %	10.5±1.3	10.0±1.2	6.4±0.5	12.5±0.8	NS	***
Total collagen %	4.3±0.5	4.8±0.4	3.6±0.2	5.2±0.4	NS	*
Soluble collagen %	21.0±3.9	22.1±3.6	9.8±0.6	28.3±2.6	NS	***
Myofibrils frag. value (MFV)	23.9±2.7	30.0±2.4	20.2±1.8	31.2±1.8	NS	**

2. The ADG resisted on pigs fed CMS diets all over the year showed a slower growth rate but also a seasonal effect.
3. The meat quality of the pigs fed this diet showed a better distribution of intramuscular fat (marbling).
4. Carcass of pigs slaugthered in summer time (spring-summer) was fatter in this country side, perhaps due to a less expenditure of ME on thermoregulation. The increment caloric (IC) in the silage diet was higher than in concentrate, so that diet was used more efficiently in winter.

References

Gerrits W.J.J, Rijnen M.J.A., Bolhuis J.E.L. and Schrama J.W. 2003. Influences of dietary fibre on physical activities and behaviour, and on its energy value for growing pigs. Perspectives in Pig Science, 23: 475-490.

Santos e Silva J., Enes, M. Figueiredo F.O., Pires da Costa J.S., José M. Abreu. 2003. Grass utilization in growing finishing Bísaro pigs (85-107 kg). Performance and carcass composition. Proceedings of 5th International Symposium on the Mediterranean Pig. 16-19 November 2004, Tarbes, France.

^{13}C stable isotope for the authentication of lamb meat

A. Janeiro[1], L. Ramalho[1], B. Henriques[1], A. Teixeira[1], C. Costa[1], M.J.M. Curto[1], J. Santos-Silva[2] & R. Bessa[2]

[1]*INETI-Departamento de Tecnologia das Indústrias Químicas, Estrada do Paço do Lumiar, 1649-038, Lisboa, Portugal*
[2]*Estação Zootécnica Nacional, Vale de Santarém, 2005-054 Santarém, Portugal.*

Summary

The $^{13}C/^{12}C$ ratio in plant material depends on isotopic fractionation, which takes place in the biosynthetic reactions of organic compounds, according to different photosynthetic cycles for CO_2 fixation by the plants, that is, the Calvin or C_3 cycle and the Hatch and Slack or C_4 cycle (Smith and Epstein, 1971). The different range of isotopic values is also transmitted to the animals in their diet and consequently to their products, providing relevant information on the composition of the diet.

In light of the above considerations we have performed a preliminary evaluation of the effectiveness of the determination of the stable isotopes ratio $^{13}C/^{12}C$ by Elemental-Analyzer/Isotope Ratio Mass Spectrometry as a method for animal feeding authentication of Portuguese lamb meat.The data obtained were studied by analysis of variance.

Keywords: food authenticity, isotope ratio mass spectrometry, ^{13}C, lamb meat.

Introduction

Food characterization is an important strategic issue for food industry (Rossmann, 2001). Authentic food information has been a major concern of many consumers and it is gaining importance, especially in the case of ovine and bovine meat. Authenticity of meat is not only linked with the geographical origin but also to the production system claimed. Organisms more or less reflect the isotopic composition of the material in their environment e.g. carbon, nitrogen or sulfur. The natural variation of the stable isotopic composition is a tool to control the authenticity of many products. However, until now only a limited kind of products such as wine, milk, juices and meat have been investigated (Renou *et al.*, 2004a, b; Piasentier *et al.*, 2003). The $\delta\ ^{13}C$ value was found to be highly dependent on diet composition, particularly with regard to maize, a C_4 plant (Boner *et al.*, 2004).

The aim of this study is to evaluate the usefulness of the $\delta\ ^{13}C$ content determination of sheep meat as an approach to establish the animal diet.

Material and methods

Animals and treatments

Thirty-two Merino Branco ram lambs were used in this trial, being reared on pasture by their dams until weaning that occurred at about 90 days of age. After that were randomly assigned to four groups of 8 each, and allotted in four units of 2 lambs of similar weight. The experimental diets were based in commercial concentrate and ground and pelleted lucerne, and soybean oil (8%) was used as supplement. Thus the four experimental diets were as follows: ground and pelleted lucerne (L); ground and pelleted lucerne plus soybean oil (LO); concentrate (C) and concentrate plus soybean oil (CO).

Lambs were slaughtered, 7 weeks after the beginning of the trial and the samples of the left *longissimus thoracis* muscle were freeze-dried, vacuum packed and stored at -80°C until the analysis was performed.

Isotope ratio mass spectrometry measurement

A quantity of about 1 mg of lyophilized muscle, for the measurement of $\delta\ ^{13}C$, was analyzed by EA-Combustion-IRMS (Elemental Analyzer from EuroVector, Italy, coupled with an Isotope Ratio Mass Spectrometer, an ISOPRIME from Micromass, U.K..

The standards used for calibration were the International Standard of Polyethylene IAEA (International Atomic Energy Agency) CH-7 and sucrose (internal standard), previously calibrated against the International Standards of Polyethylene IAEA CH-7 and Sucrose IAEA-CH-6.

Each sample was analyzed in triplicate and the values were averaged. The measurement was repeated wherever the standard deviation was higher than 0.3‰.

Participating in European proficiency tests routinely, checks the performance of the laboratory.

Results and discussion

Table 1 summarises the analysis of variance of the $\delta\ ^{13}C$ data obtained for the muscle of the animals kept under controlled feeding regimes. These results are significantly different, $P<0.0001$ which indicate that the meat submitted to both feeding regimes studied is distinguishable relatively to their $\delta\ ^{13}C$ values.

Table 1. Analysis of Variance of the $\delta^{13}C$ data obtained from lamb meat fed with lucerne(L), lucerne with oil (LO), concentrate (C) and concentrate with oil (CO).

						Significance (*P*<)		
	L	LO	C	CO	SEM	B	O	B*O
No	8	8	8	8				
$\delta\ ^{13}C$ mean values (‰)	-25.83	-25.16	-22.15	-22.38	0.111	0.0001	0.054	0.0004
St. Dev.	0.06	0.06	0.06	0.08				

No = Number of animals in each feeding regime.
SEM = Standard error of means.
B = Effect of basal diet (lucerne vs concentrate).
O = Effect of soybean oil supplementation.
B*O = Interaction between basal diet and oil supplementation.

The higher value of $\delta^{13}C$ of the animals fed with concentrate could suggest that the concentrate must have a part of maize (C_4 plant), because the $\delta^{13}C$ of maize (and for C_4 plants in general) range from -11 ‰ to -14‰.

The soybean oil supplementation of the diets did not modify the $^{13}C/^{12}C$ ratio in meat from concentrate fed lambs but increased the $^{13}C/^{12}C$ ratio in meat from lucerne fed lambs (B*O interaction, $P<0.0004$).

From this preliminary study the $^{13}C/^{12}C$ ratio seems to be a good tool to distinguish the feeding system/meat authenticity, since the isotopic composition effectively describes the kind of feeding supplied, so it is a promising way to relate meat with its production system.

We need a much more complete study and we already have an approved national project (AGRO 512 Project), where these issues are thoroughly studied.

References

Boner, M.& Forstel, H. 2004. Stable Isotope Variation as a tool to trace the authenticity of beef. Anal. Bioanal. Chem., 378: 301-310.

Piasentier E., Valusso R., Camin F. & Versini G. 2003. Stable isotope ratio analysis for authentication of lamb meat. Meat Science, 64: 239-247.

Renou J.-P., Bielicki G., Deponge C., Gachon P., Micol D. & Ritz P. 2004a. Characterization of animal products according to geographic origin and feeding diet using nuclear magnetic resonance and isotope ratio mass spectrometry. Part II: Beef meat. Food Chemistry, 86: 251-256.

Renou J.-P., Deponge C., Gachon P,. Bonnefoy J.-C., Coulon J.-B., Garel J.-P., Vérité R. & Ritz P. 2004b. Characterization of animal products according to geographic origin and feeding diet using nuclear magnetic resonance and isotope ratio mass spectrometry: cow milk. Food Chemistry, 85: 63-66.

Rossmann A. 2001. Determination of stable isotope ratios in food analysis. Food Reviews International, 17(3): 347-381.

Smith B. N. & Epstein S. 1971. Two categories of 13C/12C ratios for higher plants. Plant Physiol., 47: 380-384.

Contribution to a better definition of the production standards of the lamb "Borrego do Nordeste Alentejano - PGI"

J. Santos-Silva[1], A. Esteves[2], N. Alexandre[2], S. Alves[1], A.P. Portugal[1], I.A. Mendes[1], M. Silva Pereira[1], M. Vacas de Carvalho[2] & R.J.B. Bessa[1]

[1]Estação Zootécnica Nacional, Fonte-Boa, 2005-048 Vale de Santarém, Portugal
[2]Natur-al-carnes, Parque de Leilões de Gado- EN 246, 7301-901 Portalegre, Portugal

Summary

The objectives of this trial were to investigate the effects of: 1) the production system; 2) the increase of carcass weight from 12 to 15 kg; 3) the type of concentrate on lambs growth, carcass composition and meat quality traits. In two consecutive years, a total of 336 lambs "Borrego do Nordeste Alentejano – PGI" (BNA-PGI) were used. Two different management conditions were compared: in spring, when lambs were raised on pasture with their dams and supplemented with concentrate (120 lambs * 2 years); and in autumn when lambs were weaned and fed with concentrate afterwards (48 lambs * 2 years). Lambs were slaughtered at 28 or 33 kg live weight in order to obtain the target carcass weights. Three types of concentrate were compared: NATUR, a specific concentrate designed specially for BNA-PGI, vs. two commercial concentrates. Live weight was recorded every 14 days, from 60 days to slaughter. Carcasses were graded according to the UE grading schemes for lambs and carcass composition was estimated by partial dissection. Meat quality was evaluated by instrumental (colour and shear force) and sensorial methods. Fatty acid composition of intramuscular lipids of *Longissimus thoracis* muscle was determined. The quality of the BNA-PGI was totally independent of the type of concentrate. The increase of carcass weight had no negative effects on carcass or meat quality traits and did not compromise the requisites for certification as BNA-PGI. Shear force value was 40 % higher for lambs produced in confinement (autumn) than in pasture (spring). The panellists preferred meat from lighter lambs when raised in pasture (spring), and from heavier lambs when raised in confinement (autumn). The proportion of conjugated linoleic acid (C18:2 c9t11) was 2.65 fold higher and n-3/n-6 ratio was 2.55 higher in meat from lambs raised on pasture than from lambs raised in confinement.

Keywords: lamb, carcass quality, meat quality, fatty acid, season.

Introduction

The Borrego do Nordeste Alentejano (BNA-PGI) is a traditional product of the Northern Region of Alentejo in Portugal that is protected as a Geographical Indication. It represents an important income to the farmers of this region. In 2003, about 14 000 carcasses (169 ton) of BNA-PGI were commercialised. The lambs are the product of Merino Branco (MB) ewes, a Portuguese breed for meat production, raised by the traditional production systems, where lambs are fed with natural pastures and local forages, supplemented with concentrates. Lambs must present slaughter ages from 90 to 120 days and carcass weights from 9 to 13 kg. They show high growth rates (>300 g/day) in intensive production systems and produce carcasses up to 15 kg without excessive fat (Santos

Silva & Vaz Portugal, 2000). The possibility of certificate carcasses up to 15 kg adds flexibility to the production systems and increases the sustainability of the production chain. To improve quality, some regulations of BNA-PGI could be altered increasing the safety and the homogeneity of the product. The definition of specific formulae to produce BNA-PGI would allow a better control of the feeds, reducing variability of carcass composition and eventually the production costs, contributing to improve the image of the product, increase the market value and the farmer's profitability.

Material and methods

Three hundred and thirty six MB lambs BNA-PGI were used in 4 trials, conducted in 2 years. Two production systems were defined: in spring, when the lambs were raised in the pasture with the dams and supplemented with concentrate after 60 days of age; in autumn when lambs were weaned after 60 days, and fed in confinement with ad libitum concentrate and hay.

Three types of concentrate were tested: two commercial (A and B) and one formulated specially to BNA-PGI (NATUR). The composition of NATUR and the chemical composition of the concetrates are presented in table 1.

Two slaughter weights were considered to obtain carcasses of 12 and 15 kg. In autumn, lambs were assigned to 6 groups of 8 lambs each, according to type of concentrate (A, B, NATUR) and carcass weight (12 and 15 kg), totalising 96 lambs in the 2 years. To study carcass composition and meat quality traits, 24 carcasses were used in each year. In spring, the trials were conducted in 2 private farms of BNA-PGI. In each farm, 60 lambs were assigned to 6 groups of 10 lambs, corresponding to the type of concentrate (A, B, NATUR) and carcass weight (12 and 15 kg),

Table 1. Composition of the NATUR concentrate and chemical composition of NATUR, A and B concentrates.

	NATUR	A	B
Corn (%)	25.0		
Oats (%)	10.0		
Barley (%)	14.9		
Wheat (%)	10.0		
Corn gluten feed (%)	5.0		
Carob (%)	3.0		
Molasses (%)	2.0		
Soybean meal 44% (%)	23.8		
Magnapac (%)	1.5		
Soybean oil (%)	0.9		
Ammonium chloride (%)	0.4		
Sodium bicarbonate (%)	0.2		
Premix (%)	3.0		
Dry matter (%)	87.7	87.7	87.8
Crude protein (% DM[2])	20.9	19.6	19.2
Crude fat (% DM[2])	3.0	4.3	4.0
NDF[1] (% DM[2])	17.2	19.6	19.0
Calcium (% DM[2])	1.5	1.1	0.8
Phosphorus (% DM[2])	3.4	2.1	1.8

[1]Neutral Detergent Fiber; [2]Dry matter.

totalising 120 lambs per year. To study carcass composition and meat quality traits, 48 carcasses were used in each year. Live weight was controlled every 14 days.

The lambs were slaughtered at Sousel slaughterhouse after a fasting period of 18 hours. The carcasses assigned to be studied, were transported to the slaughter house of Estação Zootécnica Nacional (EZN) and graded according to UE grading schemes (EEC, 1992). At the 7th day, the left halves of the carcasses were divided in 8 joints (Calheiros & Neves, 1968) and the chumps and shoulders were dissected to estimate carcass composition (Santos Silva & Simões, 1999). The left loins were vacuum packed and frozen at –20°C for shear force determination (Santos-Silva *et al.*, 2002a). Colour of *Longissimus thoracis* muscle was measured using the CIE L^* a^* b^* system, after 1 hour of blooming, with a Minolta CR-300 chromometer. In this system colour is described by the coordinates L^* a^* b^*, representing lightness, redness and yellowness (CIE, 1976). Samples of *Longissimus thoracis* muscle were minced and frozen at –80°C until lipid extraction, by the Soxlet method (AOAC, 1990; method 960.39), and fatty acid determination. The fatty acid methyl esters were prepared according to Rule *et al.* (1997) and resolved in a gas chromatograph Varian 3800GC, using a Omegawax capillary column (Supelco, 30 m length and 0.25 mm of internal diameter).

The right loins and chops were collected and frozen at –20°C until the organoleptic evaluation of meat. A panel of 50 consumers of BNA-PGI was used to evaluate the effects of the type of concentrate and carcass weight on meat quality. Frozen right joints (loins and ribs) were cut transversally in chops with 1 cm of thickness. Samples were grouped according to the 6 treatments of this trial (A-12 kg, A-15 kg, B-12 kg, B-15 kg, NATUR-12 kg, NATUR-15 kg). Twenty-fours hours before the sessions, chops were thawed at 0°C, salted, grilled and served in hot plates. Each panellist was presented simultaneously with 3 or 2 chops, to compare meat from the 3 concentrates or the 2 carcass weights, respectively. Each panellist was asked to evaluate the similarity between samples and to compare them according to tenderness, taste, juiciness and overall acceptability, using a hedonic scale from 1 to 10 (1 -very tough, very dry, very bad taste, very bad overall; 10 – very tender, very juicy, very good taste; very good overall).

Data were studied using the Proc Mixed of SAS (2004), considering the fixed effects of production system, concentrate, carcass weight and their interactions, and using the year as random block. Data concerning sensorial attributes of meat were studied using a model similar to the previous, considering one fixed effect only (type of concentrate or carcass weight). The results obtained in spring and autumn were studied independently.

Results and discussion

Lambs growth and carcass and meat quality

Table 2 presents the results for lamb growth, carcass composition and meat quality traits. The columns with the results of the interactions Production system*Type of concentrate and Production system*Type of concentrate*Carcass weight are not presented because no results were significant.

Generally, the carcasses with 12 kg were graded as class 3 of 1st quality. With 15 kg, most of the carcasses were graded as R and O for conformation and 2 and 3 for fattening. As the specifications of BNA-PGI include only carcasses graded as 2 or 3, the effects studied had no implication in the certification process.

The lambs showed high growth performances (>300g), in accordance with the results obtained in other trials (Santos Silva and Vaz Portugal, 2000; Santos-Silva *et al.*, 2002a). Differences in carcass composition parameters were very narrow and probably have no implication in commercial value. Carcasses presented high proportion of muscle and low fat content independently of the production

Table 2. Effects of production system (PS), type of concentrate (C) and carcass weight (CW) on average daily weight gain (ADG), carcass composition and meat quality traits.

	PS		C			CW		Significance				
	Aut	Spr	A	B	Natur	12	15	PS	C	CW	C*CW	PS*CW
ADG (g)	317	312	322	309	313	306	323	ns	ns	*	ns	ns
CCW (kg)	13.5	13.3	13.5	13.4	13.4	12.2	14.7	ns	ns	***	ns	**
DP (%)	44.8	44.9	45.2	45.1	44.3	44.1	45.7	ns	ns	***	ns	ns
HP (%)	53.1	53.0	52.9	53.5	52.8	53.2	52.9	ns	ns	ns	**	ns
KKCF (%)	1.74	1.30	1.57	1.37	1.62	1.42	1.61	***	ns	*	ns	ns
M (%)	58.5	58.6	58.4	58.8	58.3	58.6	58.4	ns	ns	ns	ns	ns
SFat (%)	7.2	7.1	7.2	7.3	7.0	6.8	7.6	ns	ns	ns	ns	ns
DisF (%)	18.0	17.1	17.4	17.6	17.7	17.1	18.0	ns	ns	Ns	ns	ns
MB	2.58	2.42	2.47	2.52	2.51	2.43	2.56	***	ns	**	ns	ns
IF/SF	1.43	1.46	1.39	1.42	1.53	1.55	1.34	ns	ns	*	ns	ns
L*	39.9	40.3	40.0	40.6	39.9	40.5	39.8	ns	ns	ns	ns	ns
a*	16.5	16.3	16.4	16.6	16.2	16.0	16.8	ns	ns	ns	ns	ns
b*	5.20	4.77	5.34	5.03	4.58	5.10	4.87	ns	ns	ns	ns	ns
SF (kg)	5.45	3.88	4.65	4.56	4.79	4.76	4.56	***	ns	ns	ns	ns

Aut - autumn; Spr – spring; CCW – cold carcass weight; DP – dressing percentage;
HP – higher price joints (leg, chump, loin and chops); KKCF – kidney knob and channel fat;
M – muscle; SFat – subcutaneous fat;
DisF – dissectable fat (subcutaneous + intramuscular fat); MB – muscle/bone;
IF/SF - intramuscular fat/subcutaneous fat; L* a* b* - colour parameters; SF – shear force;
* - P<0.05; **- P<0.01; *** - P<0.001; ns – not significant.

Table 3. Effects of type of concentrate (C) and carcass weight (CW) on the organoleptic attributes of meat.

		C				CW		
		NATUR	A	B	Sig	12	15	Sig
Autumn								
	Tenderness	6.68	6.15	6.67	ns	6.58	7.16	**
	Juiciness	6.34^b	5.59^a	6.29^b	*	6.45	7.10	**
	Taste	6.71^b	5.90^a	6.72^b	**	6.90	7.17	ns
	Overall acceptability	6.86^b	6.10^a	6.84^b	**	6.83	7.32	ns
Spring								
	Tenderness	8.33	7.92	8.06	ns	7.84	7.02	***
	Juiciness	6.76	6.40	6.27	ns	7.73	6.58	***
	Taste	7.55^b	6.64^a	6.64^a	**	7.80	6.62	***
	Overall acceptability	7.79	6.97	7.11	**	7.91	6.58	***

Sig – significance; * - $P<0.05$; **- $P<0.01$; *** - $P<0.001$; ns – not significant.

system and carcass weight. The effects studied have not affected meat colour, and values are similar to those reported by Santos Silva *et al.* (2002a).

Lambs produced in spring showed lower shear force values, indicating a higher tenderness of meat. This was an interesting and unexpected result because the lamb's performances, carcass composition, management before and during slaughter and in storage (Sañudo *et al.*, 1998) were

Table 4. Effects of production system (PS), type of concentrate (C) and carcass weight (CW) on intramuscular fat content and fatty acid composition of Longissimus thoracis muscle adjusted to the proportion of intramuscular fat.

	PS		C			CW (kg)		Significance		
	Aut	Spr	A	B	NATUR	12	15	PS	C	CW
GI (%MS)	7.74	6.47	6.88	7.05	7.38	6.86	7.35	***	ns	ns
C14:0	2.45	4.45	3.65	3.38	3.31	3.48	3.41	***	ns	ns
C15:0	0.30	0.46	0.38	0.38	0.39	0.39	0.37	***	ns	ns
C16:0	20.4	19.8	20.3	19.9	20.0	19.8	20.3	*	ns	ns
C16:1	1.70	2.10	2.05	1.82	1.83	1.91	1.89	***	ns	ns
C17:0	1.11	0.86	0.93	1.01	1.01	0.98	0.99	***	ns	ns
C18:0	13.3	10.9	11.5	12.2	12.5	12.2	12.0	***	ns	ns
C18:1 c9	34.8	30.8	33.3	33.0	32.1	32.7	32.8	***	ns	ns
C18:2 n-6	8.94	8.10	8.25	8.52	8.79	8.52	8.52	**	ns	ns
C18:3 n-3	0.51	1.49	0.95	1.05	1.00	1.03	0.98	***	ns	ns
C18:2 c9t11	0.26	0.69	0.46	0.49	0.48	0.49	0.46	***	ns	ns
C20:2 n-6	0.44	0.32	0.40	0.38	0.37	0.40	0.36	***	ns	ns
C20:3 n-6	0.28	0.24	0.24	0.27	0.26	0.26	0.26	*	ns	ns
C20:4 n-6	3.24	2.26	2.52	2.99	2.73	2.80	2.70	***	ns	ns
C20:5 n-3	0.36	0.82	0.52	0.66	0.61	0.65	0.54	***	ns	ns
C22:4 n-6	0.22	0.06	0.14	0.15	0.13	0.14	0.14	***	ns	ns
C22:5 n-3	0.31	0.63	0.46	0.46	0.47	0.45	0.49	***	ns	ns
C22:6 n-3	0.13	0.38	0.27	0.27	0.23	0.29	0.22	***	ns	ns
hH	2.04	1.91	1.96	2.00	1.96	1.99	1.96	**	ns	ns
n-6/n-3	10.16	4.1	7.2	6.8	7.4	7.1	7.1	***	ns	ns

Aut - autumn; Spr – spring.
hH – hypocholesteremic fatty acids (C18:0+C18:1 c9 + PUFA n-6 e+ PUFA n-3) /hypercholesteremic fatty acids (C14:0+C16:0).
* - $P<0.05$; **- $P<0.01$; *** - $P<0.001$; ns – not significant.

similar. Consumer's panel have not depreciated meat of lambs fed with NATUR and preferred meat from lighter carcasses when lambs were raised in pasture (spring) and from heavier carcasses when lambs were raised in confinement (autumn) (Table 3).

Fatty acid composition of intramuscular fat

In table 4 is presented the fatty acid composition of intramuscular fat.

The production system was the main effect associated with the differences in fatty acid composition of intramuscular fat. In accordance with other reports (Rowe *et al.*, 1999; Sañudo *et al.*, 2000; Santos Silva *et al.*, 2002b), lambs raised on pasture showed a 2.5 fold higher proportion of n-3 fatty acids, reflecting a high content of linolenic acid in the grass lipids (Raes *et al.*, 2004). The lambs raised in confinement had higher proportion of n-6 fatty acids, but the differences between production systems were not as expressive as for n-3 fatty acids. The ratio between n-6/n-3 fatty acids was 2.6 fold lower for lambs raised in spring, which is advantageous considering the nutritionist recommendations (Department of Health, 1994).

The rumenic acid (C18:2 c9t11) is the main conjugated isomer of linoleic acid in ruminants fat that may have an important role in the prevention of some types of cancer (Pariza *et al.*, 2001). The proportion of the C18:2 c9t11 was 2.65 higher in lambs raised on pasture than in those raised in confinement. Other authors have reported high levels of C18:2 c9t11 in ruminants fed with diets rich in C18:3 n-3 (Raes *et al.*, 2004, Bessa *et al.*, 2005).

The ratio between hypocolesteric and hypercolesteremic fatty acids (hH) was used as and indicator of the effect of fatty acid composition on cholesterol metabolism (Santos-Silva *et al.*, 2002b). The higher value was observed for lambs raised in confinement, which is in contradiction with the results reported by Santos-Silva *et al.* (2002). The authors obtained hH values of 1.84 in lambs raised in confinement and fed with concentrate and 2.04 in lambs raised on pastured with their dams and supplemented with concentrate. The difference in the results is probably related to the higher number of n-6 isomers that were quantified in the present trial.

References

AOAC, 1990. Association of Official Analytical Chemists. Fat (crude or ether extract in meat - 960.39), Official methods of analysis. Washington DC, USA

Bessa R.J.B., P. V. Portugal, I. A. Mendes & J. Santos-Silva, 2005. Effect of lipid supplementation on growth performance, carcass and meat quality and fatty acid composition of intramuscular lipids of lambs fed dehydrated lucerne or concentrate. Livestock Production Science (In press).

Calheiros, F. & A. Neves, 1968. Rendimentos ponderais no borrego Merino Precoce. carcaça e 5° quarto. Sep. Boletim Pecuário ano XXXVI: 117-126

CIE, 1976. Commision Internationale de l'Eclairage, 18th session, London, UK, September 1975, CIE Publication 36.

Department of Health. 1994. Nutritional Aspects of Cardiovascular Disease. Report on Health and Social Subjects no. 46. HMSO, London.

EEC, 1992. Council Regulation (EEC) n° 2137/1992, concerning the Community Scale for the classification of carcasses of ovine animals. Official Journal L.214, 30/07/92, 1-5

Pariza, M.W., Y. Park & M.E. Cook, 2001. The biologically active isomers of conjugated linoleic acid. Progress in Lipids Research, 40, 283-298

Raes, K., S. de Smet & D. Demeyer, 2004. Effect of dietary fatty acids on incorporation of long chain polyunsaturated fatty acids and conjugated linoleic acid in lamb, beef and pork meat: a review. Animal Feed Science and Technology 113: 199-221

Rowe, A., F.A.F. Macedo, J.V. Visentainer, N.E. Sousa & M. Matsushita, 1999. Muscle composition and fatty acid profile in lambs fattened in drylot or pasture. Meat Science, 51: 283-288

Rule, D.C., M.D. MacNeil & R.E. Short, 1997. Influence of sire growth potential, time on feed, and growing-finishing strategy on cholesterol and fatty acids of the ground carcass and longissimus muscle of beef steers. Journal of Animal Science 75: 1525-1533

Santos-Silva, J., R.J.B. Bessa & I. Mendes, 2002a. The effect of genotype, feeding system and slaughter weight on the quality of light lambs. 1.Growth carcass composition and meat quality. Livestock Production Science, 76, 17-25.

Santos-Silva, J., R.J.B. Bessa & F. Santos-Silva, 2002b. The effect of genotype, feeding system and slaughter weight on the quality of light lambs. 2. Fatty acid composition of meat. Livestock Production Science, 77, 187-194.

Santos-Silva J. & A.Vaz Portugal, 2000. Influência do peso da carcaça na qualidade das carcaças de borregos das raças Serra da Estrela e Merino Branco produzidos em sistemas intensivos de produção. Revista Portuguesa de Zootecnia, ano VII, n° 1: 109-128

Santos-Silva, J. & J. Simões, 1999. Predição da composição da carcaça de borregos da raça Merino Branco. Revista Portuguesa de Zootecnia, VI(1), 55-62.

Sañudo, C., M. Enser, M.M. Campo, G.R. Nute, G. Maria, I. Sierra & J.D. Wood, 2000. Fatty acid composition and sensory characteristics of lamb carcasses from Britain and Spain. Meat Science, 54: 339-346.
Sañudo C., A. Sanchez & M. Afonso, 1998. Small ruminant production systems and factors affecting lamb meat quality. Meat Science, 49, Suppl. 1:S29-S64
SAS Institute Inc, 2004. SAS/STAT 9.1User's Guide, Cary, NC, USA.

Characterization of collagen and fatty acid composition of "Carne Mirandesa PDO" veal

L. Galvão[1], O.C. Moreira[2], R.J.B. Bessa[2], S.P. Alves[2], F. Sousa[1], J. Ramalho Ribeiro[2] & V. Alves[3]

[1]*ESAB, Escola Superior Agrária de Bragança, Campus de Santa Apolónia, 5301-855 Bragança, Portugal*
[2]*INIAP, Estação Zootécnica Nacional, 2005-048, Vale de Santarém, Portugal*
[3]*UTAD, Universidade de Trás-os-Montes e Alto Douro, 5000-911, Vila Real, Portugal*

Summary

The objective of this study was to evaluate chemical parameters related to meat quality, contributing to a better characterization of "Carne Mirandesa PDO" veal. This study was made in three farms, from the Northeast region of Portugal (PG, FA1 and FA2), where calves were raised permanently indoors, nurse from their dams overnight, and fed with hay and concentrate made with local ingredients and soybean meal. Twenty five calves were slaughtered at 7 months old. Twenty four hours after slaughtering, samples were collected from four different muscles: *longissimus dorsi (LD), semimembranosus (SM), gluteus biceps femoris (GB)* and *triceps brachii caput longum (TBL),* vacuum packaged and freeze-dried. Determinations of crude protein (CP), ether extract (EE), collagen and fatty acid were made. Collagen, CP and EE were significantly (P<0.001) affected by muscle type. Fatty acid composition was mostly affected by farm and by muscle type. The SM muscle was poor in saturated fatty acids (14:0, 15:0, 16:0, 17:0 and 18:0) and richer in polyunsaturated fatty acids than the other muscles. The ratio between n-3 and n-6 PUFA observed in this study are close to the usually found for pasture fed animals indicating the healthy value of "Carne Mirandesa PDO" veal.

Keywords: PDO meat, collagen, fatty acids, muscles, Mirandesa.

Introduction

Mirandesa cattle are a local meat breed from the Northeast region of Portugal, with an important role on the rural spaces maintenance, contributing to the fixation of the populations and the environmental preservation. Cows graze natural pastures (herbaceous species) during spring and early summer and afterwards are fed with hays (of natural pasture or oat) and straws (oat, barley or wheat), complemented with local feeds like squash or potatoes. The nutritive value of the feedstuffs allows the sustainability of this traditional production system, with aptitude for meat production (Galvão *et al.*, 2005).

Calves present at birth a mean live weight of 32-35 kg, are weaned at 7 months old and slaughtered with a mean live weight of 180-200 kg. Dressing percentage was about 53% and the percentage were 16.5 for fat, 62.0 for muscle and 19.3 for bone (Vaz Portugal *et al.*, 2001). The genetic characteristics of this breed, associated with the particular feed management of calf (milk complemented with hay and farm-made concentrates) contribute to particular meat characteristics like tenderness and succulence, together with tastes and flavours, very appreciated by the consumers. Its specificity, quality and genuineness, was recognized conducting to the Protected Denomination of Origin of "Carne Mirandesa". The objective of this study was to evaluate chemical parameters related to meat quality, contributing to a better characterization of "Carne Mirandesa PDO".

Material and methods

This study was made in three farms, from two villages: Prado Gatão (PG) and Fonte da Aldeia (FA1 and FA2), located in the Northeast region of Portugal. Calves were raised permanently indoors, nurse from their dams overnight, and fed with hay and concentrate made with local ingredients and soybean meal. PG concentrate presented a mean (± standard deviation) crude protein (CP) of 13.8±0.49 % dry matter (DM) and starch of 46.8±8.75 % DM. The concentrates used in FA1 and FA2 were similar, with a CP content of 15.6±0.45 % DM and a starch content of 44.8±6.94 % DM. Twenty five calves (11 males and 14 females: 6 from PG, 7 from FA1 and 12 from FA2) were slaughtered at 7 months old. Twenty four hours after slaughtering, samples were collected from four different muscles: *longissimus dorsi (LD), semimembranosus (SM), gluteus biceps femoris (GB)* and *triceps brachii caput longum (TBL),* vacuum packaged and freeze-dried.

Determinations of nitrogen (N) and ether extract (EE) without acid hydrolysis were made following the procedures described in AOAC (1990). Soluble and total collagen was calculated from the hydroxyproline content according to the method of Hill (1966). Fatty acid were extracted and methylated by the method of Rule (1997). Fatty acid methyl esters were analysed by gas chromatography using a Varian CP-3800 chromatograph (Varian Analytic Instruments, Walnut Creek, CA, USA) equipped with a 30-m fused silica capillary column OmegaWax 250 (Supelco, Bellefonte, PA, USA) with a 0.25-mm internal diameter and a 0.25-ìm film thickness. Peak identification was based on co-chromatography with known standards of fatty acid methyl esters (Sigma, St Louis, MO, USA). All results are presented as weight percentage of identified fatty acids.

Data were analysed using the "proc mixed" from SAS (SAS, 2004). The model considered the fixed effects of sex, farm and muscle. Muscles were treated as repeated measures.

Results and discussion

The effect of sex and farm were not significant for N, EE and collagen content of muscles. All these parameters were significantly (P<0.001) affected by muscle type (Table 1). SM muscle was the

Table 1. Nitrogen, ether extract content and collagen (% of dry matter) of gluteus biceps femoris (GB), longissimus dorsi (LD), semimembranosus (SM), and triceps brachii caput longum (TBL) muscles from "Carne Mirandesa PDO" veal.

	GB	LD	SM	TBP	s.e.m.	Sig
N	14.1^{a}	14.3^{a}	14.7^{b}	13.7^{a}	0.19	***
EE	10.0^{b}	10.3^{bc}	6.4^{a}	12.6^{c}	1.07	***
Sol. Col.	3.1^{a}	4.3^{b}	2.8^{a}	6.2^{c}	0.43	***
Tot. Col.	7.4^{a}	9.0^{a}	9.5^{a}	14.1^{b}	0.79	***
Sol. Col %	41.7^{b}	44.0^{b}	28.9^{a}	44.2^{b}	1.58	***

N- nitrogen, EE- ether extract, Sol. Col. – soluble collagen, Tot. Col. – total collagen, Sol. Col. % - soluble collagen in percentage of total collagen. Means with different superscripts are significantly different (P<0.05).

Table 2. Effect of farm (PG, FA1 and FA2) and mucle type (gluteus biceps femoris-GB, longissimus dorsi-LD, semimembranosus- SM, and triceps brachii caput longum- TBL) on fatty acid composition of "Carne Mirandesa PDO" veal.

	Farm				Muscle				
	Al	PG	VG	s.e.m.	GB	LD	SM	TBP	s.e.
12:0 [F]	0.14[a]	0.22[b]	0.12[a]	0.022	0.17	0.16	0.15	0.17	0.00
14:0 [F,M]	3.71[a]	4.98[b]	3.63[a]	0.307	4.24[a]	4.32[a]	3.69[b]	4.16[a]	0.20
15:0 [F,M]	0.50[a]	0.62[b]	0.47[a]	0.038	0.55[a]	0.57[a]	0.49[b]	0.52[a]	0.02
16:0 [F,M]	25.1[a]	26.8[b]	25.4[a]	0.49	26.1[a]	26.7[a]	24.8[b]	25.5[a]	0.36
16:1 [M]	4.24	4.42	4.45	0.275	4.11[a]	4.23[a]	4.90[b]	4.24[a]	0.19
17:0 [M]	0.96	0.88	0.96	0.045	0.99[a]	0.98	0.82[b]	0.94[a]	0.02
17:1	1.00	0.85	0.96	0.046	0.94	0.91	0.97	0.93	0.02
18:0 [F,M]	14.3[a]	12.6[b]	14.3[a]	0.45	14.2[a]	14.2[a]	12.1[b]	14.3[a]	0.33
18:1 [F]	38.0[a]	32.1[b]	36.0[a]	1.04	35.0	35.8	34.9	35.7	0.65
18:2n-6 [M]	5.62	7.17	6.60	0.572	6.31[a]	5.51[b]	7.70[a]	6.34[a]	0.57
18:3n-3 [F,M]	0.71[a]	1.54[b]	0.71[a]	0.053	0.99[a]	0.90[a]	1.14[b]	0.93[a]	0.04
20:1 [F]	0.17[a]	0.12[b]	0.16[a]	0.011	0.14	0.15	0.15	0.15	0.00
20:2n-6 [M]	0.28	0.26	0.31	0.030	0.26[a]	0.23[a]	0.36[b]	0.27[a]	0.02
20:3n-6 [M]	0.46	0.58	0.58	0.066	0.51[a]	0.44[a]	0.70[b]	0.52[a]	0.04
20:4n-6 [M]	1.80	2.43	2.06	0.232	1.90[a]	1.64[a]	2.77[b]	2.09[a]	0.18
20:3n-3 [F]	0.03[a]	0.00[b]	0.02[a]	0.005	0.01	0.01	0.02	0.01	0.00
20:5n-3 [F,M]	0.47[a]	1.28[b]	0.54[a]	0.080	0.75[a]	0.63[a]	1.09[b]	0.58[a]	0.07
22:5n-3 [F,M]	0.82[a]	1.50[b]	0.89[a]	0.097	0.97[a]	0.81[a]	1.49[b]	1.02[a]	0.09
22:6n-3 [F,M]	0.14[a]	0.28[b]	0.22[a]	0.031	0.23[a]	0.16[a]	0.32[b]	0.16[a]	0.02
Σ Odd-br	3.99	3.70	4.02	0.127	4.06[b]	4.02[b]	3.65[a]	3.88[b]	0.09
Σ n-3 [F,M]	2.19[a]	4.60[b]	2.38[a]	0.242	2.96[a]	2.51[a]	4.05[b]	2.71[a]	0.22
Σ n-6 [M]	8.16	10.4	9.56	0.883	8.98[a]	7.82[a]	11.5[b]	9.23[a]	0.68
n6/n3 [F,M]	3.89[b]	2.30[a]	4.16[b]	0.222	3.43[b]	3.41[b]	3.13[a]	3.83[c]	0.14
PS [F,M]	0.23[a]	0.32[b]	0.26[a]	0.025	0.25[a]	0.22[a]	0.36[b]	0.26[a]	0.02

F – significant effect for farm ($P<0.05$), M – significant effect for muscle type ($P<0.05$), Σ Odd-br – sum of odd chain and branched chain fatty acids, n6/n3 – ratio between n-6 and n-3 polyunsaturated fatty acids, PS – ratio between polyunsaturated fatty acids and saturated fatty acids. Means with different superscripts in farm and in muscle type sections are significantly different ($P<0.05$).

leanest and with higher N content, whereas the TBP was the fatter muscle. These muscles presented also the extreme values of soluble collagen, which in TBP was significantly higher compared to the other muscles. Differences in composition between muscles are well known.

Fatty acid composition was mostly affected by farm and by muscle type (Table 2). Few fatty acids were affected by sex with 20:2 n-6, 22:6 n-3 and sum of n-3 PUFA higher ($P<0.05$) in males than in females and 22:5 n-3 higher ($p<0.05$) in females. However, 18:1 cis-9 (oleic acid) tended ($P<0.06$) to be higher in females, whereas 18:2 n-6 (linoleic acid) ($P<0.07$) tended to be higher in males.

Farm effects were significant for most of the fatty acids studied. The animals from PG farm had higher 12:0, 14:0, 15:0, 16:0, 18:3 and n-3 polyunsaturated fatty acids (PUFA) and lower 18:0, 18:1 and 20:1 than animals from FA1 and FA2 farms. Veal from FA1 and FA2 farms had no differences in fatty acid composition. The differences between farms could be related with the reported differences between concentrates and geographical factors.

The effects of muscle type on fatty acid composition were mainly attributed to the SM muscle and only slight differences were observed between GB, LD and TBP. The SM muscle was poor in saturated fatty acids (14:0, 15:0, 16:0, 17:0 and 18:0) and richer in polyunsaturated fatty acids than the other muscles. This is consistent with the fact that SM muscle is leaner than the other muscles (Table 1) and, consequently, with higher relative proportions of phospholipids and PUFA. Muscle differences in fatty acid composition are well known and related to total fat content and proportions of muscle fiber types (Wood *et al.*, 2004).

The ratio between n-3 and n-6 PUFA is an important index of health nutritional quality of foods. Meat from pasture fed animals have a low n3/n6 PUFA ratio, ranging between 2.0 and 2.3 (Nuernberg *et al.*, 2002; Raes *et al.*, 2004) and are considered healthier than concentrate fed beef, ranging from 16 to 20 (Enser *et al.*, 1998). The values observed in this study are close to those reported for pasture fed animals indicating the healthy value of "Carne Mirandesa PDO" veal.

References

Association of Official Analytical Chemist (AOAC). 1990 Official Methods of Analysis. Ed. S. Kenneth Helrich. USA.

Enser, M., B. Hallet, G. Hewitt, J. Fursey, D. Wood & G. Harrington, 1998. Fatty acid content and composition of UK beef and lamb muscle in relation to production system and implications to Human nutrition. Meat Sci. 49: 329-341

Galvão, L., O.C. Moreira, R. Valentim, J. Ramalho Ribeiro & V. Alves. 2005. Dynamics of pastures and fodder crops for Mirandesa cattle breed. In: Book of Abstracts of 56th Annual Meeting of the European Association for Animal Production, Uppsala, Sweden. Wageningen Academic Publishers, Wageningen, The Netherlands, 356.

Hill, E., 1966. The solubility of intramuscular collagen in meat animals of various ages. J. Food Sci., 37: 161-166.

Nuernberg, K., G. Nuernberg, K. Ender, S. Lorenz, K. Winkler, R. Rickert & H. Steinhart, 2002. N-3 fatty acids and conjugated linoleic acids of longissimus muscle in beef cattle. Eur. J. Lipid Sci. Technol. 104: 463-471

Raes, K., S. De Smet, & D. Demeyer, 2004. Effect of dietary fatty acids on incorporation of long chain polyunsaturated fatty acids and conjugated linoleic fatty acid in lamb, beef and pork meat. A review. Anim. Feed Sci. Technol. 113: 199-121

Rule D.C., 1997. Direct transesterification of total fatty acids of adipose tissue, and of freeze-dried muscle and liver with boron –trifluoride in methanol. Meat Sci. 46: 23-32

SAS, 2004. SAS/Stat user guide, SAS Inst., Cary, NC, USA.

Vaz Portugal, A., D.R. Navas & J.F.F. Mira, 2001. Relatório Final de execução do projecto PAMAF 3047 - Raças Bovinas Autóctones, contribuição para a sua produção nos ecossistemas tradicionais e seu reflexo sócio-económico. INIA -Estação Zootécnica Nacional, Vale de Santarém, Portugal

Wood, J.D., R.I. Richardson, G.R. Nute, A.V. Fisher, M.M. Campo, P.R. Kasapidou & M. Enser, 2004. Effects of fatty acids in meat quality: a review. Meat Sci. 66: 21-32.

Session 6. Round table and Conclusions

Round table
From traditional to certified animal products. Products perception by consumers, marketing systems and interaction with producers: what future role for organic animal farming in the Mediterranean?

J.M.C. Ramalho Ribeiro

Estação Zootécnica Nacional (EZN-INIAP), Santarem, Portugal

Introduction by J.M.C. Ramalho Ribeiro (Portugal)

In the Mediterranean there are many typical animal products as a result of the importance of domestic animals into the food diet. The pigs used to be fed at the doors step, to be slaughtered in winter, and its meat processed in order to become the best protein food of the family diet. Latter the same happens with the chickens, lambs, goat kids and beef. Based on local resources (animals and feeds), the animal production systems obtained high quality products with which the different communities using their imagination, tradition and ingredients transformed into delicious products and dishes. Today the gastronomy of Mediterranean area is very rich and diversified as well as healthy and very much appreciated by those who enjoyed the pleasure of the food.

This diversity means low quantities which bring a problem of dimension and a difficulty for the marketing and promotion of those products. Beside, most of these procedures are not achieved according to adequate safety recommendations. The consumer becomes more and more concerned with that he eats and the governments, following those concerns, regulate the market and define the rules to protect the consumer health. Therefore there is the need to maintain the quality and the tradition procedures in one hand and in the other one to respect the hygienic recommendations and the market rules.

Is it possible to maintain quality and traditional procedures together with the improvement of the animal production systems namely feeding, reproduction, genetic improvement etc.? Has this problem the same importance in the Northern and in the South Mediterranean? Those are the questions behind the idea of promoting this Round Table with the participation of five invited colleagues which presented the following comments.

G. Zervas (Greece)

From the papers presented during the two days of the Symposium it is obvious that the Mediterranean countries have a great number of traditional speciality products, mainly coming from extensive sheep and goats farming systems. In addition to that, it could be seen that there is also a large diversity among these countries as processing techniques, sanitary conditions, adaptation to legislation, safety control of products, typicity, market's organization, etc. concerns. Today's market, which has been affected by globalization, demands - apart from quality and typicity - safety and traceability which the traditional Mediterranean products of animal origin cannot always provide or guarantee. Thus, their comparative advantages in many cases become very limited or are restricted to local markets.

At present, Mediterranean people, in particular those aged over 35 years, have closed links with the countryside and can appreciate much more the quality of traditional products and are prepared to pay extra price for them. People of the younger generation don't have such experience and they are much less linked with the gastronomic and cultural traditions, due to American model of "fast food" to which they are more and more adapted.

Thus, the typical traditional Mediterranean products of animal origin, with their specific originality and authenticity which is due to animal breeds, to natural feed resources these animals use, to limiting routine medication and to specific technology used for their processing, should be produced and marketed according to consumers' demands and expectations which include safety, quality, typicity and traceability.

J.-P. Boutonnet (France)

The notion of "comparative advantage" is pertinent for nations. Each nation produces and exports the product for which it is more efficient (e.g. Portugal: wine; England: wool). Comparative advantage is related to trade between nations, and to commodities. Competitiveness relates to "capacity (of a nation or a region, or a firm) to maintain or increase its market share". It can be obtained by price (production cost) or by other means (monopolistic competition). In this last case, the product is not a commodity but a specific product.

Mediterranean animal husbandry does not produce any commodity (staple food, raw material for industry) for which it cannot be price-competitive, but specialities, specific products. These typical Mediterranean animal products are of two types:

- Processed products: cured ham and cheese. Their characteristics are the result of both farmers and processors know how. Building of the added value depends on the good co-ordination between all the actors of the chain.
- Meat from young (not weaned) animals, fed with grain. Consumers of the region know and appreciate this type of meat, which creates a strong barrier to imported meat.

All these products can be sold at high prices only because of a specific, regional or local knowledge or know-how. Not only farmers and processors knowledge, but also consumers' knowledge. Consumers have to be "trained" to appreciate these special products.

In the region of origin, products are well known by consumers and producers. Local market can accept personal trust as a mean of guaranteeing the quality of the product. Their typicity is not labelled as there is no need for that. When they are sold on a larger market, they need to be identified by a formal label. This certification is the main characteristic of typical products on a global market. This formal control is expensive and can only be made if the market allows a value added for such products.

F. Casabianca (France)

First of all, traditional products and organic ones show deep differences. Organic product is a very recent conception based on the forbidding of chemical inputs and has nothing to do with tradition.

Second point is designing the relevant framework of our research on product certification: we must consider it as a whole system, how to put together i) technical aspects of production and resource management; ii) logistical dimensions referred to the relationship among actors along the supply chains; and iii) institutional features which arrangement identify what we now call the "*gouvernance*" of the system. As an example, the question of assessment and control has implications in these three directions.

Third point concerns the choices made by local actors when they decide to build up a product certification. They must put a part of the technical certification and they must also put a part of the technical culture they share into the code of practices justifying the certification. But, they have also to anticipate the required competencies of the targeted clients supposed to appreciate their product. And they must reflect on the internal rules in order to insure a correct repartition of added value among the various actors of the economic chain. In these perspectives, the movement from traditional product to certification has to be considered as a major innovation process at a local level, producing more questions than effective solutions.

D. Crespo (Portugal)

I would like to call your attention for some economic aspects of producing and marketing traditional food products of high quality, such as local specialities or organic products. Indeed some high quality traditional products which are available in the market in limited quantities have a demand much superior to the offer and therefore attain very rewarding prices for the producers.

However, some products of superior quality face a great competition from products with similar aspect which are very abundant in the market but of lower quality. The consumer is not always prepared to differentiate quality and associated prices, therefore giving preference to the cheaper products. As to produce high quality traditional food products involves an extra cost, if the consumer is not prepared to distinguish quality and accept the respective higher prices, then the producer may be tempted to abandon this quality production.

The above considerations may apply particularly to organic products, which may not be easily distinguished from non organic ones by the consumer. Our experience as producers of organic sheep milk cheese of high quality has been a demonstration of the difficulties encountered. Indeed we have been the first to produce organic sheep milk cheese in Portugal and we thought that its introduction in the market at a rewarding price was an easy task; indeed it was not! We had to find special niches in the market in a long time taking process to get our production consumed by people able to differentiate quality and pay the respective price. We had even sent cheese to the Organic Food Exhibition of Nuremberg (Germany), where it had a great success during the tasting sessions, and this was confirmed by a couple of orders coming from selected shops of organic food. Unfortunately, as the organic cheese production is not compatible with the use of fungicide products (as it is common with non organic cheese), and the transport took a bit longer than expected, those orders of cheese arrived in Germany covered with a thin layer of blue fungus, which resulted in their rejection!

G. Piredda (Italy)

A brief overview on the situation of the sheep milk products manufactured in Sardinia may help us to understand the problems that producers have to face to satisfy the demand of the modern market.

The local production of cheese, from 2 800 000 milking ewes, is the core of the Sardinian agricultural production. The cheese-making is usually done using traditional methods in small cheese plants or in the farm. The industrial companies, and, of course, the market, demand milk with low price and constant characteristics. This is, however, not compatible with the local production system. The best way to valorise the sheep milk production may need a further step into the certification as PDO (Protected Designation of Origin) cheeses. As a matter of fact the PDO "Pecorino Romano",

"Pecorino Sardo" and "Fiore Sardo" (respectively 27 000 tons, 1 400 tons and 300 tons in the year 2003) constitute the majority of the produced sheep cheese.

The new rules for the production of agricultural food and the evolution of the consumer's request have forced the production system to remarkable changes. At the same time it is necessary to emphasize the characteristics and the quality of the products. Therefore all subjects involved in the cheese production, like cheese-makers, breeders, supplement producers and even the rennet producers, find highly convenient to obtain process and product certifications. This activity is well supported by the Sardinian Government, which promotes the cooperation between public organizations and private companies to improve the quality and the image of local productions. Nevertheless there is a problem to tackle: the cost of the controls that, according to the European rules, must be paid by the producers and they are often too expensive for the small productions. Moreover the control systems, and consequently the costs of the controls, are different from country to country.

In the future, the same control system for all European Countries which should also convey, hopefully, a reduction of the costs could help the local production to evolve towards the certified production that the modern market requires.

Final considerations and conclusion by J.M.C. Ramalho Ribeiro

The Round Table participants look into these problems from a different view.

Summarising, George Zervas considered the importance of those products for the people over 35 recognising that the young generation have today different food habits where traditional products are less important.

Jean-Pierre Boutonnet reminded us the need to be very precise on what we are discussing, clarifying what we should understand by competitive advantage (mainly market related) as well as certified products (characterization and identification), concluding by the cost of it and the required recognition by the consumers in order to be prepared to pay the added value.

François Casabianca introduced the idea that we should separate organic and traditional products as they should be treated in different ways. Then he focussed on the certification as a process and the need to look at it as an innovation but also clearly identifying its technical and cultural components.

David Crespo described his own experience on the certification of organic products (ewe milk cheese) and the involved costs, as well as the promotion required to have the market informed and prepared to pay the difference. This should not be an individual effort, but a collective task.

Finally Giovanni Piredda presented a successful case with the Sardinia cheese which has been supported by the local government but should in future be supported by the farmers.

The main problems are the costs of controls which must be harmonized and reduced in order to be reasonable.

In conclusion, organic is not traditional, and traditional is usually a local or regional recognised product which needs to be promoted into the global markets. To do so, the certification must include the different characteristics of the product (identification, technical and cultural contents) which should be used in order to promote its added value and its acceptance by the consumers.

Conclusions of the Symposium: Modernisation of the Mediterranean animal products through certification?

J.-C. Flamant

Mission Agrobiosciences, Toulouse, France

At the end of these two days, after having heard this very intensive succession of reports, I should say that we are probably more able to built that could be the ideal programme of such a Symposium if we should plan to repeat it. I mean that with such a Symposium we enlarge our perspective and we obtain a better understanding of its subject.

Specialty products versus commodity products

"*Comparative advantages*"... In choosing these key words for our meeting, we were referred to the theory of Ricardo as mentioned by J.P. Boutonnet in the final Round Table. In fact, I was really impressed that some of the main speakers needed to make reference to the History (with a great "H", from the domestication of the animals, Greek and Roman civilization, and Islam, etc.) and to the large world market in the framework of globalisation with its negative consequences for the local specificities. In fact, our colleagues helped arguments from the historical and present dynamics to support their own vision of that they consider to be the animal production systems in the Mediterranean region. In this respect I could say that the subject of this Symposium introduced trouble into the certitudes.

May be we obtained a first key at the beginning of our meeting with the interesting report of Daniele Rama who has proposed to distinguish specialty products and commodity products. In progressing in the course of the sessions, we obtained a first list of the factors which participates to these specialty products, typical of the animal production systems in the Mediterranean: extensive animal production systems, local breeds, traditional processing and practices, feeding resources from local territory, low animal productivity... These characteristics contrast with those of the commodity products, mainly intensive animal production systems, cosmopolitan breeds, industrialised processing, feeding resources from purchase raw material, high level of animal productivity with the help of research and technology.

In fact, this classification between specialty and commodity products has to be completed by other analysis for understanding the situation. For instance, many of the speakers pointed out that diversity is the main characteristic of the typical Mediterranean products. In this respect, several papers and posters have designed like a mosaic of the Mediterranean region, an impressionist mosaic as the impressionist painters, or may be also a puzzle... We have to find the good vision of this mosaic, the significant figure of this apparent puzzle, the right meaning. It could mean that if Mediterranean countries have a comparative advantage on the world market it could come from its global capacity in marketing speciality products more than from commodities products. But, at this stage, my feeling is that this obvious and interesting diversity could hamper the possibility for a generalised and unique conclusion of our Symposium in this way.

Typical animal products in the economy

"*Typical*", it is the other key word of our Symposium. I remember that nine years ago, one of the previous symposium of this Mediterranean series was hold in Badajoz and Zafra, in Spain closed to here, dedicated to the specific quality of the animal products of Mediterranean origin[1]. If we push away the information produced during this present Symposium, we can say that "*typicity*" takes its roots from:

- the scarcity of the resources provided by the territory;
- the livestock material with low productive potential but with a good adaptation to the constraints of the systems;
- the seasonal variations of the weather associated to climatic uncertainty;
- various particular processing systems linked or not linked to the characteristics of the raw material, milk or meat, and in connexion with special and local cooking practices.

And I am asking if these difficult and uncontrolled conditions are really a chance for obtaining products with a comparative advantage on the large market... George Zervas discussed the possibility for local typical products to be involved in organized market and commodity chains, reasoned in the framework of globalisation (Cf. the EAAP Round Table in Cairo in 2002[2]). I note that François Casabianca provided a very useful contribution in proposing to make a distinction between "*archetype typical product*" and "*stereotype typical product*" about the case of the ham production. "*archetype typical production*" is closely related to the living local tradition characterised by a large diversity and small quantities, and "*stereotype typical production*" makes a choice within this diversity for retaining traits which characterise products for larger marketing in commodity chains.

I also note that we have not considered the possible comparative advantage of Mediterranean products in the commodity chains for mass consumption. But, is it possible to produce other things than specialty products in the Mediterranean conditions? So, what it is the dimension of the market niche of the specialty products for supporting the economy of the region? Is the society willing to pay extra cost for them and for providing the justified income to the producers, by direct payment to the producers or by tax? Or, in other words, could the consumers buy these products in enough quantities for supporting numbers of farmers in the rural territories?

In complement to this economic aspect, we need to consider the contribution of agricultural and animal production to land management and rural development, a very important aspect in the difficult conditions of the Mediterranean mountains and plateaux. We had very eloquent reports about this aspect with, for instance the contribution of Javier Calatrava and Samir Sayadi in south-eastern Spain. But considering the market, we have to measure the fact that a large part of the Mediterranean typical products are bought and consumed in the Mediterranean countries, and only a small part of them are involved in large international market as commodity products and being mainly "*stereo typical products*".

---

[1]"Basis of the quality of typical Mediterranean animal products", EAAP Publication n° 90, 1998.

[2]Cairo Round Table on EAAP website : http://www.eaap.org/content/round_table_cairo_english.htm

What happens with tradition?

We have a session about the improvement of the traditional production systems. It seems to me the most crucial issue of our Symposium. We heard arguments about typical products as having their roots in tradition, in the heritage of past generations - and it is why probably some of us evoked the words of Homer and his Odyssey - but I remark that, may be, it is as if a better efficiency could be attained with traditional products as far as the tradition is death, as far as if it is death it can be revisited. This issue needs to come back to the discussion about "archetype traditional products" and "stereotypical traditional products": may be people involved in that it is call "archetype tradition" are not well prepared to pass to "stereotype tradition". And it is really an issue to debate. It is really more difficult if the tradition is living as it is demonstrated in the very interesting case of the Anti-Taurus in the East Mediterranean Turkey. How do we consider the relation to the world market of the products which are processed from still living tradition, with a to-day human reality of the link with the previous generations?

I should like to conclude by mentioning something about what I call our own cultural attitude, that of the animal scientists are here mainly. Could I point out that for fifty years and even one century, we and our predecessors we were pushing the development of meat and milk production for the mass food market through the modernisation of the traditional animal production systems, by increasing the individual productivity of the farm animals which were till there only a limited part of the daily food of the population mainly depending from local grain and vegetable production. May be we have now to change our vision of what means modernisation, mainly in the Mediterranean region in reference to specialty products. I again say that the subject of this Symposium has provided trouble into our certitudes. So I propose to consider that the new way for modernity and true innovation is that we call... "*certification*"!

www.ingramcontent.com/pod-product-compliance
Lightning Source LLC
LaVergne TN
LVHW060616300926
849162LV00008B/1635
9789076998862